EARLY PHYSICS AND ASTRONOMY

A HISTORICAL INTRODUCTION

History of Science Library

Editor: MICHAEL A. HOSKIN
Lecturer in the History of Science, Cambridge University

EARLY PHYSICS
AND
ASTRONOMY

A Historical Introduction

OLAF PEDERSEN

Professor of
History of Science, Aarhus University

and

MOGENS PIHL

Professor of Physics, University of Copenhagen

MACDONALD and JANES: LONDON

AND

AMERICAN ELSEVIER INC: NEW YORK

Sole distributors for the British Isles and Commonwealth
Macdonald & Co. (Publishers) Ltd.
49–50 Poland Street
London w1A 2LG

Sole distributors for the United States and Dependencies
American Elsevier Publishing Company, Inc.
52 Vanderbilt Avenue
New York, N.Y. 10017

All remaining areas
Elsevier Publishing Company
P.O. Box 211
Jan van Galenstraat 335
Amsterdam
The Netherlands

Standard Book Numbers
Macdonald ISBN 0356 04122 0
American Elsevier ISBN 0444 19575 0
Library of Congress Catalog Card Number 72 181849

Printed in Great Britain
by Unwin Brothers Limited
The Gresham Press, Old Woking, Surrey
A member of the Staples Printing Group

Contents

Preface

TODAY, books on the early history of science are far from rare, and to present still another work on this subject is a venture which, perhaps, calls for some explanation. Long teaching experience has convinced us of the need for a book which might serve not only as a first introduction but also as a practical guide to a more detailed study. This book thus contains a general and, we hope, easily comprehensible account of some major topics in Ancient and Mediæval science, followed by a biographical and bibliographical appendix which will enable the reader to begin a more searching study of details.

The text itself does not pretend to survey the whole field. Much is left out, and each of the twenty chapters is nothing more than an essay illustrating one important theme in the complex pattern of the whole field. Among the inevitable limitations the most regrettable is, perhaps, that the book omits any detailed account of Egyptian and Babylonian science before the Greeks, as well as any reference to the rich development in Mediæval India or China. On the one hand we feel unable to deal competently with Indian or Chinese contributions; on the other hand we deem it unnecessary to offer a new account of Egyptian ·or Babylonian science, considering the number of introductory works on these topics which have recently been presented by eminent scholars. For the same reason the history of pure mathematics has also been omitted, while the main emphasis of the book is on the mathematical description of nature as it gradually emerged in astronomy and also, to a lesser extent, in mechanics and physics. Again, many interesting but more special subjects have been ignored, such as time-reckoning, cartography, and others.

The second part of the book contains, besides a selected bibliography of general works, a dictionary of all the scientists and other authors mentioned in the text. In it we have collected such biographical information as seems useful in a work of a purely introductory character. But the main purpose of this section is to provide the reader with references to original sources, translations, and a certain number of secondary works. It goes without saying that it has been impossible to achieve any kind of completeness: we have merely aimed to give enough references to allow the start of a more elaborate study of any of the subjects mentioned in the first part of the book.

For some years this book has been available in its original Danish form. In preparing the English translation most chapters have been re-worked or given a new shape. The bibliographical references have been extended and, as far as possible, updated to 1970.

We are indebted to many colleagues who have pointed out errors or suggested improvements and would also like to express our gratitude to Dr. Michael A. Hoskin of Churchill College who took the initiative of having the book published in an English version. We also feel indebted to Mrs. Christina Jones, M.Sc., and Mrs. Ruth Gardiner, M.A., who most kindly took upon themselves the burden of rendering our draft translation into readable English. Finally, we are most grateful to Mrs. Kate Larsen who has followed this book at all stages from the first draft of the Danish version to the completion and printing of the English translation.

Mogens Pihl   Olaf Pedersen

Chapter One

Science Before the Greeks

The Origin of Science

IN previous centuries it has often been maintained that science in general, and natural science in particular, were created by the Greeks during the fifth, fourth, and third centuries B.C., when classical Greek culture enjoyed its most flourishing period. Although this opinion is supported by a certain number of facts, it is now regarded as true only with essential modifications. With this in mind, it is worth noticing that the Greeks themselves did not conceal the fact that their science presupposed results achieved by others, a fact well illustrated by the historian Herodotos (c. 450 B.C.), who credits the Egyptians with the discovery of geometry just as he thinks that astronomy emerged among the Babylonians. To him these new sciences came into being for purely practical reasons, so that Egyptian geometry is supposed to have been created because of the need to reconstruct the boundaries between fields each time the inundations of the Nile had removed them.

Aristotle (384–322 B.C.) also places the cradle of mathematics in Egypt. For him, science is one result of a fundamental human desire for acquiring knowledge for its own sake. This desire can be satisfied only if society contains at least one class with sufficient leisure to study. According to Aristotle this was not the case until the priestly caste emerged in Egyptian society.

Modern archaeology has done much to confirm the suspicion that at least some systematic knowledge of nature existed long before the Greeks. Thus, French and Spanish cave paintings from the Stone Age

I

reveal a reasonably well-developed knowledge of the natural history of a number of animals. Rock carvings from Palaeolithic times also prove that, perhaps about 15,000 years ago, people had already begun to observe the stars and to see them as constellations. Of course examples like these do not prove a Stone Age existence of any natural science in the modern sense, but they are testimonies to the fact that men possessed the desire to observe and describe natural phenomena which is the core of any scientific activity. This supposition holds true whether this interest was motivated by practical reasons springing from the desire to master nature and to exploit its resources, as Herodotos supposed, or by a more detached love of abstract knowledge for its own sake, as maintained by Aristotle.

The Fertile Crescent

The prehistoric civilizations in the Fertile Crescent in the Nile, Tigris, and Euphrates valleys have provided the fullest archaeological material for an evaluation of the development of science and technology in the pre-Greek world. A detailed account of the contributions of both the Egyptian and Babylonian culture to the history of science falls, however, outside the scope of this book and there is space to mention only a few very general features important to understanding the later achievements of the Greeks.

Some of the oldest archaeological finds show that chemical technology was well developed in both the great civilizations of the Near East. Several pieces of clay apparatus from Mesopotamia reveal an unmistakable familiarity with such procedures as distillation (Fig. 1.1), sublimation, and extraction, as far back as the fourth millenium B.C. Cuneiform texts from a somewhat later period describe the manufacture of mercury from cinnabar and of sal ammoniac from dung. Recipes for the extraction of perfumes are also known. In the second millenium B.C., the Egyptians had methods for making glass, many different dye stuffs, soda, potash, alum, nitre and a few antiseptics. The connection between ores and metals seems to have been discovered in the fourth millenium B.C. By about 2,000 B.C., the primitive metallurgy of gold, silver, copper, lead and tin, besides that of the alloys electron (gold and silver) and bronze (copper and tin),

Fig. 1.1.A. *Chemical apparatus from about 3,500 B.C. found at Tepe Gawra in Northern Mesopotamia and shaped like a pot with a double rim forming a groove around the top which is provided with a tight-fitting lid. The apparatus can be used for distillation, the vapour from the boiling liquid being condensed under the lid and collected in the groove.*

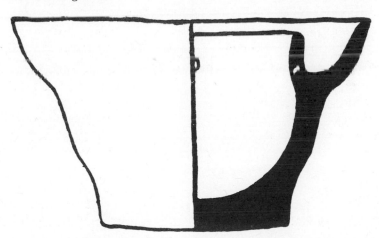

Fig. 1.1.B. *A variation of the apparatus shown in Fig. 1.1.A. The inner rim is pierced with small holes so that the distillate returns to the pot to be evaporated once more. In this form the apparatus can be used for extraction.*

3

was understood throughout the Near East. Iron was known, but was useless until the method of hardening it was discovered towards the end of the second millenium B.C.

As far as astronomy is concerned, the achievements of the early civilizations are very considerable. Astronomical papyrus texts are of a fairly recent date (mostly from Hellenistic times), but even around the middle of the second millenium sundials and water clocks were already being used in the Nile valley to count the hours of day and night. Here, as everywhere, the changing seasons could be connected with the heliacal rising of selected fixed stars, that is the time of the year when these stars reappear immediately before dawn after their periods of invisibility. In addition, the direct eastwards motion of the planets among the fixed stars was known, together with their stationary points and retrograde periods, whereas there are no Egyptian records of eclipses until Hellenistic times. On the other hand, eclipse observations played an enormous role in Babylonian astronomy, and planetary positions were systematically observed and recorded from about the beginning of the second millenium. From about 600 B.C. Babylonian astronomers developed various highly ingenious mathematical methods, based on arithmetic series, of describing planetary movements accurately enough to make possible predictions of eclipses, oppositions, and so on.

The existence of a few papyri written at the beginning of the second millenium, but containing much older material, shows that in the field of mathematics the Egyptians were able to solve a great variety of practical problems by mathematical methods as early as 2,000 B.C. They deal, for example, with determinations of areas and volumes, and computation of the amount of raw materials such as grain necessary for the manufacture of a given amount of the finished product like bread or beer. The best known of these texts is the lengthy *Papyrus Rhind*, which is generally believed to be a manual for the education of the royal scribes, or other members of the administrative class. Only a few mathematical texts have survived in Egypt, but from Mesopotamia several thousands of cuneiform clay tablets with mathematical contents have been discovered. Some of these are tables for multiplication, reciprocals, squares and square roots, cubes and cubic roots and so on, in the sexagesimal system of numbers (Fig. 1.2). Other tablets contain worked examples showing how mathematics was applied to astronomy and time reckoning, to the

Fig. 1.2. Clay tablet from the old Babylonian period (c. 1700 B.C.) showing a square with its diagonals. The number 30 is inscribed along the side of the square. Along the diagonal is written 42, 25, 35 which in the sexagesimal system means $42 + 25/60 + 35/60^2$. If this number is divided by 30 one has 1, 24, 51, 10 which is the sexagesimal value of $\sqrt{2}$. This number is also seen along the diagonal. In decimal notation it is equal to 1.414213 (c.f. the modern value 1.414214).

computation of interest, in probate cases, and in public administration, for the digging of canals, fortification, and in the distribution of land. Sometimes, more theoretical problems are dealt with. The old Babylonian Plimpton text, for example contains relations between Pythagorean numbers, that is, sets of three numbers a, b, c, satisfying the condition $a^2 = b^2 + c^2$. Quadratic equations with one or two unknown quantities were studied in a rather abstract way, and with obvious interest in their purely mathematical properties.

5

The Empirical Character of Science

Archaeology has shown the extent to which pre-Greek civilizations were dependent upon technology and mathematics. This seems to prove that exact science came into being long before the Greeks. In a sense, this is true, but both Egyptian and Babylonian science and mathematics were, as will be seen, very different from those of the Greeks. A finer investigation reveals that the achievements of the Egyptians, and of successive peoples in Mesopotamia, were very closely related to the practical demands of everyday life, and involved none of the elements considered today as essential to science: the evidence so far suggests that these peoples knew nothing of logical proof or of natural laws.

This statement, although rather sweeping, is borne out in many different fields. Even if, for example, the metallurgists of the Bronze Age knew how to reduce ores to metals, there is no indication that they had any fundamental understanding of why the methods employed led to the desired results. Such methods were presumably developed by trial and error, and were not based on any chemical theory of the nature and interconnection of the substances involved in the process.

Arithmetical methods for the computation of planetary positions were similarly developed to amazing perfection by the Babylonians during the last six centuries B.C. But these were not, as far as is known, accompanied by any connected ideas of the physical structure of the universe. Here Babylonian astronomy was strictly phenomenological although equally successful as the geometrico-physical astronomy of the Greeks. Nevertheless, the art of developing theories based upon physical models seems to have been unknown. Even more significant is the fact that most of the pre-Greek mathematics had a similar empirical foundation. The mathematical methods employed were imparted through the study of numerical examples, but without any attempt at general proofs. Without doubt, general theorems and procedures were known, but they were extracted from mathematical experience and not deduced from explicitly stated presuppositions.

It must not be forgotten that Egyptian mathematics, for example, contains at least some features pointing to the existence of more theoretical methods. In the so-called *Papyrus Moscow*, dating from the beginning of the second millenium B.C., a correct computation can be found for the volume of a truncated pyramid (Fig. 1.3). The

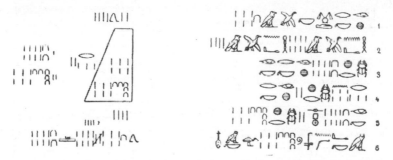

Fig. 1.3. Section of Papyrus Moscow showing a truncated pyramid with given dimensions. The computation of its volume is shown underneath (right) in hieratic script, and in a modern transcription using hieroglyphs (left).

computation corresponds closely to the well-known formula $V = \frac{1}{3}h(a^2 + ab + b^2)$, where h is the height and a and b are the sides of the end squares. However, it would be possible to find the volume of a pyramid in an empirical way. A wooden pyramid could be submerged in a jar filled with water and having the same height and base, when a third of the water in the jar would be seen to overflow. This purely practical method fails, however, in the case of a truncated pyramid, and one might therefore suspect that, after all, the Egyptians possessed more advanced algebraic methods, enabling them to deduce the volume of the truncated from that of the complete pyramid in much the same way as in later elementary stereometry. But nothing else in Egyptian mathematics indicates that this was the case, and until more evidence is discovered, the only reasonable conclusion is that if the Egyptians really possessed such theoretical methods they did not consider them sufficiently important to be included in their mathematical writings.

The opinion commonly held, therefore, is that neither the Egyptians nor the Babylonians had any theoretical proofs, and the level of their scientific culture has accordingly been described by some authors as pre-logical. This does not mean that it was illogical. In fact, most of the mathematical methods employed are quite correct, showing that the human mind was able, at that time, to think in a consequent way, and to distinguish between true and the false in the same way as later generations. The term pre-logical is therefore misleading, and it would be better to stress that the idea of natural causality seems to be missing and with it the possibility of a theoretical understanding of empirical methods. For this reason, it is more convenient to discard the word pre-logical, and to use the more explicit term pre-causal to describe this mode of thought.

The Mythological Explanation of Nature

It is difficult to believe that the various pre-Greek civilizations could have been completely devoid of any intellectual structure equivalent to the theoretical science of later times, for it is impossible to visualize a people who never pondered on the origin and development of the material world. How would the Egyptians have replied if we could have asked them for their opinion on how the universe came into being? The question can be answered from a wealth of Egyptian literature. The Egyptians would have told a tale tracing the changes in nature back to individual acts of the Gods. Thus the famous Memphis cosmogony tells that there was, in the beginning, only one freshwater ocean by the name of Nun, from which the Sun God Atum arose (Fig. 1.4). He saw that he was alone and begat with himself the God Schu (identified with the air) and the Goddess Tephnut. This pair begat Geb and Nut (Earth and Heaven), whose progeny in turn were the two pairs Isis and Osiris, and Set and Nephtys.

Similar speculations are known from the Mesopotamian area where, for instance, the high gods of the Babylonian Parnassos were identified with the planets. In fact, all ancient civilizations seem to have had

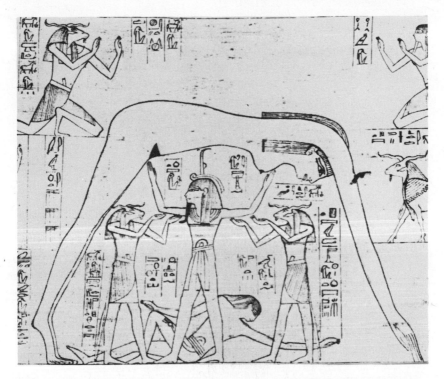

Fig. 1.4. *Illustration from the Egyptian Book of the Dead after a copy from about 970* B.C. *found at Thebes. The Goddess Nut of the Heavens rests on her hands and feet and thus makes up the whole universe. The reclining figure is the Earth God Qub (or Geb).*

their own mythologies, as have the 'primitive' peoples of today. Despite their legendary form, it would be wrong to discard such mythological explanations as arbitrary fantasy. Very often they are merely a thin veil, covering a structure of relations between natural phenomena, and expressed in a manner well adapted to both the natural and religious experience of the human mind. Often it is only the idea of causality which is lacking: individual natural phenomena are thought to be caused directly by the free will of the God presiding over that part of nature. From this stems the idea of influencing the course of nature through religious intercession or invocation of the God, respecting his essential liberty, or—in a quite opposite way— through magical practices forcing his will to conform with the desires of the magician.

9

Chapter Two

The New Concept of Nature

The Attack on Mythology

As the sixth century B.C. approaches, the general conception of nature begins to change, with far-reaching consequences in the way nature is described. First and foremost, this is apparent through the many attacks on the mythological explanation of nature, both from religious and from philosophical points of view, an evolution taking place outside the older cultures, and concerning peoples who had previously occupied a more obscure position.

A very strong movement to maintain a purely monotheistic religion is thus discernible in Israel, and carried by the prophets from about the seventh century B.C. onwards. The universe is considered to have been created by a transcendental God who has established universal laws governing the course of nature. This idea leaves no place for the earlier Gods of nature, nor for the conception of the will of supernatural Gods or demons as the cause of natural phenomena. The new monotheistic religion is therefore equivalent to the demythologization of knowledge about nature. But the victory of this movement was not complete. The Old Testament contains a highly mythological story of the Creation in *Genesis* 2, following the almost completely unmythological account in *Genesis* 1.

Nature and the Greeks in Homeric Times

Long after the peak period of the Near Eastern cultures, the Greeks in Hellas and Ionia were still fairly ignorant of anything comparable to Egyptian or Babylonian science. Thus in the eighth century B.C. the Homeric Epics reveal a very primitive view of natural phenomena. At the time when Babylonian astronomers collected planetary observations in a systematic way only a very rudimentary knowledge of a few celestial phenomena, mainly of importance to navigation, can be found in Greek astronomy. Thus Odysseus steers an eastern course by keeping the Great Bear to his left. Amongst other constellations mentioned by Homer were Orion and the Pleiades. He knew also of the fixed stars Arcturus and Sirius. Among the planets, he was acquainted with the morning and the evening star, but was unaware that both are identical with the planet Venus. There is a vague conception of the annual motion of the Sun; when it shines only faintly upon the Hellenes, Homer's explanation is that the Sun God has gone away to the land of the Ethiopians, from whence he will return. The turning point is an island called Syrie.

In the poem *Works and Days* by Hesiodos there are references to a similar small range of astronomical facts, used mainly for calendar purposes. The farmer must sow, for instance, when the Pleiades disappear from the night sky (at their achronitic setting) and reap when

Fig. 2.1. The forge of Hefaistos. Attic vase painting from the 6th century B.C. showing the furnace and tools of the blacksmith.

the Pleiades reappear (at their heliacal rising). Another poem, the *Theogony*, by the same author, contains an expression of the mythological conception of nature even stronger than that of Homer, with a long genealogy of Gods: Eros, Gaia (the Earth), Uranos (the Heavens), and so on, ending with the well-known figures of Olympos. But even Hesiodos was somewhat influenced by the new ideas and, like *Genesis*, his poem also includes a less mythological account.

Of the metals, Homer knew of gold, silver, copper, tin and iron, and the alloys electron and bronze. He describes a number of technical processes in some detail, such as forging (Fig. 2.1), casting, the hardening of steel, and sheet metal work, but says nothing about the extraction of metals from ores. In Hesiodos's *Work and Days*, ancient metallurgy forms the basis of an interesting division of history into golden, silver, copper, and iron ages, which succeed each other in this order, and which correspond fairly well to the order in which these metals actually came into use in pre-historic times.

The Ionian Philosophers

In time, the Greek knowledge of nature began to develop presum ably because of influences from more advanced civilizations. Towards 600 B.C., scientific and technological lore penetrated from Egypt and Babylonia into the Greek world, through commercial relations between Egypt and Phoenicia and the Ionian colonies which had already existed for centuries along the western coast of Minor Asia. But, as Plato said, *whatever the Greeks adopted from foreign parts they carried to a greater perfection*. This does contain a grain of truth as far as the Ionians—placed as they were on the border line of two different cultures—were concerned, for they soon developed a new type of philosophy and a new attitude to nature destined to mark all later science.

Unfortunately, the history of these Ionian philosophers is rather obscure, and none of their own writings has survived to the present. As far as is known, the first step was taken by Thales from Miletos (*c.* 625–545 B.C.) who is first mentioned by Herodotos a century and a half later. The most remarkable elements of the Thales tradition are that he predicted an eclipse of the Sun, that he considered water as the fundamental substance from which everything else had developed, and that he regarded the Earth as a flat disc floating on water. Furthermore, Thales is credited with the proof of some geometrical theorems, for instance, the proposition that 'a circle is halved by its diameter', or the theorem that 'in an isosceles triangle the angles at the base are equal'.

The contents of these propositions was, of course, known to mathematicians long before Thales, so that his achievement was not merely one of understanding. The important fact is that he is said to have proved them. If true, this is the first reference to the use of logical proofs in mathematics, and thus reveals a completely new status of philosophy, plus a distinction from the purely intuitive methods used in Egypt and Mesopotamia. It is very probable that the extent of Thales' knowledge of geometry was much more limited than that of his oriental contemporaries, but this does not preclude the fact that he may have given geometry a new character by initiating its transformation from an empirical set of rules to a logical structure. But apart from the particular legend of Thales, it is undisputed that the Greeks both started and concluded this significant process, not only in mathematics but also, though less widely, in a number of other disciplines.

The movement away from the mythological towards a more rational explanation of nature is very clearly seen both in the ideas of Thales himself and in those of his successors in the Miletian school, Anaximander (*c.* 610– *c.* 547 B.C.) and Anaximenes (*c.* 585– *c.* 525 B.C.). In the work of these philosophers the word *physis* emerges for the first time as a technical term of immense scientific importance. Later, it was usually translated as *nature*, so that physics meant natural science in general. For the early Greek philosophers, the word had a slightly different meaning, as it used to denote features common to material substances. The theory was that all existing things have been produced from a kind of prime matter, or fundamental principle (*physis*), as a result of the action of various natural forces. To Thales this prime matter was water, an opinion he may have acquired during his travels in Egypt, where the Nile deposits delta mud from which plant and animal life arise.

Primitive as it is, such an idea must be considered a physical hypothesis expressed in logical terms, and quite different from the mythological explanations of former times. How fast this new rational approach to nature grew, even outside philosophical circles, can be clearly seen in Herodotos' account of the inundations of the Nile. Thales had explained the phenomenon as an effect of the Etesian winds blowing in summer from the north and thus hindering the waters of the Nile from flowing out through the delta, so that they spread over the country. Herodotos shows that this explanation is false, because the inundation also occurs in years when the Etesian winds are not blowing. This argument shows how only one century after the Ionian philosophers Herodotos makes use, without hesitation, of their fundamental principle of causality—that similar effects must be traced back to similar causes.

Anaximander

The most important of the Ionian philosophers was Thales' pupil Anaximander, who wrote a now lost work *On nature*, said to be the first Greek treatise in prose. Traditionally, details of his life and opinions are perpetuated not only by Aristotle and Theophrastos, but also by a great number of secondary authors. The main points are as follows:

1. Anaximander from Miletos was a pupil of Thales. He was born in 610 and died shortly after 547 B.C. [Simplikios, sixth century A.D., after Theophrastos, fourth century B.C.].

2. He wrote the first Greek treatise in prose, the now lost Περι φυσωσ or *On physis* [Apollodoros, second century B.C.].

3. He assumed that the *physis* or *arche* was το απειρου, that is, the infinite or the unlimited [Simplikios, after Theophrastos].

4. He believed the *physis* to be eternal and without age, and to comprise the whole universe [Hippolytos, third century A.D.].

5. He assumed the existence of innumerable worlds, emerging and disappearing again [Simplikios].

6. He believed that there is an eternal motion during which the emergence of the worlds takes place [Hippolytos].

7. He thought that *something which is able to produce heat and cold* was separated from the eternal when the world began. From this arose a sphere of flames growing around the air surrounding the Earth just as bark grows round a tree. When this sphere was subdivided and confined to certain rings, the Sun, the Moon and the stars came into being [Pseudo-Plutarch, second century A.D.].

8. Anaximander regarded the stars as wheel-like condensations of air filled with fire, and provided at certain places with openings through which flames are discharged [Aëtios, c. 100 A.D.].

9. He described the order, the sizes and the distances of the planets [Simplikios, after Theophrastos].

10. He assumed that the Sun is the highest, the Moon below it, then the fixed stars, and the planets the nearest of the heavenly bodies [Aëtios].

11. He declared the Sun to be a ring 28 times as big as the Earth and shaped like a cart-wheel. It has a hollow rim, filled with fire which at a certain place is seen through an aperture as in a pair of bellows [Aëtios].

12. He said the Sun itself (that is, the aperture in the sun-wheel) is as big as the Earth [Aëtios].

13. At the time of the 58th Olympiade (548–544 B.C.) Anaximander discovered the obliquity of the ecliptic [Plinius, first century A.D.].

14. He assumed the Moon to be a wheel like the Sun, 19 times as big as the Earth, and in an oblique position like the Sun [Aëtios].

15. He thought that eclipses of the Sun and Moon occur because of a stoppage of the apertures [Aëtios].

16. He held that the Earth is floating in the middle of the universe [Theon of Smyrna, c. 130 A.D.].

17. He said that the Earth remains where it is because it has the same distance from everything. Its shape is convex and round like a stone pillar. We live on one of the end surfaces, and the other is at the opposite side [Hippolytos, after Aristotle, fourth century B.C.].

18. He considered the Earth to be shaped like a cylinder, the height of which is one third of the width [Pseudo-Plutarch].

19. Anaximander drew a map of the inhabited part of the world, which Hekataios later made the basis of his geography [Diogenes Laërtios, third century A.D.].

20. He maintained that thunder and lightning are caused by the wind. When this has been compressed inside a thick cloud and suddenly breaks through, the sound is heard when the cloud is bursting, and the fissure looks like a spark because of the contrast with the dark cloud [Aëtios].

21. He said that rain is due to moisture drawn up from the Earth by the Sun [Hippolytos].

22. Anaximander thought living beings arose from the moist

element when this was evaporated by the Sun. To begin with man was like other animals, that is a fish [Hippolytos].

23. He believed that in the beginning man was born from various other species. His reason for this is that whereas other animals find their own food before long, man alone needs a longer period of suckling. Had man, therefore, originally been as he is now, he would never have survived [Pseudo-Plutarch].

24. He declared that the first living beings were produced in the moist elements, and were covered with thorns. As time went on they emerged onto dry land, and changed their manner of living, because the thorns broke off soon afterwards [Aëtios].

25. Anaximander is said to have invented the gnomon and placed it as a sundial in Sparta [Diogenes Laërtios].

26. He constructed a 'sphere' (perhaps a kind of celestial globe) [Diogenes Laërtios].

Even without a detailed commentary, these elements of the Anaximander tradition give a strong impression of an original and courageous thinker making conscious efforts towards producing a rational explanation of fundamental physical principles (Points 3–7), the nature and motion of the heavenly bodies (8–15), the shape of the Earth, its place in the universe, and its geography (16–19), meteorological phenomena (20–21), and the evolution of living creatures (22–24), as well as inventing and constructing astronomical instruments (25–26). All these points reveal the general trend of early Greek thought, and every early philosopher tries his hand at them, although not always as clearly and lucidly as Anaximander.

The Pythagoreans

With the political unrest in Asia Minor caused by the Persian expansion, many Ionian philosophers went to other Greek colonies, or to Hellas itself. Thus the first philosopher in Athens was Anaxagoras (c. 500–428 B.C.) from Klazomenae, and from Kolophon Xenophanes (c. 530 B.C.) brought the new philosophy from Ionia to Syracuse in Sicily. Most famous among these emigrants was Pythagoras from Samos who—according to a very obscure tradition—left his island in the second part of the 6th century B.C. in order to travel through the near east before he settled down at Crotone in southern Italy. In the next 150 years, this region became the centre of the Pythagorean school of philosophy, and a representative of one of the most consistent attempts to develop a unified mathematical description of nature.

Like most Greek schools, the Pythagorean institution was organized as a religious community, of both men and women, based on a common set of ideas of which many were oriental in origin. Pythagorean anthropology was marked by the conception of man as a divine soul entombed in a material body, enslaved by corporeal baseness and impurity as well as by intellectual darkness and ignorance. The main preoccupation was thus with the liberation of the soul. On the material side, this was attempted through ascetic excercises such as vegetarianism, whereas the spiritual delivery had to be achieved through the intellectual education which took place inside the community. It is known that the members were divided in two classes—a large body of 'acousmatics' who received instruction in the more general ideas of the sect, and a smaller, rather esoteric set of 'mathematicians' who participated in the more advanced studies continued in great secrecy until, towards the end of the fifth century, Philolaos published part of the doctrine in three books which are now lost.

Fortunately the main outlines of this advanced study, or the Pythagorean 'mathematics', are not lost. The word comes from the participle $\mu\acute{\alpha}\vartheta\eta\mu\alpha$ meaning 'that which has been learned', through study, that is, and in contrast to more popular subjects like rhetorics, poetry, and music needing only elementary instruction. Plato still used the word occasionally in this sense. Aristotle said of the Pythagoreans that

> they devoted themselves to mathematics and were the first to further this study. And since they were educated in it they thought that the principles of mathematics were the origins of all things . . .

It is clear that 'mathematics' has now acquired more of its usual meaning. How that came about is easy to understand from the contents of this esoteric doctrine. The philosopher Archytas (first half of the fourth century B.C.), himself a late member of the School, said that

> they have given us a clear knowledge of the course of the stars and their risings and settings, of geometry, arithmetic, and spherics, and not least of music, for these studies prove to be sisters.

This is the set of disciplines which became known later as 'the Pythagorean *quadrivium*'—arithmetic, geometry, music theory, and astronomy as distinct branches of pure and applied mathematics in the modern sense. Their interconnection appears from the following scheme, which reveals an obvious interest in the mathematical description of nature.

	Theory of discrete quantities	*Theory of continuous quantities*
Pure discipline	Arithmetic	Geometry
Applied to nature	Harmonics	Astronomy

As far as astronomy was concerned the process of mathematization was well under way before the Pythagoreans; there is a tradition although it may not be very reliable—that it was Pythagoras himself who discovered the mathematical ratios of the musical intervals. Simple experiments with a monochord will show that the string will produce the fifth or the octave when its length is reduced to ⅔ or ½. The discovery that simple numbers are sufficient to describe acoustic phenomena may well have led the Pythagoreans on to a more ambitious programme of research. Aristotle continues (from above)

> . . . since they realized that the properties and relations of the musical scale could be expressed in numbers, and since all other things seemed in their nature to be modelled after numbers, and the heavens to be a musical scale and number.

In other words, the natural sciences, like harmonics and astronomy, are comprehensible in terms of arithmetic and geometry respectively; but in the end the Pythagoreans tried to reduce everything to a theory of numbers of the kind still extant in Book 7 of Euclid's *Elements*. In it is expounded their classification of numbers into even and uneven, prime and compound, quadratic, rectangular, triangular, cubic, etc., and their curious doctrine of 'friendly' and 'perfect' numbers.

In Antiquity, number theory became the common interpretation of the original Pythagorean doctrine. Thus a late author, Diogenes Laërtios (third century A.D.), states that, from unity and indeterminate duality come numbers. Numbers give rise to points, points to lines from which plane figures are produced. From the latter spatial bodies emerge, and from these material bodies the elements which are four in number—fire, water, earth, and air. These elements interchange their positions and are completely transformed into each other. From them arises a world which has a soul, is rational and spherical, with its centre in the Earth, which is also spherical and inhabited all the way round.

Detailed interpretation of such texts can be difficult, but the main features of Pythagorean philosophy of science are clearly discernible. There is a mathematical structure behind the visible universe; the description of nature must therefore be expressed in terms of mathematics. From now on, this connection between physics and mathematics takes a progressively stronger hold upon the minds of natural philosophers, and must be thought of as the most important contribution to the advancement of science made by the Pythagoreans. It retained its fascination, and its inspiration to scientists persisted even after the specific Pythagorean doctrines had been abandoned as naïve or as obscure manifestations of an arbitrary number mysticism.

Greek Mathematics

During the fifth century B.C. new ideas such as those of the Pythagoreans were held by mathematicians and philosophers throughout the Greek world. In mathematics, a very important discovery was made about 450 B.C., presumably in Pythagorean circles, namely that there exist incommensurable lines, such as the side and the diagonal of the same square, which means that these quantities cannot be expressed by integral numbers, found from measuring them by the same unit, however small that unit is. No integer or fraction represents the ratio of the two lines, a discovery which meant that the fundamental Pythagorean concept of number as the nature of things had to be discarded, involving, in turn, a strengthening of geometry at the cost of number theory and algebra. In geometry, such incommensurable quantities were perfectly acceptable, and Greek mathematicians concluded that geometrical entities are of a more general nature than those which can be represented by numbers or ratios of numbers.

Against this background mathematicians gradually turned their attention more and more to geometry, and the result of the development was a very considerable advance within the field.

The logical methods introduced by the Ionian philosophers were first carried on to perfection, and the logical proof became the basis of geometry. At the same time geometry acquired an increasingly abstract character. It had been an empirical or intuitive doctrine concerning concrete material bodies studied from a particular point of view. Now, it became an abstract discipline concerned with the properties of ideal geometrical entities (the point, the straight line, the circle, and so on) which have no real material existence outside the imagination of mathematicians. To a great extent, geometry acquired a strictly formal character because of the claim, now generally accepted and presumably formulated by Oinopides of Chios (about 450 B.C.), that all geometrical proofs must be based on the properties of the straight line and the circle alone, and that all geometrical constructions shall be made by means of two mathematical instruments only, the ruler and the pair of compasses. This quite arbitrary condition gave rise to a number of so-called insoluble problems such as the trisection of the angle, the duplication of the cube, and the quadrature of the circle which have no exact solutions in the sole terms of the straight line and the circle.

The intensification of work on geometrical questions led to an enormous increase in geometrical knowledge as such, and the great profusion of results made it difficult to survey the whole field. This is the reason why mathematicians like Archytas of Tarentum (c. 375 B.C.) and others during the fourth century tried to comprise the whole of geometry into a single connected structure of propositions. These early attempts have been lost, but Euclid's *Elements* or *Stoicheia* (the name means 'root') of about 300 B.C. is an existing work which amasses the results of the preceding centuries in one enormous handbook.

This famous work should not be thought of as a textbook or manual. It ignores any appeal to mathematical imagination and is a bad guide to research into new fields. Essentially, it is an encyclopedia, arranging all established mathematical knowledge in a sequence of clearly stated propositions, theorems, and methods of construction, and giving the logical proofs of all the statements. As such, it became inestimably valuable to the mathematician who worked with analytical methods and was in constant need of verifying his presuppositions. Moreover, the *Elements* has, to a great extent, the character of a logical deductive system in which the main body of the statements (propositions and theorems) is deduced from a small number of fundamental axioms and its postulates accepted without proof. Thus Euclid's work made geometry the most developed branch of Greek science, inside which new, logical thought demonstrated its ability to handle a large domain of knowledge in a transparent and unified way.

Results of the Scientific Awakening

It is impossible to overestimate the importance of scientific progress during these early centuries. Whole fields of knowledge were conquered by logic and the principle of causality. Mathematics became completely dominated by logical proof. The attitude to nature began its divorce from myth, and the idea of causal relations between natural phenomena penetrated to such an extent that Aristotle, without question, defined science in general as the study of causes. Finally, the absolute necessity of mathematics as a tool for the description of nature, had been realized, at least inside the Pythagorean tradition.

This scientific revolution gave rise, however, to a number of serious problems never previously considered. Geometry had become an abstract body of knowledge of so-called geometrical entities, but how were these related to the material objects which alone are capable of perception? Similarly, the Greeks had to discuss whether it were possible to acquire knowledge solely by means of the rational activity of the human mind and outside mathematics. These and other difficulties stemming from the development of science occupy the great philosophers of the fourth century B.C., stimulating them to a systematic investigation of this whole complex of ideas. These philosophical efforts influenced in turn the development of science itself, both in Antiquity and in the Middle Ages.

Chapter Three

Plato and Greek Mathematics

Plato's Academy

ELEMENTARY schools were known in the various Greek city-states at an early date, but in the fourth century B.C. Athens became the seat of a number of institutions for more advanced study and education. Here, philosophers and scientists tried to solve the many problems created by the new scientific methods and the change in attitude towards both nature and man himself. The first of these schools was the so-called Academy founded and directed by the philosopher Plato (c. 429–347 B.C.). Initially the school was, like the Pythagorean schools in Italy, a religious community for the honouring of the muses. Its main achievement, however, was the work done by Plato and his collaborators whose aim was to elucidate all philosophical problems and to develop a consistent doctrine of the nature of things based on a coherent theory of knowledge.

Plato's direct teaching was embodied in a number of works which, unfortunately, have been lost. However, his more literary writings are extant in the form of a long series of dialogues in which Plato develops his ideas in the form of conversations between his own teacher Socrates (c. 470–399 B.C.) and a number of historical or fictitious persons. These works present a detailed, although somewhat unsystematic investigation, with a careful discussion of ethical problems as their climax.

It seems that both Socrates and Plato received an overwhelming impression of the development of Greek mathematics, and the certainty with which mathematicians produced their results and established undeniable mathematical proofs. Mathematics had therefore to live up to a unique status as the science before all others. This was illustrated also by the inscription, 'Let no one unversed in geometry come under my roof', which according to Tzetzes, was placed above the entrance of the Academy. The same preoccupation with mathematics is illustrated in the merging, in about 368 B.C. of the Academy with a mathematical school founded by the mathematician Eudoxos from Cnidos (c. 408–355 B.C.), who was for a number of years, Plato's collaborator. In fact, many of the investigations in the *Dialogues* seem to have been made for the purpose of solving epistemological problems raised because of the changes in mathematical thought.

Epistemology

In the dialogue *Menon* Plato describes Socrates in the market discussing with Menon the definition of goodness. During their conversation, Socrates beckons Menon's slave and makes this completely uneducated man carry a mathematical argument to its correct conclusion. The slave is quite unversed in mathematics, but Socrates makes him answer correctly and realize the logic of the proof mainly by asking him well chosen questions. The result is that the slave has now acquired a knowledge which he did not previously possess. On the other hand, Socrates has told him nothing essential, apart from a couple of definitions. From where does this new knowledge stem?

Socrates (or Plato) tries to answer this question by means of a theory assuming that the slave has been brought to remember the geometrical truth as a result of his being questioned by Socrates. This implies that sometime in the past he did possess this knowledge, but has forgotten it. Since the slave never knew the proposition during his lifetime it must be assumed that he came to know it during a previous existence before his material life on earth. The conclusion is that knowledge is acquired through remembrance of truths which were manifest to the soul before it was united with the body.

Certain Pythagorean ideas of the body as the tomb of the soul are, without doubt, at the root of this remarkable theory of reminiscence, a view supported by another theory, and which can be illustrated, for example, by the famous 'Parable of the Cave' in the dialogue of the *Republic*. The fundamental idea is that changing and perishable things in the material world of the senses are imperfect, material copies of pure, immaterial, unchanging, and eternal models called 'ideas' or 'archetypes'. In its previous existence the soul had a clear vision of these ideas, and they are not so long forgotten that they may not be brought to mind when sensible images of their imperfect material imitations awaken the memory, or when the soul is subject to adequate teaching or education.

This theory of knowledge is now extended beyond the realm of mathematics. The purely intellectual means of acquiring knowledge of mathematical truth is only one particular instance of the general procedure of the human mind. Also, the knowledge of nature is a re-awakening of memories of something in the ideal world. Sensory experience therefore plays a minor role for Platonic philosophers. What our senses tell us of material phenomena is not the source of knowledge, but at most the occasion through which the memories of ideas appear and are viewed by the soul, a point very clearly expressed in the *Republic*:

> Although they [the mathematicians] make use of visible figures and reason about them they do not think of these figures, but of what they represent. Thus it is the square as such, and the diagonal as such, which are the object of their discussion, but not the lines they are drawing. When they make models or draw objects—which perhaps in turn are reproduced by shadows or reflected in water—they use those things in a similar way as pictures, striving to contemplate those absolute objects which can be seen in no other way than by thought alone.

Metaphysics

The Platonic doctrine of ideas is not only a theory of knowledge, but also a metaphysics which, among other things, must account for the abstract character of mathematics. To Plato, the ideal world is fundamental and is, in fact, the only world which really exists in the accepted sense, since the only things that exist are those without change. Material objects exist only because of a certain relation to their ideal type. This relation is sometimes described by the word 'participation'. Plato also says that material things are imitations of their ideas or that they are mixtures of the definite and the indefinite, that is, of the completely defined type and the shapeless matter. Thus the relation between idea and object is expressed in more or less obscure

25

metaphors. But certainly their relation implies no identity. Objects are different from their ideas, and ideas exist separately in a particular, immaterial world. Furthermore, the relationship maintains the perfection of the ideal types compared with the imperfection of the world of the senses. A wheel is round because it participates in the idea of a wheel. This idea is the mathematical form of the round, which again is the same as the abstract geometrical concept of the circle.

Mathematical forms do not, however, have the same ontological status as the pure archetypes. They exist in a region between the ideal and the material world. Like the ideal they are unchanging and eternal, but like the material they are multiple, in contrast to ideas which are unique. Yet the mathematical object known as a circle exists only in an immaterial world where the intellect is able to see it in its essential perfection. This is what mathematicians do in geometry. No material circle has a similar perfection and is, by its material nature, an inferior object.

The same metaphysical structure is also presupposed in the other sciences. The ideal world is primary and perfect, the material secondary and imperfect. Since knowledge is a vision or recollection of the perfect world of ideas, it follows that, strictly speaking, we have no knowledge of the material world as such. First, this is because knowledge of this kind would be impossible, since the soul in its previous existence never had any vision of material things and has therefore no memory of them now. Second, the soul would never be satisfied with the ever-changing testimony of unstable and perishable things revealed by the senses. If it has tried, but once, to look into the world of ideas, the soul will never again content itself with anything less perfect. Therefore, any science must try to transcend the material world and to regard immaterial ideas as its proper subject of study. On the one hand, the abstract character of mathematics is only a special case complying with the general characteristics of all real knowledge. On the other, the very nature of such mathematics makes it the prototype on which any other science ought to model itself.

Plato on Astronomy

In Plato's time only one science, mathematics, or, more correctly, geometry, could be said to have reached this ideal stage. Plato therefore exhorts other sciences, but particularly astronomy, to make an effort to acquire a similar status. In the *Republic* Socrates says that,

> we ought to learn astronomy in a different way from the present one if we shall profit by it to the purpose of which we are talking.

This is explained in more detail as follows:

> Thus we must pursue astronomy in the same way as geometry, dealing with its fundamental questions. But what is seen in the Heavens must be ignored if we truly want to have our share in astronomy [. . .] Although celestial phenomena must be regarded as the most beautiful and perfect of that which exists in the visible world (since they are formed of something visible), we must, nevertheless, consider them as far inferior to the true, that is to the motions [. . .] really existing behind them. This can be seen by reason and thought, but not perceived with the eyes.

These texts present many problems. Perhaps the most immediate interpretation is that Plato wants to change astronomy into a purely speculative science unconnected to observable phenomena which must be disregarded by the astronomer. In this case, the proper subject of astronomy would be the ideal background of the phenomena.

Another possible interpretation is that astronomers ought to do what mathematicians are already doing: instead of studying individual material phenomena in a purely empirical way, their task should be to construct an abstract doctrine reflecting the structure of the ideal astronomical world accessible to thought and reason alone. In this way, the observed phenomena would have found their theoretical explanation.

A third, although less likely, interpretation has been proposed, namely that here Plato advocates the creation of a general system of mechanics able to account for the motion of bodies in general. Through such a system, the motion of the celestial bodies would be explained from general principles, that is general mechanical laws.

However, the Greek tradition has more to say on Plato and astronomy. Thus the commentator Simplikios (sixth century A.D.) wrote that,

> Just as Eudemos narrates in the second book of his *History of Astronomy*, like Sosigenes, who follows Eudemos here, Eudoxos from Cnidos was the first among the Greeks to take up this kind of hypotheses since

Plato, as Sosigenes says, had proposed as a problem to all who consider these things seriously to investigate the uniform and regular motions by means of which planetary phenomena may be saved.

This is clearly another conception of astronomy. To use an expression much in vogue in later years, the problem is to save celestial phenomena from their apparent irregularity by showing how they might result from a system of regular circular motions. In modern phraseology, it could be said that the task of the astronomer is to formulate a mathematical theory which, from certain presuppositions (in this case a number of uniform circular motions) makes possible a deduction of the more irregular movements which the planets are seen to perform. Astronomy ought, in other words, to be reconstructed in the form of a deductive system, just as geometry had been.

Perhaps this difficult problem of interpretation admits of no unique solution. Neither can there be any certainty that Plato's thinking was consistent and completely uncontradictory. The best course is to follow what the first Greek astronomers actually did (see Chapter Six) in order to understand how Plato's astronomical programme was conceived by contemporary professional astronomers, who were for the most part Plato's own pupils, and as such acquainted with the scientific ideas of their master.

The Philosophy of Nature in Timaios

Plato's own conception of the universe appears in one of his last dialogues, *Timaios*. Formally, *Timaios* is not a scientific work. Its purpose is to outline the regular order of the world as a pattern of the order which ought to rule human society. The book gives a sketch of Plato's scientific and astronomical opinions by whose means he attempts to synthesize a number of earlier, essentially Pythagorean, ideas (Timaios is the name of an earlier Pythagorean philosopher). The fundamental idea is that the world order has been worked out by a sort of creative God (Demiurge) on the basis of an organizing plan or thought which presupposes that human reason is able to understand the universe as a regular unity. The Ionian philosophers had already made this attempt, but on several occasions Plato dissociates himself from these earliest philosophers because, according to him (Socrates) they have lost themselves in the problems of the outer world, thus neglecting the questions most essential to man.

In many ways, *Timaios* reflects an animistic and anthropomorphic philosophy foreign to the Ionian philosophers' fundamental attitude, although they have a few similar ideas. It includes, for instance, the peculiar conception of a similarity between macrocosmos and microcosmos, between the universe as a whole, and the individual. Not only is the world, like man, composed of the four elements. It is animated by a world soul, who organizes it into a whole just as the human soul is the uniting principle of the organism. Also the individual celestial bodies are animated and must, in other words, have appeared to Plato as living beings.

Timaios gives furthermore, an exposition of the planetary system which will be dealt with later (p. 65), a survey of the doctrine of the four elements and their geometrical forms (p. 144), a series of physiological speculations and attempts to explain the cause of a number of diseases, as well as of sensory impressions like colours and sounds. In short it resembles a draft of the all-embracing philosophy of nature implied in Plato's other works.

The fact that *Timaios* was Plato's only work available before the twelfth century in a Latin translation, meant however that Plato's cosmology had an immense influence on the conception of the universe in the early Middle Ages, whereas the great triumphs of the Greek mathematical astronomy fell into oblivion. In particular, the doctrine of the similarity between macrocosmos and microcosmos acquired a momentous importance, becoming fertile ground to astrology when, with its doctrine of the influence of the celestial bodies on man, this study spread from the Orient all over the Hellenistic world, and, in the Middle Ages, over Europe.

The Aftermath of Platonism

The following observations can be made in an attempt to estimate Plato's role in the history of science. Being under the influence of the explosive development of geometry from an empirical to an abstract, deductive science, Plato urged other sciences to proceed in the same way. In doing so, he furthered without doubt, the advance of scientific theories. But as the axioms of geometry appeared self-evident,

29

unconnected with the empirical geometry developed through observation, Plato considered geometry, as well as other sciences, to be related to an ideal world which could only be reached by way of thought. In doing so, he minimized the role of sensory experience and, as a result did not maintain that all natural sciences must draw their data from the world of experience. The result was that Plato's disciples often made too little of experience and were not very careful about their assumptions. Their work on scientific problems was often characterized, therefore, by elegant but far flown speculations on the consequences of arbitrary and unrealistic assumptions.

With the neo-Platonic movement arising in the last period of Antiquity, Platonism experienced a renaissance while at the same time Platonic philosophy established an increasingly intimate relation with the religious currents of the time. Following the great Christian teacher Augustine (A.D. 354–430) Platonism became one of the most important conditions of thought and science in Latin Europe up to the 13th century, when interest once more began to centre around Aristotelian views (p. 191).

Chapter Four

Aristotle and the Knowledge of Nature

Aristotle's Research Institute

IN the middle of the fourth century B.C., at the Academy in Athens the second great name in Greek science in this period arose, namely that of Aristotle of Stagira (384–322 B.C.), who for twenty years was Plato's most outstanding pupil, and thereafter one of his most thorough critics. In Athens, about the year 334 B.C., Aristotle founded his own school, quite a new one of its kind, called the Lyceum or the Peripatetic, and containing in particular a library and a collection of objects for teaching purposes. His teaching was embodied in a long series of systematic works on natural science, logic, epistemology, metaphysics, social problems, and so on. At the Lyceum he guided many of his pupils into further research in botany, the history of philosophy, and other disciplines. It is believed that he also wrote some more popular works, of which all but one are lost.

Even if Aristotle was, in many ways, influenced by Plato, his approach to the problems was obviously different from that of his teacher. Plato considered mathematics the science before all others. Aristotle, too, valued it greatly and made an extensive use of mathematical arguments and examples in his works on the philosophy of nature, which also contain theorems and proofs which are not found in Euclid, as well as valuable reflections on the basis of mathematics. But he considered mathematics to be a sort of auxiliary science, or a special language, by whose aid one may readily express oneself in other sciences.

Aristotle's work shows a strong interest in naturalhisr toy. He was a keen observer of living things, and is remembered as the founder of zoology through his great work *The natural history of animals*, in which he classifies and describes more than five hundred different species. This is yet another instance of the Greek urge to find order and regularity in the phenomena of nature, and is in itself an outstanding scientific deed. This familiarity with at least some aspects of the phenomena of the material world is probably the most important condition of Aristotle's apprehension of scientific knowledge as such. Furthermore, it motivates the vivid interest which he takes especially in the classification problems of formal logic, a discipline of which he was also the founder.

Form and Matter

Another condition of Aristotle's epistemology is Plato's doctrine of ideas which is, however, thoroughly remodelled. Aristotle considered all material objects to be constituted of two essentially different elements. The first, called matter, is in itself something shapeless and indefinite. The second was later given the Latin name *'forma'*, and is immaterial, but always fixed, well-defined and unchangeable. Things come into being when some form becomes united with a sort of matter, and shapes matter according to its own nature. The object then acquires all the attributes of the form: a particular geometrical figure, a certain gravity or lightness, the faculty of reacting in a definite and regular way to other objects, and so on. This doctrine of considering things to be constituted of form and matter is called hylemorphism, from the Greek ὕλη = substance (in fact, wood) and μορφή = form.

Being eternal, unchangeable, and immaterial, these 'forms' largely recall the Platonic 'ideas'. On one essential point, however, they differ from these ideas: they have no separate existence in a world of their own. Aristotle rejects the doctrine of participation as incomprehensible; things are not poor imitations of ideas—they have an existence of their own, and the 'forms' exist only inside the concrete things, not outside them.

Cognition

Aristotle's rejection of 'participation' has several important consequences for the knowledge of the physical world. As Aristotle does

32

not admit any separate world of ideas, obviously he cannot conceive of cognition as a recollection (nor of any previous existence of the soul). To Plato, the cognition of a wheel was the same as the recollection of its idea, caused possibly by a fundamentally unimportant visual impression of a wheel. To Aristotle, the cognition occurs directly at the sight of the actual wheel: the eye receives a sensory impression which proceeds to reason, there to be transformed into an intellectual picture of the form of the wheel. It is as if the form has separated from its material connection and has insinuated itself into the soul which is also the form' of the body.

For Aristotle, therefore, sensory impressions are not occasions incidental to cognition but its necessary condition. The direct observation of material phenomena thus becomes the only way to acquire knowledge of the physical world. In other words, experience, not recollection, is the core of cognition. It is easy to see that to a scientist oriented towards zoology, as Aristotle was, a view of this kind is more natural than the theory of Plato. Calling attention to the principle of experience as the foundation of all knowledge must be considered as Aristotle's main philosophical achievement. But this does not imply any refusal of theoretical knowledge. On the contrary, although experience is the foundation of any science, the task of the scientist is not exhausted by the collection of individual observations, which must be processed by reason in a particular way, and to a specific purpose, two activities characteristic of scientific knowledge.

The Process of Abstraction

For Aristotle, scientific knowledge is characterized by various degrees of abstraction. Thus he always abstracts from individual features of a complex phenomenon: a scientific investigation of a rose begins with the description of a certain number of individual specimens; but the botanist is concerned only with the properties roses have in common, and disregards their individual peculiarities. The scientist tries to find the 'form' of an object which is identical in all individual roses. Their individuality springs from their matter alone and is of no scientific interest.

33

Classification of the Sciences

The process of abstraction may be more or less far-reaching. Even if an abstraction is performed from the individual features of a number of specimens of the same kind, they may be considered as material entities, that is, natural phenomena can be examined as they appear in the material world. This is known as the first degree of abstraction, comprising all natural science, and designated by Aristotle simply as *physics*. A more precise definition of physics is the science of motion. But the word motion has a double meaning. It denotes local motion in which the body changes its position in space, but it is also used to describe qualitative changes from one state to another.

At the second level of abstraction the things of the material world are regarded as separated from their material properties. One can then examine, for instance, their geometrical size or figure, numerical relations, and so on. This degree of abstraction thus comprises *mathematics* in the most general sense, and mathematics is conceived as an empirical science of the real world, but abstract and general because of the universal character of mathematical forms. The material and mathematical properties do not exhaust the fullness of being. When matter, shape, and number have been removed, something general remains. At this third degree of abstraction one can ask only very fundamental questions concerning, for instance existence or non-existence, or causal connections. The kind of knowledge acquired at this level is *metaphysics*.

The classification of scientific knowledge into physics (that is natural science), mathematics, and metaphysics is one of the main tenets of all later Aristotelian philosophy. It may be developed by more refined subdivisions in which, for example, the individual branches of physics are distinguished from each other. Here, a special problem arises concerning sciences such as astronomy, which describe the material world using mathematical means. Properly speaking, a science of this kind belongs to the first degree of abstraction, but its mathematical language links it with the second.

Many later authors, such as Ptolemy, changed the Aristotelian classification, and considered astronomy and mathematical physics (optics, mechanics, harmonics) as branches of mathematics. Mediæval philosophers often adopted a middle position, making these sciences intermediate between the first and second degrees of abstraction (see p. 34).

Science as Causal Explanation

For Aristotle, the purpose of scientific method is to demonstrate a fundamental order and regularity behind the complex of natural phenomena perceived by the human senses. In this case, 'order' means causal connection, and the main purpose of science is to find causes for the various phenomena so that we can see how they condition and produce each other. Thus a scientific explanation is, to Aristotle, the same as a causal explanation. Aristotle distinguishes between four types of causes, all of which are necessary to produce a certain pheno-menon, so that science must explore them all. The *material* cause is the material necessary to produce the phenomenon. The *formal* cause is the form introduced into this matter so that an independent substance emerges. The *efficient* cause is the force necessary to unite matter and form. And lastly, the *final* cause is the purpose of the whole process.

It is easy to apply these causal categories to many human activities. When a house is constructed, the material cause is bricks, and timber. The formal cause is the architect's design. The efficient cause is the work of the artisans, and the final cause the builder's wish for a house. As a matter of course, Aristotle supposed the same scheme to be valid in the knowledge of the physical world. Later, this hypothesis was to prove one of the weaker parts of his natural philosophy and the cause of many difficulties. In particular, the introduction of final causes into physical science became a stumbling block to the advancement of physics. Galileo, for instance was told that his newly discovered moons of Jupiter were nonexistent because they were invisible to the naked eye, and thus could serve no purpose.

Aristotelianism

It would be unfair both to Aristotle and to the history of science to let his decisive role be obscured by hind-sight. Certainly his particular physical opinions had to be revised or discarded at a much later date.

Nevertheless, it was Aristotle who laid the groundwork for natural science through his stubborn maintenance of the principle of experience and his rational classification of, and distinction between various types of science. The immediate consequence of his clarifying activity was a theoretical separation between the various parts of natural philosophy, with a subsequent freedom of development for the individual disciplines. This is the reason why soon after his death it is possible to observe outspoken tendencies for specialization, keeping the problems of the individual sciences apart, even if single scientists or philosophers continued to work in more than one field of experience.

Hellenistic Science

The geographical extension of Greek science which took place in the decades after Aristotle's death was most important. The fourth century B.C. had been dominated by the great schools in Athens. Now, a new centre of learning came to light in the new Egyptian city Alexandria, where, shortly after 300 B.C., king Ptolemy Soter created a new scientific institution with schools, a library and a museum. This was done at the insistance of Demetrios of Phaleron, who was a disciple of Aristotle, and took the Lyceum in Athens as his immediate pattern. In the course of time a long line of the very best scholars worked there, supported by the government and provided with facilities in the form of books, instruments, and so on to an extent never surpassed by any other Greek school. Even though the ancient schools of Athens continued their activity, both Greek and Oriental scholars from all parts of the Mediterranean world preferred to go to Alexandria to study with teachers like Euclid, Eratosthenes, and later Heron and Ptolemy.

Through this development, Greek civilization and science passed into its true Hellenistic phase. Historians have often stressed the more negative aspects of this period, and it seems true that Hellenistic thought gradually became formalistic and diluted in a way which compares unfavourably with the great creative efforts of the fifth and fourth centuries B.C. Many Alexandrian scholars were more systematic than imaginative: they wrote compilations and commentaries, occupying themselves more with the keeping and systematizing of results already achieved than with new research. Nevertheless, Hellenistic science was an impressive and far-reaching movement. Archimedes, the greatest genius of the exact sciences in Antiquity, had his scientific background in Alexandria, and here one of the most influential and closely reasoned works in astronomy, Ptolemy's *Almagest*, was written.

In the following chapters, some characteristic aspects of the Greek achievement in the individual sciences will be considered. Sometimes it will be necessary to refer back to pre-Alexandrian eras, but it will be repeatedly obvious how initial developments found their final fulfillments in scientists connected in some way with the school of Alexandria.

Chapter Five

Practical Astronomy

THE attempt to give a rational description of the universe was one of the main interests of the Ionian philosophers whose cosmological speculations form the beginning of the Greek astronomy. In this chapter some of the more important Greek contributions to the knowledge of celestial phenomena as such will be considered, although a definite exposition is a difficult matter: the sources are very fragmentary; it is often hard to distinguish between what the Greeks achieved themselves and what they took over from Egypt and Babylonia, and finally it is difficult to appreciate what kind of observations were used as the basis for their more speculative explanations of the phenomena. The previous section on Anaximander (Chapter 2 p. 15) was a good illustration of such interpretative problems. However, one thing is clear, it was no simple matter to come to terms with even the most fundamental celestial phenomena.

The Identity of the Heavenly Bodies

Homer was, as has been mentioned, unaware that the morning and evening star are identical with the planet Venus. Parmenides of Elea (c. 450 B.C.) seems to have been the first to realize the truth. Nor was the identity of the Sun clearly established. To Heraclitos of Ephesos (c. 500 B.C.) the Sun was new every morning, and a similar opinion was ventured by Xenophanes of Kolophon (c. 570–450 B.C.), for whom the Sun was a collection of fiery particles which assemble in the morning to form a radiant cloud. This cloud travels across the Earth, regarded by

Xenophanes as flat and infinitely extended, only to dissolve again at night. Most peculiar is the theory of Empedocles of Agrigentum (c. 493–433 B.C.), for whom the Sun did not exist at all as a material entity. He explained day and night by the assumption that a bright and a dark hemisphere revolve around the Earth, the light from the bright hemisphere being reflected from the Earth back onto the heavens as a strongly illuminated spot which we call the Sun.

The Nature of the Stars

The conception of the heavenly bodies as permanent material substances in space does not seem to have been commonly accepted until about 400 B.C. The theory probably gained weight from the great meteor which fell near Aigos Potamoi in 467 B.C. The records of this event led the first philosopher in Athens, Anaxagoras of Klazomenae (c. 500 B.C.) to maintain that the Sun is a red-hot stone bigger than Peleponnes. Gradually, the fixed stars were conceived of in the same way, and finally Democritos of Abdera (c. 456–370 B.C.) declared the Milky Way to be a collection of faint stars, and not a cloud or reflection as previously supposed.

As for the nature of the heavenly bodies, Leucippos of Miletos (c. 440 B.C.), one of the originators of the atomic theory, assumed that the Sun and the stars glow only because of their fast motion through the air. The most common assumption, however, was that the heavenly bodies were fiery and thus luminous. Anaximander and Anaximenes even supposed the moon to be self-luminous, although Thales is said to have declared before this that it receives its light from the Sun, an idea clearly expressed for the first time by Parmenides.

The Motion of the Planets

It must have been known that all heavenly bodies participate in the diurnal rotation of the heavens from east to west as long as man had noticed the course of the stars. That the wandering stars—the Sun, Moon and planets—also have their individual course in the opposite direction was known to Babylonian astronomers at a very early stage, whereas the Greeks seem to have realized this fact only much later.

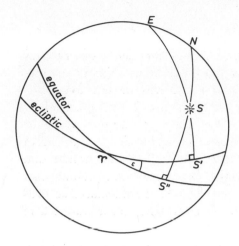

Fig. 5.1. The heavenly sphere with the celestial equator and its pole N, together with the ecliptic with the pole E. ♈ is the vernal equinoctial point. The position of a star S is given by the ecliptic coordinates

$$longitude \; \lambda = ♈S'$$
$$latitude \; \beta = S'S$$

or by the equatorial coordinates

$$right \; ascension = ♈S''$$
$$declination = S''S$$

However, when Parmenides says of the Moon that 'she always looks behind her towards the rays of the Sun', this implies some understanding of the phases of the Moon, and accordingly of its motion in relation to the Sun. Anaxagoras must have realized the existence of these motions, since he was able to give the proper explanation of eclipses:

> The Moon is darkened when the Earth, and sometimes heavenly bodies below the Moon, come between it and the Sun. The Sun (is darkened) when at New Moon the Moon comes between the Sun and the Earth.

Empedocles put forward the same explanation, so that the direct motion towards the east of the Sun and Moon must have been well-known by his time. On the other hand, he seems to have held that the planets are able to move about freely and irregularly. The Pythagorean, Alcmaion of Croton (*c.* 500 B.C.), is often mentioned as the first Greek philosopher to realize that the planets, too, move according to laws, and perform direct motions like the Sun and Moon. But it is not known who was the first to discover that this direct motion is sometimes followed by a transitory retrograde (indirect or western) motion, with stationary points in between. This was known to Democritos, but probably the Greeks first heard of the phenomenon through contacts with Babylonian astronomy. All the main features of planetary movements were known at Plato's time, if not before, a fact obvious from the *Timaios*.

The First Instruments

All the celestial phenomena mentioned above can either be observed directly or deduced from simple qualitative observations without any special aids; the regular art of observation started with the introduction of the first astronomical instruments although this cannot be accurately pinpointed. In any event, Anaximander (see p. 17) did not invent the so-called *gnomon*. This fallacy was corrected by Herodotos, who tells of how the Babylonians taught the Greeks to use both the gnomon and the *polos*. Herodotos is probably right, just as one must presume that the water clock mentioned by Empedocles had been introduced according to the eastern pattern. As early as about 1500 B.C. it was used in Egypt as well as in Mesopotamia.

These first instruments were very simple. The gnomon was merely a vertical pole erected on a horizontal plane, while the polos consisted of a hemispherical bowl placed with its rim parallel to the horizon. Originally, a small ball was suspended in the centre of the bowl, but later this was replaced by a gnomon which rose from the bottom of the bowl to its centre. The position of the Sun was registered in both instruments by the shadow cast by the gnomon or the ball. In its simplest form the water clock was merely a vessel filled with water and provided with a narrow aperture in the bottom through which the water streamed out, the level of the water marking the lapse of time.

Step by step these instruments were improved and in the Hellenistic period of astronomy both Hipparch of Nicaea (*c.* 190–120 B.C.), and Ptolemy (*c.* A.D. 90–160) developed new types, some of which are shown in Figs. 7.14 and 7.16. But before such improvements Greek astronomy had already achieved some of its most important results. With admirable boldness and ingenuity the Greeks had acquired enough data to advance mathematical planetary theories of great elegance, and to attempt a determination of the dimensions of the Earth and the solar system by means of quite simple aids.

The Length of the Year

Even with a simple gnomon it is possible to perform a large number of measurements fundamental to astronomy. First, it is easy to find the meridian of the locality by bisecting the angle between the two positions of the shadow in which the shadows are of equal length. In doing so, the cardinal points are also determined. Next, the moment of true noon when the Sun is at its highest point in the sky, can be established, for then the shadow of the gnomon passes the meridian (Fig. 5.2). The date for the shortest noon shadow of the gnomon will

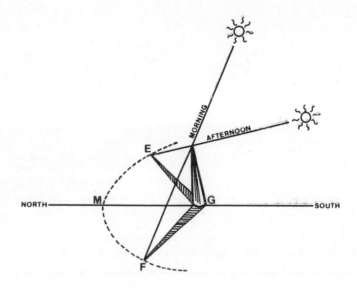

Fig. 5.2. *Determination of the meridian by means of a gnomon G. The dotted line is a curve (usually a hyperbola) with the base of the gnomon as centre. Two equally long shadows GF and GE are found, GF before, noon, GE after noon. The meridian line GM bisects the angle between the shadows.*

then determine the moment of summer solstice. Winter solstice is correspondingly established as the day when the noon shadow has its greatest length. The length of the year (that is the tropical year) is found as the interval between two succeeding solstitial dates. Thales had stated the length of the year to be 365 days, presumably on the basis of a certain knowledge of the Egyptian calendar. Later it became a major concern of Greek astronomy to determine this fundamental constant as exactly as possible.

43

The length of the tropical year, i.e. the interval between two successive Summer solstices

Thales c. 600 B.C.)	365^d
Meton c. 430 B.C.	$365\frac{5}{19}^d$
Calippos 4th century	$365\frac{1}{4}^d$
Aristarchos 3rd century	$365\frac{1}{4}^d + \frac{1}{1623}^d$
Hipparchos 2nd century	$365\frac{1}{4}^d - \frac{1}{300}^d = 365^d 5^h 55^m 12^s$
Modern	$365^d 5^h 48^m 46^s$

The table shows how the results were gradually improved until Hipparchos stated the value which came to be generally accepted in Antiquity and the Middle Ages. For the synodic month Hipparchos used the value $29^d 12^h 44^m 3\frac{1}{3}^s$ which is well known from Babylonian astronomy—a proof that the Greek astronomers were acquainted with some of the results of their eastern colleagues.

The Length of the Seasons

By means of a gnomon it is also possible to establish the dates of the equinoxes (when day and night are of equal length) (Fig. 5.3). The best procedure is to determine the moment at which the shadow at noon bisects the angle formed by the positions of the noon shadow at winter and summer solstice, or to find the date on which the morning and evening shadow are on a straight line. Such measurements were performed in 432 B.C. by two astronomers from Athens, Meton and Euctemon, and at approximately the same time by Democritos. The seasons proved to be of different length, which even Thales is said to have known. These numerical values, too, were gradually improved, as is shown in the table. Later, they were used as a starting point for Hipparchos's theory of the Sun (see p. 84) which was one of the first important theoretical victories of Greek astronomy.

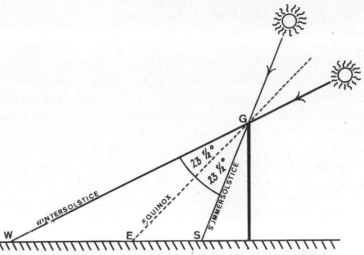

Fig. 5.3. Determination of the seasons and the obliquity of the ecliptic by a gnomon. The figure is drawn in the plane of the meridian. The shortest shadow at noon GS marks the summer solstice, the longest GW the winter solstice. The angle WGS is bisected by GE which represents the noon shadow at the equinoxes when the Sun is on the celestial equator. The angle WGE is the obliquity of the ecliptic relative to the equator.

The Length of the Seasons

Season	Length in Natural Days According to			
	Euctemon and Meton 5th century	Democritos 5th century	Calippos 3rd century	Hipparchos 2nd century
Spring (vernal equinox to summer solstice)	93	94	94	$94\frac{1}{2}$
Summer (summer solstice to autumnal equinox)	90	92	92	$92\frac{1}{2}$
Autumn (autumnal equinox to winter solstice)	90	89	89	$88\frac{1}{8}$
Winter (winter solstice to vernal equinox)	92	90	90	$90\frac{1}{8}$

45

The Annual Motion of the Sun

Contradictory legends make it impossible to decide the identity of the first Greek to discover that the Sun performs its annual (direct) motion among the fixed stars in the plane of the ecliptic. One legend recounts that Cleostratos of Tenedos (second half of the sixth century B.C.) divided the zodiac into the twelve signs well-known today: Aries, Taurus, Gemini, and so on. Eudemos maintains, however, that Oinopides of Chios (*c.* 450 B.C.) was the first to find the zodiac, while Aëtios says that Oinopides had stolen the discovery from Pythagoras, passing it off as his own. Finally, Diodor has it that

> Oinopides associated with the Egyptian priests, and from them he learned that the Sun has an oblique orbit and moves in the opposite direction to the other stars.

This last explanation has at least a ring of truth in so far as the Greeks' insight was most probably of eastern origin.

Certainly the polos had been used to study the motion of the Sun. Being hemispherical, the bowl of the instrument was similar to the celestial sphere. Its centre of similarity was the style (the ball or the point of the gnomon) seated in the centre of the bowl. This means that in the course of the day the shadow point on the inner side of the bowl forms a curve similar to the diurnal arc of the Sun on the sky, and thus almost a circle. If, therefore, the direction of the shadow at the equinox is marked the celestial equator will be delineated as a circle in the polos, whereas at solstice one will see the tropics marked as two small circles parallel to the equator at a distance of $23\frac{1}{2}°$. Used in this way the apparatus showed the Sun's motion graphically, just as it yielded two scientific results of the utmost importance.

Sundials

The first of these results was the invention of the sundial, which in its Greek form was simply a polos graduated in a particular way, known both from archaeological finds and from a description by the Roman engineer Vitruvius (second half of the first century B.C.) of the so-called *analemma* construction, and of an apparatus called the *scaphae* (Fig. 5.5).

Fig. 5.4. *Greek sundial from the 4th century* B.C. *found on the isle of Delos. The hemispherical bowl is ordinary painted pottery. The gnomon is a bronze pin.*

Vitruvius credits Aristarchos with this invention, but a specimen from the fourth century B.C. found on the island of Delos proves that it is much older (Fig. 5.4).

To understand the principle of this instrument the situation at the equinoxes (cf. the analemma in Fig. 5.6) must first be considered. Here the Sun is on the celestial equator so that the point of the style casts a shadow describing, on the inner surface of the bowl, one half of a great circle corresponding to the equator. When this arc is divided into 12 equal parts the instrument can be used for dividing the day into 12 equal intervals identical to our present hours. At other times of the

47

Fig. 5.5. Roman scaphae found at Civitá Lavinia outside Rome. The curved surface is part of a hemisphere, and the instrument is essentially the same as that shown in Fig. 5.4 except that the unnecessary parts of the hemisphere are removed. The gnomon was a thin rod placed horizontally at the centre of the hemisphere where all the hour lines meet.

year the construction is a little more intricate. Fig. 5.6 shows the situation at the summer solstice when the Sun is on the Tropic of Cancer 23½° north of the equator. The point of the shadow will then describe in the polos a small circle 23½° below the equator, intersecting the rim (representing the horizon) at two points on a line through *F*. This circle is drawn in the plane of the paper as the arc *SPG*, and *FG* is drawn perpendicularly to *SF*. Then the arc *SPG* is half of the diurnal arc of the Sun.

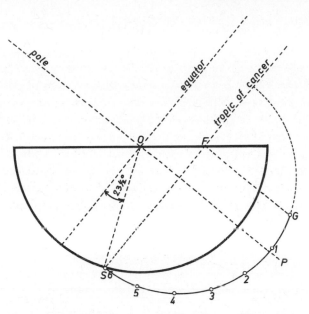

Fig. 5.6. Analemma for the construction of seasonal hours in the hemispherical sundial.

For everyday use this arc was divided into six equal parts. The shadow will pass the dividing points at regular intervals of time of one twelfth the length of the day at summer solstice. These intervals are called 'hours', so that the length of one hour will depend on the season, the hours of the day being long in summer time and short in winter time, while the hours of the night vary in the opposite way. This curious division of the day into unequal hours was used throughout the Middle Ages. For astronomical purposes, however, the invariable equinoctial hours were always used and have now completely replaced the variable kind.

The division of the day into hours, a practice already employed in Egyptian and Babylonian astronomy meant an enormous progress in time reckoning, because imprecise statements like 'at cockcrow', or 'when the market is full' (that is late in the morning), were no longer needed.

49

Gnomonics and Conic Sections

Dividing the day into hours had another important, but theoretical result. It is a very reasonable hypothesis that the conic sections in geometry may well have been discovered as a by-product of gnomonics—the doctrine of sundials (Fig. 5.7). These important curves are usually ascribed to a pupil of Eudoxos called Menaechmos (*c.* 350 B.C.), who is said to have invented them during his purely mathematical studies of the famous Delian problem of the duplication of the cube. The connection between conic sections and sundials becomes apparent as soon as a dial is constructed with a plane surface instead of a hollow hemisphere (Fig. 5.8). Then the point of a shadow will no longer describe a circle, but another curve.

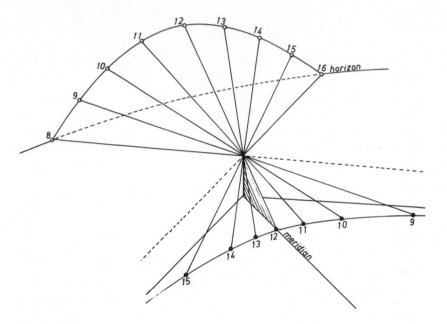

Fig. 5.7. How the shadow of a gnomon moves upon a horizontal plane. The figure corresponds with a place in the northern hemisphere and with a date on which the day length is 8 hours. Above is the diurnal arc of the Sun rising above the horizon and divided in 8 equal parts from sunrise to sunset. Only the noon shadow is shown in full, but the positions of the other shadows are marked by one hour intervals from 9ʰ to 15ʰ. Their end points are seen to lie upon a hyperbola with the meridian line as axis and the sunrise and sunset shadows as asymptotes. At another date the shadow will describe a different hyperbola which at the equinoxes degenerate into a straight line through the foot of the gnomon.

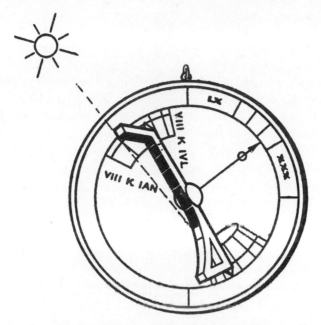

Fig. 5.8. Roman equatorial sundial from the third century A.D. (diameter about six centi-metres). The instrument has three parts: 1, a circular disc of brass of which the first quadrant is divided into degrees; 2, another circular disc turning about the centre of the first one and, by means of the pointer, adjustable to the geographical latitude. Two further divisions are engraved opposite each other. They serve to adjust the third part of the instrument, which consists of a gnomon attached to a curve scale with hour divisions, according to the date, that is to the declin-ation of the Sun. The instrument is suspended on a ring and turned about the vertical until the shadow of the gnomon falls precisely along the scale. In this position the discs of the instrument will be in the plane of the meridian while the gnomon and the scale are in the plane of the ecliptic. The instrument can therefore be used without the cardinal points being known if only the latitude is determined beforehand. It was found during an excavation of a Roman military camp. Reproduced from A History of Technology III, 598, *Oxford, 1954.*

If all the rays passing the point of the style in the course of one day, are considered, it is obvious that together these rays will form a circular cone around the axis of the world, the angle between the axis and one of the rays being equal to the complement of the Sun's de-clination. Therefore, the path of the shadow on an arbitrary plane can be conceived as the intersection between the plane and the cone, that is, a conic section. If the plane is perpendicular to the axis of the world (that is parallel to the equator) the path of the shadow is a circle. On a horizontal plane it will be hyperbolic except at places where the Sun is

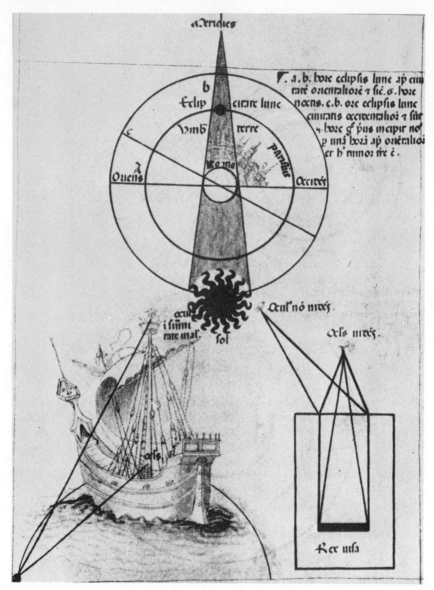

Fig. 5.9. The two Aristotelian proofs of the spherical shape of the Earth. At the top is a drawing of a lunar eclipse. Below and to the left is an illustration explaining why the mast is the first you see of a ship approaching the shore. From a 13th-century manuscript of Sacrobosco's Tractatus de sphera, *now in the Biblioteca Estense at Modena.*

circumpolar. Other positions of the place will give a straight line or part of an ellipse or a parabola. Vitruvius also describes these kinds of sundials and explains how the curves can be divided into hours.

Geodesy

The size and shape of the Earth were problems which occupied Greek philosophers from the earliest times. Thales thought that the Earth was flat, for Anaximander it was shaped like a drum, but even initially the Pythagoreans thought that the Earth was spherical. This assumption was presumably made chiefly for philosophical reasons, the sphere being the perfect geometrical figure. The first real proofs of the spherical shape of the Earth are recorded in Aristotle, who showed that the Earth cannot be flat, for if one travels north or south the stars vary in their altitude above the horizon, which would not be the case on a flat Earth. Furthermore, during a lunar eclipse the Earth casts a round shadow on the Moon's surface, regardless of the position of the Sun, which is another proof of the Earth's spherical shape. A third proof is the well-known observation of a ship approaching the shore, when the top of the mast is seen before the hull (Fig. 5.9).

The determination of the size of the Earth was one of the greatest achievements of Alexandrian science, still impressive in its ingenious simplicity. In his work on the *Circular motions of the heavenly bodies* the astronomer Cleomedes (*c.* A.D. 130–200) reports that Eratosthenes of Alexandria (*c.* 275–194 B.C.) found the circumference of the Earth from the following hypotheses:

1. The town of Syene is directly south of Alexandria, that is on the same meridian (Fig. 5.10).

2. Their distance apart is 5,000 *stadii*.

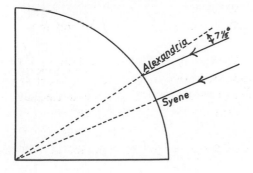

Fig. 5.10. The principle of Eratosthenes' measurement of the Earth.

3. Syene lies under the Tropic of Cancer, since the gnomon casts no shadow at noon at summer solstice.

4. The shadow of the gnomon (in the polos) in Alexandria at the same time forms an angle with the vertical of $\frac{1}{50}$ of a complete circle.

From this Eratosthenes deduced that the circumference of the Earth measured along the meridian through the two cities is

$$50 \text{ times } 5,000 = 250,000 \text{ } stadii$$

The exact length of a *stadium* is unknown but is usually assumed to be 157.7 metres. This leads to the astonishingly precise value of 39,370 kilometres.

Later, Poseidonios of Rhodes (*c.* 135–50 B.C.) tried to repeat this determination, but this time with Alexandria and Rhodes as endpoints of the base-line. The difference in latitude between the two localities was found as the difference in culmination height of the star Canopus. The distance was also estimated in this case to be 5,000 stadii, but with much greater uncertainty because the straight line between the two places crossed the Mediterranean. The result was slightly less accurate than that of Eratosthenes, namely 240,000 *stadii*.

Distances in the Solar System

Aristarchos of Samos made an even more courageous undertaking. His only extant treatise deals with his determination of the distances of the Sun and the Moon (Fig. 5.11). The treatise is written in the axiomatic style which shows clearly that scientists were now trying to follow the method of the mathematicians. Like Euclid's *Elements*, therefore, the treatise begins with a number of presuppositions or hypotheses, as follows:

1. The Moon gets its light from the Sun.

2. The Earth is like a point and a centre relative to the sphere in which the Moon is moving.

3. When we look at the half Moon, the great circle dividing the dark and the bright part points directly at our eyes.

4. The half Moon has an angular distance from the Sun equal to one quadrant minus $\frac{1}{30}$ of a quadrant (that is 87°. Actually the value is 89°; 50).

5. The width of the Earth's shadow is twice the diameter of the Moon.

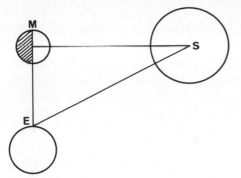

Fig. 5.11. *How Aristarchos tried to find the relative distances of the Sun and the Moon. The observer at E sees the Moon in the first quarter (half Moon) and measures the angle MES, thus determining the ratio EM:ES.*

6. The Moon comprises $\frac{1}{15}$ of a zodiacal sign (that is 2°. The true value is $\frac{1}{2}$°).

From these axioms Aristarchos was able to prove the following three propositions or theorems:

1. The Sun has a distance from the Earth greater than 18, but less than 20, times the Moon's distance from the Earth.

2. The true diameters of the Sun and the Moon have about the same ratio (that is between 18 and 20).

3. The ratio between the diameters of the Sun and the Earth is greater than 19:3 and less than 43:6.

From these values it is easy to deduce the distances of the Sun and Moon, expressed in diameters of the Earth.

Aristarchos's treatise is important as the first attempt at determining distances in the solar system from astronomical measurements. In view of the poor quality of existing instruments for measuring angular distances it is not surprising that the results were far from accurate. A number of other determinations are listed in the table below, from which it can be seen that the best astronomers of Antiquity came very close to the true distance of the Moon, whereas they greatly underestimated the distance of the Sun.

The mean distances of the Sun and the Moon in diameters of the Earth

	Moon	Sun
Aristarchos	$9\frac{1}{2}$	180
Hipparchos	$33\frac{2}{3}$	1245
Poseidonios	$26\frac{1}{2}$	6550
Ptolemy	$29\frac{1}{2}$	605
Modern	30	11740

The Astronomy of Hipparchos

About a century after Aristarchos, Greek astronomy reached its summit through Hipparchos of Nicaea in Bithynia (*c.* 190–120 B.C.), who worked partly in his native town, and partly on the island of Rhodes. Hipparchos was equally eminent as an observer and a theoretical astronomer. Moreover, he was acquainted with at least some essential features of Babylonian astronomy and frequently made use of its numerical results. It was Hipparchos who introduced into Greek mathematics the Babylonian division of the circle into 360°, with sexagesimal subdivisions of the degree into minutes and seconds. The result was that after his time Greek astronomers usually made their computations in the Babylonian sexagesimal system.

Even more remarkable were Hipparchos's own achievements. One of these was his great *Catalogue of fixed stars.* If we are to believe Pliny, Hipparchos obtained the idea of this Catalogue from the observation of a new star. It is known from Chinese records that a bright nova appeared in 134 B.C., and this may well be the year in which he began the catalogue. It differed from earlier and more imprecise descriptions of the constellations in giving exact values of the ecliptic longitude and latitude of about 850 stars.

Closely connected with this work was Hipparchos's discovery of the precession of the equinoxes. He found that the period in which the Sun returns to the same fixed star (the sidereal year) is a little longer than the period of return to the vernal equinox. This discrepancy was correctly interpreted as a consequence of a motion of the fixed stars and the equinoxes relative to each other, usually described as an eastward motion of the starry heavens relative to a fixed equinox. The annual value of precession was found to be 36″, corresponding to 1° per century. This became the commonly accepted value of the precessional constant in Antiquity.

Finally, Hipparchos's geographical efforts deserve mention. Even before 300 B.C., geographical latitudes and longitudes had come into use for defining positions on the Earth. Hipparchos advocated that such co-ordinates should be determined by astronomical means and not from vague information on the number of days' journey between two localities. Now it is easy to determine the geographical latitude which is equal to the polar altitude of the locality, found by means of a gnomon combined with the declination of the Sun. It is more difficult to find the geographical longitude between two places. Hipparchos drew attention to the fact that this difference can be found if the same event can be observed from the two places. The reason is that this event will occur at different moments of local time, the time difference being equal to the difference in longitude. There are not many events visible to observers far away from each other, so Hipparchos proposed to make use of eclipses of the Moon. The entrance of the Moon into the shadow of the Earth is an objective phenomenon happening at the same moment for all observers. The only trouble is that eclipses are rare so that the method is slow when there are geographical co-ordinates of many localities to be determined. So as not to overlook any useful eclipses, Hipparchos is said to have computed a list of future eclipses for a span of 600 years. It did not survive and seems to have been lost at an early stage, since nothing came of this proposal.

This brief and incomplete sketch of the history of practical astronomy shows how incorrect it is to say that the Greeks disregarded experience or were not acquainted with empirical methods. In fact, they possessed extraordinary gifts of making observations of the utmost importance by means of very primitive instruments. Also admirable is the scientific spirit behind, for example, Hipparchos' long-term project of eclipse observations.

Chapter Six

Early Greek Cosmology

The Universe of Spheres

T HE discussion of the practical astronomy of the Greeks as something
disconnected from astronomical theories in the previous chapter
was slightly out of historical context. As always in Greek science,
the practical knowledge of phenomena and theoretical speculations on
their causes go hand in hand. In the thinking of Anaximander, for
example, a fairly complete theory of the solar system had already
emerged. His conception of the planets as wheels filled with fire
rotating about the Earth is perhaps the first Greek attempt to explain
astronomical phenomena by means of a mechanical model. This model
is developed a little further by Anaximenes. It is true that he took a
retrograde step, considering, like Thales, that the Earth was as flat as a
table top. Yet he maintained that 'the stars do not go beneath the Earth,
but turn round it like a cap revolving on our head', showing that Anaximenes conceived of the heavenly bodies as spherical shells, or, as may
be inferred from his statement that the stars are fastened like nails to the
firmament, as bodies attached to revolving spheres. If this interpretation is correct, then Anaximenes is the first of the Greeks to assume
the existence of the heavenly spheres which later became the predominant feature of both the Greek and the Mediæval universe.

Pythagorean Astronomy

The Pythagorean philosophers, it will be remembered, assumed the Earth to be spherical in shape just as they thought the universe as a whole to be of this perfect geometrical form. Beyond this, it is difficult to get a clear picture of Pythagoras's astronomical views. Aristotle (who never mentions Pythagoras by name and possibly did not believe in his existence) says that the so-called Pythagoreans assumed the existence of a central fire in the middle of the spherical universe. This 'fire of Hestia', may be argued, is invisible to man because it is always hidden by the ground beneath our feet. Hestia was the goddess of the sacred fireplace of homes and public buildings. Her cult was very widespread and connected with old and primitive conceptions of fire as being life-giving and purifying. The fire, the Pythagoreans believed, is surrounded by ten concentric spherical shells or spheres. The inmost but one of these spheres takes the Earth round the central fire in the course of one day. The inmost sphere moves a globe invisible to us, the 'anti-Earth', which also performs a revolution about the centre in the course of one day, always opposite to the Earth. Outside the Earth are the planets in the following order: Moon, Sun, Venus, Mercury, Mars, Jupiter, Saturn. Each of these spheres turns from west to east, completing a revolution in a period characteristic of the planet. Furthest out is the tenth sphere containing the fixed stars, which also moves, but so slowly that it is imperceptible to the eye.

In this form the theory is usually connected with the rather obscure Pythagorean Philolaos, who may have lived at the end of the fifth century B.C. It appears as a conscious attempt to save or explain a number of astronomical phenomena by a mechanical model, and is thus one of the theories of a mechanical universe developed since Anaximander's time. Yet the theory is strongly marked by the philosophical ideas of the Pythagoreans. Thus the sphere of the fixed stars is assumed to move in order to raise the number of moving spheres to ten, which was the sacred number in Pythagorean number mysticism: from an astronomical point of view this motion is unnecessary before the discovery of the precession. Perhaps the curious anti-Earth was introduced in order to explain that eclipses of the Moon are more frequent than those of the Sun. A modern conjecture that the anti-Earth was invented as a kind of counterweight, to balance the real Earth in their mutual motion around the central fire, has no foundation in the extant sources.

Leaving the anti-Earth and the central fire out of account, the system must be evaluated by the extent to which it is able to explain the phenomena known to Greek astronomers in the fifth century B.C. Firstly, the daily motion of the heavens appears as a result of the Earth's diurnal circulation around the central fire. Secondly, the model roughly explains the direct motions of the planets, as the planets are continually moving towards the east. Retrograde motions remain unexplained but were possibly not known at the time when the model was developed. Finally, the model gives the previously known and correct explanation of eclipses apart from the consequence that the anti-Earth, as well as the Earth is able to eclipse the Moon.

The most peculiar feature of the Pythagorean universe is, without doubt, the doctrine of the harmony of the spheres. This idea can best

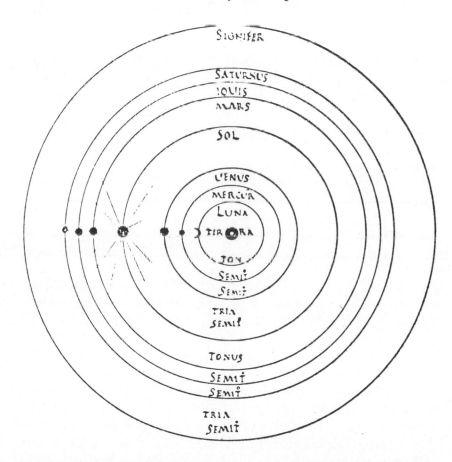

Fig. 6.1. The harmony of the spheres. Ancient representation of a concentric planetary system showing the musical interval corresponding to each pair of spheres.

be understood as a result of the general striving to discover mathematical structures in nature, leading to a mathematical theory of
the planets founded upon an analogy between planetary distances and
musical intervals, for the mathematical ratios in music were known
to the Pythagoreans (see p. 137) (Fig. 6.1.). Aristotle relates that the
Pythagoreans assumed the movements of the spheres to produce
powerful tones whose pitch depended upon the velocities of the spheres;
these velocities, in turn, were conditioned by the distances from their
centre. The outer spheres, it was argued, are the faster and produce the
highest notes; together they create a harmony. Our inability to hear
this celestial music was explained by the fact that since our birth we
have been accustomed to its uniform sound and therefore disregard it.

Unsophisticated though this Pythagorean astronomy may appear,
it had a marked effect on the future development of theoretical
astronomy, not least by laying down the principle of uniform circular
motion as the prime mathematical tool of theoretical astronomy.

Geocentric and Heliocentric Systems

The astronomy of the Pythagoreans proves that in these early times
neither the idea of a moving Earth nor the assumption of its being
outside the centre of the universe was foreign to astronomers. In fact,
a whole series of more or less obscure philosophers nourished similar
ideas. The Pythagorean Hicetas of Syracuse (fifth century B.C.) modified the Philolaic system, giving the Earth a daily rotation about its
own axis. Another Pythagorean, Ecphantos from Syracuse, (c. 400 B.C.)
discarded the central fire, replacing it by the Earth, so that the diurnal
motion of the heavens is explained by the daily rotation of the Earth
positioned at the centre of the universe.

In the history of astronomy, Hicetas and Ecphantos are little more
than mere names. Better known by far is the chief supporter of the
theory of the daily rotation of the Earth, namely Heracleides of Pontos,
who was a disciple of Plato and a famous academic philosopher in his
own right. According to Aëtius he taught like Ecphantos, that the
Earth turns round its own axis. Furthermore, in his commentary to the
Timaios, Chalcidius maintains that Heracleides developed a new theory
of the motion of Venus in order to explain the fact that this planet keeps
within a maximum distance from the Sun (Fig. 6.2). Heracleides

III. Syſtema Ægyptiorum, Vitruuij, Capellæ, Macrobij, Bedæ &c.

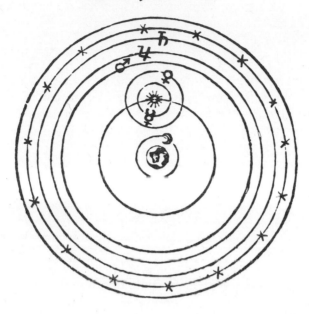

Fig. 6.2.   The geo-heliocentric system of Heracleides from Pontus (after Riccioli, Almagestum novum, Bologna 1651).

supposed this greatest elongation to be 50° (in fact it is 46°) and explained it on the assumption that Venus moves on a circle having its centre in the Sun and a radius of 50° as seen from the Earth. When it is found on the eastern part of this circle the planet is seen as an evening star, while it appears as a morning star on the western part. It seems that Heracleides used a similar hypothesis to explain the motion of Mercury, which exhibits very much the same phenomenon. His planetary system can therefore be described as geo-heliocentric: geocentric for the Moon, Sun, Mars, Jupiter, and Saturn, but heliocentric in the case of Venus and Mercury. Almost two thousand years later another variant on the geo-heliocentric theory was developed by Tycho Brahe.

In the third century B.C. a consistent heliocentric theory was created by Aristarchos of Samos, already known from his attempt to determine the distances of the Sun and Moon (p. 54). In the preface to his

63

treatise *The Sand Reckoner* Archimedes wrote to King Gelon:

> You are not unaware that by the universe most astronomers understand a sphere the centre of which is at the centre of the Earth [. . .]. However, Aristarchos of Samos has published certain writings on the (astronomical) hypotheses. The presuppositions found in these writings imply that the universe is much greater than we mentioned above. Actually, he begins with the hypothesis that the fixed stars and the Sun remain without motion. As for the Earth, it moves around the Sun on the circumference of a circle with centre in the Sun.

To save the phenomena Aristarchos has to make two further assumptions. First, the sphere of the fixed stars must be immensely distant, or the apparent positions of the stars would change as a result of the annual motion of the Earth. Secondly, Aristarchos must have supposed that the Earth has a diurnal rotation about its own axis. This is not confirmed in any surviving text, but because the diurnal rotation of the Earth is the only hypothesis able to explain the apparent daily motion of the heavens, it goes without saying that it must have been part of Aristarchos's theory. His system became, in time, a source of inspiration to Copernicus, and the Greek astronomer may justly be named the Copernicus of Antiquity.

The Fate of Heliocentric Astronomy

Only one ancient astronomer is known as a supporter of the new theory. According to Plutarchus, the Babylonian Seleucus (*c*. 150 B.C.), regarded Aristarchos's theory as true in the sense that it described the real physical structure of the universe; yet Aristarchos himself, if we are to believe Archimedes, considered it merely a mathematical hypothesis. But as a physical hypothesis it presented a sharp contrast with the physics of Aristotle, which cited several 'physical' reasons why the Earth, as the most heavy element, has its 'natural' place at the centre of the world. From the scientific point of view this seems to be the main reason why Aristarchos acquired so few followers in Antiquity.

Much more serious was the opposition Aristarchos met from the traditional Greek religion. It is known from Plutarchos that the Stoic Cleanthes (*c* 331– *c*. 232 B.C.) led a campaign of popular resentment against the astronomer, accusing him of sacrilege for having 'displaced the hearth of the world'. The affair is reminiscent of the persecution suffered by Galileo because of his adherence to the Copernican system, but was essentially much more serious. When Galileo was brought before the court of the Inquisition it was more as a result of ecclesiastical and academic intrigues than from any fundamental incompatibility between the Catholic faith and modern astronomy. But in the eyes of the ordinary Greek, Aristarchos had sinned against deep-rooted ideas about Hestia's Fire, and the Earth as a Divine Being. Such religious tenets could not be shaken by abstract astronomical theories incomprehensible to the ordinary man. The intellectual climate of Greece was not prepared for theories of this kind, and from the era of Plato and Aristotle onwards all other planetary systems adhere to strictly geocentric principles.

Plato's Universe

Plato's astronomy also owed a debt to the Pythagoreans. The dialogues *Timaios*, the *Republic*, and *Epinomis* contain many references to a universe of concentric spheres around the Earth, which is supposed to remain motionless at the common centre of the spheres (Fig. 6.3). Around the terrestrial globe, Plato maintains, are found first the element of water contained in a spherical shell with a thickness of 2 radii of the Earth. It is surrounded by the atmosphere, or a sphere of air, with a thickness of 5 radii. Next comes the outer sphere of fire with a thickness of 10 radii, in the superior part of which the fixed stars are found. Their distance from the centre is seen to be 18 radii. Between the atmosphere and the fixed stars the seven planets are moving. In the *Timaios* their order, reckoned from the Earth is: Moon, Sun, Mercury, Venus, Mars, Jupiter, and Saturn. The later dialogues give a slightly different order; in the *Republic* the Sun, Venus, and Mercury are all supposed to have the same velocity and the same distance from the Earth. In *Epinomis*, a new sphere of a fifth element called the *ether* is introduced between the air and the fire, which increases the dimensions of the universe. The stars are imagined to be much greater than before; thus the Sun appeared larger than the Earth, which was impossible with the data given in the *Timaios*.

In Platonic astronomy each planet is ascribed two uniform, circular motions. First there is a daily rotation from east to west about the poles

65

II. Syſtema Platonis, aut Platonicorum.

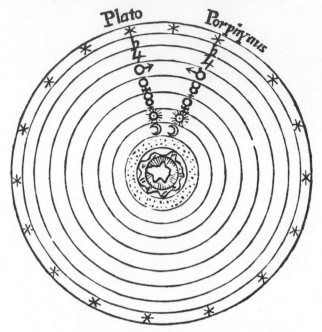

Non deſunt tamē rationes & authoritates,quibus pro-
babile fieri poſſet Eudoxum, Calippum & Ariſtotelem
ſecutos fuiſſe ſyſtema Pythagoræ , de quibus ſuo loco.
Interim falſum eſt,quod ait *Clauius* in ſphæra, ſolum eſ-
ſe Authorem libelli de Mundo ad Alexādrum, qui Mer-
curium ſub Ioue ac ſupra Venerem ponat. Demum Plato concedit in Timæo, Terram verti circa ſuum centrum,quod negat Ariſtoteles.

*Fig. 6.3. The order of the planets according to Plato and the neo-
Platonists (Riccioli, Almagestum novum, Bologna 1651).*

of the equator, that is, about the axis of the world. Secondly, each
planet rotates about the poles of the ecliptic from west to east, completing its revolutions in a sidereal period (that is the period of return
to the same fixed star). A model like this explains only the crudest
astronomical phenomena, namely the participation of the planets in
the diurnal motion of the heavens (their rising and setting), together
with their individual slow movements in the opposite direction relative
to the fixed stars. But several phenomena well known in Plato's time
remain unexplained, notably the retrograde motions of the five planets,

as well as their varying latitudes (distances from the ecliptic). In the case of the Sun the theory is unable to account for the different lengths of the seasons, as the direct motion is supposed to take place with a uniform velocity (p. 85).

It was, perhaps, difficulties of this kind that allegedly led Plato to ask the philosophers how a theory accounting for all known phenomena could be created on the basis of uniform, circular motions. The result was a series of geometrical models of planetary motions which—with Archimedes's statics—form the most elegant developmental contribution of the Greeks to a mathematical description of nature.

Chapter Seven

Mathematical Astronomy

I F Plato was first to challenge philosophers with the question of how the irregular motions of the heavens might be saved by uniform circular movements (p. 27), his mathematical colleague at the Academy, Eudoxos of Cnidos (*c.* 405–355 B.C.), was first to meet the challenge. Apart from his contribution to pure mathematics—the famous theory of proportions and incommensurable quantities conserved in the fifth book of Euclid's *Elements*—Eudoxos's work was soon abandoned. But his answer to the problem of planetary motions has considerable historical importance. It showed how the problem was attacked by the most brilliant mathematician of the fourth century B.C. who was, moreover, a personal acquaintance of Plato's, so that his solution can legitimately be accepted as an indication of how Plato wanted theoretical astronomy to proceed.

The four main phenomena that an astronomical theory had to take into account, were, it will be remembered, as follows:

Firstly, it is obvious even to the most casual observer that the planets partake of the *diurnal rotation* of the starry heavens from east to west with a period of revolution of one sidereal day (about $23^h 56^m$) (Fig. 7.1).

Secondly, more careful observations reveal that besides the diurnal motion all planets perform a much slower *motion in longitude* from west to east relative to the fixed stars.

Thirdly, apart from the Moon and the Sun the five other planets will, from time to time, stop their direct (eastward) motion among the fixed stars in order to take up an indirect, or *retrograde* (westward) *motion* (Fig. 7.1). After a while the direct motion is resumed. At the beginning (first station) and end (second station) of the retrograde phase the planet seems to be stationary.

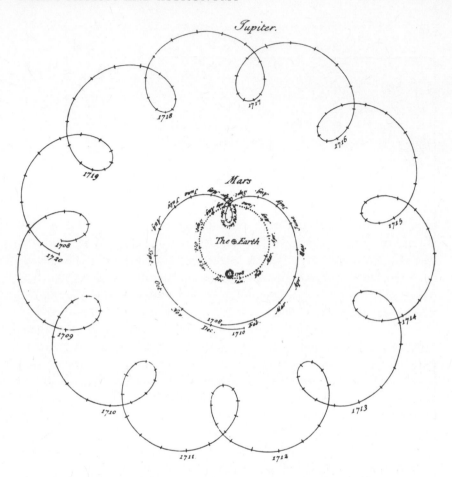

Fig. 7.1. *The motions of Mars and Jupiter relative to the Earth during the years 1708–10 (Mars) and 1708–20 (Jupiter). The two geocentric orbits exhibit the varying distances of the planets from the Earth and the characteristic loops around the points of minimum distance (From Roger Long,* Astronomy, *Cambridge 1742).*

Fourthly, while, by definition, the Sun moves on the ecliptic, the path of the Moon and any other planet deviate a little from it, causing the latitude of the planet to vary between a northern maximum and a southern minimum of the same numerical value. The point on the ecliptic where the planet is passing from southern to northern latitudes is called its *ascending node.* The opposite point of the ecliptic is the *descending node.* The *motion in latitude* combined with the motion in longitude

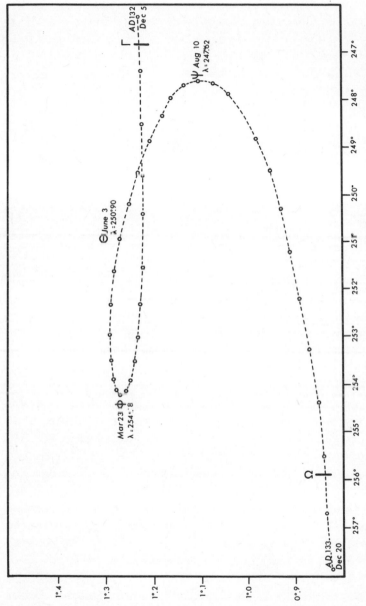

Fig. 7.2. The motion of Saturn in A.D. 133 (when it was observed by Ptolemy) showing its first visibility Γ, the first stationary point φ, the opposition θ, the second stationary point ψ, and the last visibility Ω. The motions in longitude and latitude produce a loop whose appearance may vary greatly from one opposition to the next.

produces the characteristic loops around the retrograde phases (Fig. 7.2).

Several supplementary phenomena support these main observations, among them the phases and eclipses of the Moon, the eclipses of the Sun, the period of invisibility during which the planet is too near to the Sun to be visible against the clear sky, the heliacal rising and setting (the end and beginning of the invisibility period), conjunction and opposition with the Sun.

The Anomalies of Planetary Motion

The great stumbling-blocks to Plato and other philosophers were the irregularities connected with each of these phenomena, apart from the daily revolution of the heavens. This is the case even with the Sun. It is true that the Sun has a constant period of revolution in that it returns to the same fixed star within a regular interval of time called the *sidereal year* or to the vernal equinox in one *tropical year*. But the irregularity arises from the fact that it moves with a changing velocity during a complete revolution. This was discovered as soon as the solstitial and equinoctial dates could be determined from observations. For while the Sun moves exactly 90° upon the ecliptic during any of the seasons, the latter comprise unequal numbers of days (see p. 86). The angular velocity of the Sun is therefore not constant, but varies about a mean value called the *mean daily motion*.

The irregular motion is even more apparent in the Moon. Since the Moon can be seen at night against a background of fixed stars, fairly crude observations will show that its true daily motion varies between a minimum of 10° and a maximum of 14°. But the Moon is even more irregular, as it has no constant period of return neither to the same fixed star, nor to the equinox, that is the *sidereal and tropical months* are variable periods with a mean value. The five other planets all resemble the Moon in having irregular daily motions and irregular sidereal periods with definite mean values.

It follows from this that all the seven planets exhibit the same irregularity in that their true daily motions vary with the position of the planet relative to the fixed stars. This deviation from uniform circular motion depends, therefore, on where the planet is on its path around the heavens. In ancient astronomy the deviation was usually called the *first anomaly* and considered to be what would be called nowadays a function of the ecliptic longitude of the planet. This concept makes it possible to define an *anomalistic* period of the planet as the period of return to the same daily motion (angular velocity). In the Sun this anomalistic period is constant, whereas, in the other planets it is variable, but with a definite mean value.

Among the phenomena are irregularities of a different type. Simple observation shows that the retrograde motions of the planets Saturn, Jupiter, and Mars always take place around the points where the planets are in opposition to the Sun. In fact, the opposition is at the middle of the retrograde arc. Venus and Mercury are never in opposition, but here a conjunction with the Sun takes place near the middle of the retrograde arc. This shows that in the five planets the deviation from uniform circular motion does not depend only on the position of the planet, but also on that of the Sun. In other words, the first anomaly is supplemented by a *second anomaly* depending on the position of the Sun relative to the planet.

In modern terms, the total deviation from a uniform, circular revolution in longitude of the planet must be a function of (at least) two variables, namely the position of the planet, and the position of the Sun. A basic aim must be to find this function, but the task was often stated in other terms by Greek astronomers and, before them, by the Babylonians.

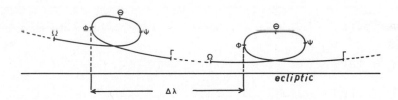

Fig. 7.3.   Two successive occurrences of the main synodic phenomena. The arc Δλ is called the synodic arc. It corresponds to a synodic period (see p. 74) and varies around a mean value.

The second anomaly reveals a coupling between the motion of the planet and that of the Sun. This is inferred not only from oppositions, but from a whole class of so-called *synodic phenomena* (Fig. 7.3) including stationary points, first and last visibility and, of course, conjunctions.

73

The Aim of Planetary Theories

As they were easily observable, synodic phenomena played an important role in ancient astronomy, particularly among the Babylonians. The main problem of Babylonian planetary theory could thus be stated: Given the time t_0 and the longitude λ_0 of a certain synodic phenomenon: to find times t_1, t_2, . . . and longitudes λ_1, λ_2, . . . of the subsequent occurrencies of the same phenomenon. This is a rather difficult problem, since neither the time interval (the synodic period) nor the longitude interval (the synodic arc) are constant, but vary about mean values. However, the Babylonians succeeded in developing abstract mathematical methods for solving the problem, jumping, as it were, from one synodic phenomenon to the next following of the same kind without being interested in the intermediary motion of the planet.

In the course of time, Greek astronomers acquired a much more general view of the aim of planetary theory. Not satisfying themselves with considering discrete sets of synodic phenomena they tried to find methods enabling them to determine both the longitude $\lambda = \lambda(t)$ and the latitude $\beta = \beta(t)$ of each of the planets at any given time t. This new way of presenting the problem required new mathematical tools. This is why the Greeks abandoned the purely algebraic methods of the Babylonians in favour of geometrical models whose configurations could be varied in a continuous way and examined for any given moment of time. The first of these models of which we have a more than fragmentary knowledge is the famous system of Eudoxos which marks the beginning of a long development fulfilled only five centuries later by the achievements of Ptolemy.

The Concentric Spheres of Eudoxos

Eudoxos described his system in a treatise *On the velocities* (that is of the planets). The work is lost, but the general features of the theories contait ined can be reconstructed from a brief summary in Aristotle's *Metaphysics* and from remarks by later commentators like Simplikios, who built upon the work of Sosigenes, and ultimately upon the lost history of astronomy by Eudemos. And as Eudemos lived only a generation later than Eudoxos, the legend is reasonably reliable.

The most conspicuous feature of Eudoxos's system is that it does not make use of any geometrical tools other than a number of spheres concentric with the Earth and performing uniform rotations around axes through its centre. This is, without doubt, an inheritance from Pythagorean astronomy. In his lunar theory Eudoxos used three such spheres, of which the outermost accounts for the diurnal rotation of the Moon from west to east around a first axis running through the poles of the celestial equator.

The second of Eudoxos's spheres rotates from east to west around a second axis fastened to the first sphere and passing through the poles of the ecliptic. It completes a revolution relative to the first sphere in 223 synodic months or 6,585 days, the mean synodic month being $T_s = 29.53$ days.

The third sphere rotates from west to east around the third axis fastened to the second sphere and forming an angle of 5° with the second axis. This third sphere carries the Moon and completes a revolution in a *draconitic month* $T_d = 27.21$ days, defined as the period of return to the same node.

It is easy to deduce the qualitative behaviour of this system. Because of the coupling between any one sphere and the axis of the next (reckoned inwards) the motion of any sphere will enter as a component into the motion of all interior spheres. Thus the diurnal motion of the first sphere will be transmitted to the second and the third, regardless of the proper motion of the third. The Moon will therefore take part in the diurnal rotation of the heavens.

Now, if the third axis had been identical with the second the Moon would have moved along the ecliptic with a period of revolution relative to the fixed stars of one sidereal month T_f determined by

$$360° : T_f = 360° : T_d - 360° : 6{,}585$$

or $T_f = 27.32$ days. It would then be in conjunction with the Sun at regular intervals of 1 synodic month T_s determined by

$$360° : T_s = 360° : T_f - 360° : 365\tfrac{1}{4}$$

But in that case an eclipse of the Sun would occur at the beginning of each synodic month, that is, at every new Moon, and an eclipse of the Moon at every full Moon in between.

To avoid this unacceptable clash with experience Eudoxos gave the third axis an inclination relative to the first. This, he said, causes the

Moon to move most of the time outside the ecliptic with a maximum latitude of 5°, so that most of the conjunctions or oppositions with the Sun will occur without any eclipse.

The slow retrograde motion of the second sphere assures, he argued, that the Moon will be at the same point in the heavens at regular intervals of 223 synodic months, or 18.6 years. This means that if an eclipse is visible from a certain place at a certain time, there will be another eclipse of the same kind 18.6 years later (the *Saros period*).

This lunar theory is a really admirable part of Eudoxos's system. It saves all the phenomena of the Moon's motion known in his time, even the Saros period which was discovered (or introduced from Babylonian astronomy) not long before. In fact, in the hands of the mathematicians, this model could be expected to lead to quantitative predictions of a fairly reliable character.

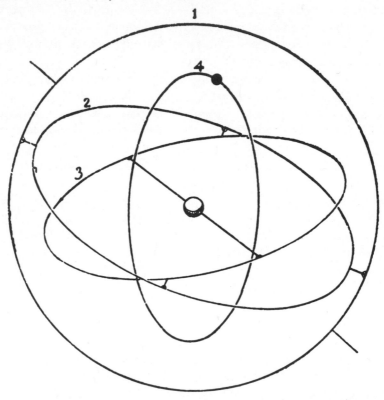

Fig. 7.4. Planetary mechanism on Eudoxos's model with four concentric spheres. After Künnsberg, Eudoxus, Dinkelsbühl, 1888.

Compared with his lunar theory, Eudoxos's theory of the Sun seems much less fortunate. It was developed in strict analogy with the lunar model, comprising three spheres of which the axis of the third had an inclination of 30′ relative to that of the second. As a result the Sun will have a motion in latitude, a belief held by several ancient astronomers. The most probable explanation is that they did not define the ecliptic as the path of the Sun but, more vaguely, as a great circle somewhere along the middle of the zodiac. Hipparchos seems to have been the first to discard the belief in the latitude of the Sun.

In his theories of the five planets Eudoxos has to introduce an extra sphere to account for the retrograde motions. As before there is, for each planet, a first sphere producing the diurnal motion from east to west. The second sphere rotates from west to east about an axis through the poles of the ecliptic with a zodiacal (sidereal) period characteristic of the planet. A point on the equator of the second sphere then represents the mean position of the projection of the planet on the ecliptic. The axis of the third sphere passes through this point and the centre, producing a component motion perpendicular to the ecliptic. Finally, a fourth sphere is introduced (Fig. 7.4). It carries the planet at a point on its equator and its axis is inclined to a small angle relative to the third axis. The third and fourth spheres have the same period of revolution, equal to the synodic period of the planet, but they move in opposite directions.

The Hippopede

To see how the model of the five planets saves the phenomena of retrogradation and motion in latitude the third and fourth spheres are first projected on the second as shown in Fig. 7.5. Here A represents the planet which at a given time is at the intersection between the equators of the third and fourth spheres. At a given time later, A would have moved with the fourth sphere to Q, but due to the opposite motion of the third sphere it arrives at the point P which must now be determined. It is easy to see that QP is part of a circle (with centre O') parallel to the equator on the third sphere. The angles AOQ and $QO'P = \beta$ are equal. B is a point on the equator of the fourth sphere 90° from A. Finally the point P_0 is marked on the equator of the third sphere at the distance β from A.

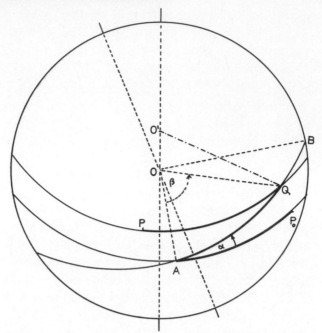

Fig. 7.5.   Figure showing the motions of the third and fourth spheres
in the system of Eudoxos (see p. 79)   according to O. Neugebauer.

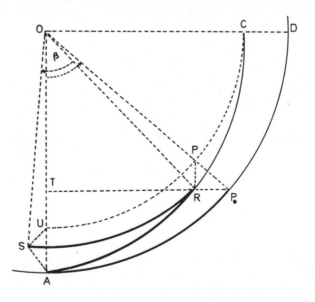

Fig. 7.6.   The determination of the hippopede.

By projection on the equatorial plane of the third sphere Fig. 7.6 is obtained. O' is projected to O, Q to R, and P to S where $OR = OS$. The point B is projected to C through which a circle is drawn with centre at O. By this projection the circle AQB is transformed into an ellipse with OA as half of the major, and OC as half of the minor axis.

If P_0 is rotated around OA it will come to Q. Therefore P_0, Q, and R are in the same plane perpendicular to OA. Therefore the line $P_0 T$ perpendicular to OA must contain the point R. Since OP_0 intersects the circle through C at P', the point R must lie on the line through P' parallel to OA, which follows by a known affinity construction of the points of an ellipse.

The projection of the small circle PQ is a circle through R on which the point S is marked so that the angle ROS is equal to β. Being the projection of the planet P, S will give its position.

The same result can be interpreted geometrically. Let the point U be the intersection between OA and the circle through C. Then the triangle AUS will be congruent with the triangle $P_0 P' R$. Therefore the angle USA is $90°$, from which it follows that, due to the motion of the planet, S will describe a circle with AU as its diameter. This circle

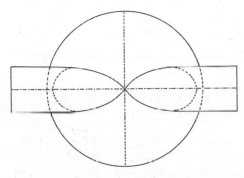

Fig. 7.7. *The hippopede as the intersection between a sphere and a cylinder.*

defines a cylinder which intersects the sphere in the curve along which the planet P will move. Eudoxos named this curve the *hippopede*, or 'horseshoe-curve'; its general form is shown in Fig. 7.7 and is a kind of spherical lemniscate.

So far, it can be seen that, because of the motions of the third and fourth spheres, the planet will move a little forwards and backwards, and up and down, relative to its mean position. If this motion is combined with the direct movement of the third sphere, the result will be a curve similar to that shown in Fig. 7.8. It appears that the planet now

Fig. 7.8. Loop-shaped path of
a planet in the system of Eudoxos.

has an overall movement in longitude towards the east, but also a
motion in latitude. Together they produce a loop reminiscent of Fig.
7.8.

Eudoxos succeeded through this mechanism in describing the main
phenomena of the planets in a qualitative way. His complete system
included a total of 27 revolving, concentric spheres—three for the Sun,
three for the Moon, four for each of the five planets, and a single sphere
outside the others carrying the fixed stars.

Criticism of Eudoxos

In the generation following his death Eudoxos's system proved vulner-
able to criticism in several ways. For instance, it implied that the Sun
moves with a constant apparent velocity among the fixed stars, from
which it follows that the four seasons are equal in length, contrary to
the knowledge of experience. For this reason the astronomer Calippus
tried to 'save' the inequality of the seasons by adding two more spheres
to the mechanism of the Sun, and at the same time the mechanisms of
several other planets were provided with additional spheres.

Aristotle also augmented the number of spheres, but for quite
different reasons. He regarded the spheres as real bodies of celestial
ether (it is not known how Eudoxos conceived them) and was unable to
imagine how they could move without disturbing each other. It was
also contrary to his physical principles to assume an extended empty
space between the spheres. Aristotle tried to satisfy these physical presuppositions by interposing a large number of 'compensating' spheres
to create a kind of dynamical isolation between the various 'revolving'
spheres of the system. This increased the total number of spheres, but
Aristotle thought one single sphere was sufficient to produce the diurnal
motion of all the planetary mechanisms, a simplification of the original
theory in which Eudoxos had placed a separate sphere inside each
mechanism to account for the daily rotation.

Despite these modifications, the system of Eudoxos proved unable
to satisfy astronomers. A disciple of Eudoxos, Polemarchos of Cyzicos

(c. 340 B.C.) realized that the apparent diameters of the Sun and Moon must vary. Therefore the system of Eudoxos, and other concentric theories were dealt a death blow when Autolycos of Pitane (c. 310 B.C.) pointed out that no modification of the system would allow changes in the distance of any planet from the Earth, since all spheres have the same centre. These, and perhaps other discrepancies formed the background of the theoretical investigations made by Greek astronomers in the Hellenistic period which followed Eudoxos and whose aim was to find acceptable alternations to Eudoxos's solution.

But even if Eudoxos's brilliant attempt proved to be a failure it had far-reaching consequences. In its Aristotelian version, the system survived for a very long time among scholars whose veneration of Aristotle, combined with their scanty mathematical equipment, made them ignorant of the obvious deficiencies of the concentric system. However, the models of Eudoxos encouraged the study of revolving spheres from a mathematical point of view, and it is not mere chance that the fiercest critic of the concentric system, Autolycos of Pitane, was the author of the first mathematical expositions of spherical astronomy which have come down to us under the titles *On the moving sphere* and *On risings and settings*. Later, this doctrine of 'Sphaerics' became a standard branch of Greek mathematics.

The Investigations of Apollonios

Among the scholars searching for new openings in theoretical astronomy was one of the very best mathematicians of the ancient world, Apollonios of Perga (c. 262—c. 190 B.C.). Little information persists about Apollonios as a practical astronomer; no astronomical treatise of his has been conserved, and the *Almagest* of Ptolemy gives only a brief exposition of Apollonios's results. However, it seems that Apollonios first concentrated on finding a model which was able to account for the varying distances of the Sun and Moon, upholding the general principle of uniform circular motions as the basic component of all planetary movements.

The first geometrical model investigated by Apollonios seems to have been the *eccentric circle*, that is (see Fig. 7.9) a circle whose centre, D lies outside the centre T of the Earth. Suppose a planet P is moving with constant velocity on this circle in the direction west to east. There will

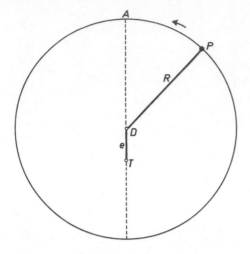

Fig. 7.9. The eccentric model of Apollonios.

then be a position A (the *apogee*) where P has a maximum distance $(R + e)$, and another position Π (the *perigee*) where the distance has a minimum value $(R - e)$. Here, R is the radius of the eccentric circle and e is the eccentricity TD.

Furthermore, suppose that the motion of P is uniform when seen from the centre D. Expressed in modern terms, this means that P has a constant angular velocity ω. It follows that seen from T this motion will be non-uniform, with a minimum apparent velocity at the apogee, and a maximum at the perigee. This eccentric model seems to simulate at least two of the features of planetary motion: the varying distance TP, and the varying apparent velocity, that is, the first anomaly.

Apollonios realized that the same results can be obtained by means of another model, whose fundamentals he may well have derived from the geo-heliocentric system of Heracleides of Pontos. This second model employs a circle of radius R concentric with the Earth, called the *deferent*, and another circle of radius r called the *epicycle*. The centre C of the epicycle moves on the deferent with a constant angular velocity ω_d (Fig. 7.10). The planet P moves on the epicycle with constant angular velocity ω_e.

In the eccentric model, the position of the planet relative to the Earth is given by the vector

$$\overrightarrow{TP} = \overrightarrow{TD} + \overrightarrow{DP}.$$

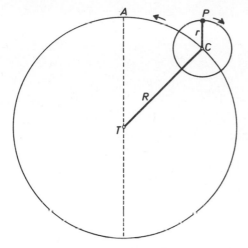

Fig. 7.10. The epicyclic model of Apollonios.

In the epicyclic model the corresponding relation is

$$\overrightarrow{TP} = \overrightarrow{TC} + \overrightarrow{CP}.$$

It follows that the two models are equivalent if firstly the vector $\overrightarrow{TC}$ is equal and parallel to $\overrightarrow{TD}$, and second that the vector $\overrightarrow{CP}$ is equal and parallel to $\overrightarrow{DP}$.

This second condition implies that the vector $\overrightarrow{CP}$ rotates around C from east to west with the same numerical angular velocity as the epicycle centre has on the concentric deferent, and as the planet has on the eccentric circle. This is an important theorem on the equivalence of the two models.

Apollonios then realized that the epicyclic model has further possibilities, and that it can be made to simulate retrograde motions by a suitable choice of the angular velocity of the planet on the epicycle. Fig. 7.11 shows a model in which the planet has a direct motion on the epicycle. This means that the apparent velocity of P, seen from T, has a maximum when P is in the neighbourhood of A on the epicycle, but a minimum in the neighbourhood of Π. This minimum may be negative if the angular velocity ω_e of P on the epicycle is large enough compared with the angular velocity ω_d of C on the deferent. In that case, P will be retrograde around Π. In order to find the first stationary point S_1 the velocity of P is resolved into two components. One is the vector

83

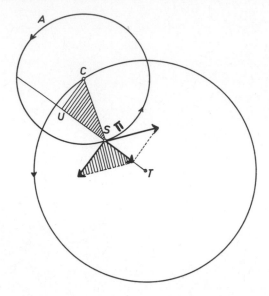

Fig. 7.11. Determination of the stationary points on Apollonios's model.

$\vec{v}_e$, due to the motion of P on the epicycle. It is perpendicular to $\overrightarrow{CS_1}$ and has the value $\omega_e.r$; the other is the vector $\vec{v}_d$ perpendicular to $\overrightarrow{TS_1}$ and with the value $\omega_d.TS_1$. The condition that S_1 is a stationary point is, that $\vec{v}_1 + \vec{v}_2$ is a vector through T. From the similar triangles shaded in the figure we then have,

$$\frac{v_e}{v_d} = \frac{CS_1}{S_1 U}.$$

But $CS_1 = r$, $v_e = \omega_e . r$ and $v_d = \omega_d . TS$. This gives the condition,

$$\frac{TS_1}{S_1 U} = \frac{\omega_e}{\omega_d}.$$

It should thus be possible to account for the second anomaly by a model of the last type.

The Solar Theory of Hipparchos

Apollonios succeeded in creating several new geometrical models of interest to theoretical astronomers. As far as is known, Hipparchos was the first to try adapting such models to actual observations.

Although Hipparchos had some knowledge of the elegant linear, algebraic methods of Babylonian astronomy and simply took over several Babylonian parameters, it is a remarkable fact that in his own work he chose to continue along the path outlined by Apollonios.

Hipparchos's first theoretical result was the creation of a theory of the motion of the Sun. To this end he chose the eccentric model. In Fig. 7.12 the outer circle represents the ecliptic with the centre T; the

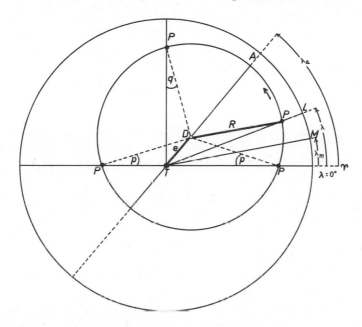

Fig. 7.12.   The theory of the motion of the Sun according to Hipparchos.

inner circle is the deferent with the centre D, placed on the apogeum line TA forming an angle λ_a with the direction from T to the vernal equinox ♈. On this deferent the Sun is supposed to move with a constant angular velocity $\omega = 360° : 365\frac{1}{4} = 59'8''$. Thus the angle $ADP = a(t)$ is a linear function of time.

The angle $\lambda = ♈ TP$ is the *true longitude* of the Sun. The *mean longitude* $\lambda_m = \lambda_a + a(t)$ is the angle $♈ TM$ where TM is parallel to DP. Also λ_m is a linear function of time.

Hipparchos first showed how the basic parameters R, e, and λ_a of this model could be derived from the length of the four seasons.

In the figure, P_1 is the position $\lambda = 0°$ of the Sun at the vernal equinox; P_2 is the position $\lambda = 90°$ at summer solstice $94\frac{1}{2}$ days later, and P_3 is the position $\lambda = 180°$ at the autumnal equinox $92\frac{1}{2}$ days later again.

In a slightly modernized form the method used by Hipparchos is as follows. From the triangles TDP_1 and TDP_2 we have

$$\frac{R}{\sin \lambda_a} = \frac{e}{\sin p} \quad \text{and} \quad \frac{R}{\cos \lambda_a} = \frac{e}{\sin q},$$

from which follows

$$tg\lambda_a = \frac{\sin p}{\sin q} \quad \text{and} \quad \frac{e}{R} = \frac{\sin p}{\sin \lambda_a}.$$

The small angles p and q can be found from the lengths of the seasons. Since the daily motion of the Sun on the eccentric is $\omega = 59'8''$ the arc $P_1P_2P_3$ traversed by the Sun from the vernal to the autumnal equinox is

$$p + 180° + p = (94\tfrac{1}{2} + 92\tfrac{1}{2}) \cdot \omega.$$

In the same way, the arc P_1P_2 from the vernal equinox to the summer solstice is

$$p + 90° + q = 94\tfrac{1}{2} \cdot \omega.$$

According to Hipparchos the solutions of these equations are

$$\lambda_a = 65°30' \quad \text{and} \quad \frac{e}{R} = \frac{1}{24}.$$

By this method, only the ratio e/R can be found but not the absolute values of e and R which would demand that the parallax of the Sun could be measured, which was impossible in ancient astronomy.

Hipparchos's Theory of the Moon

It seems that Hipparchos's ultimate aim was a theory enabling him to predict eclipses (see p. 57), which also demanded a theory of the motion of the Moon. The *Almagest* reveals that Hipparchos made use of the epicyclic model to this end (see Fig. 7.13). Here the deferent vector

86

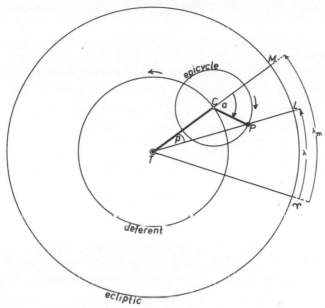

Fig. 7.13.   *The lunar theory of Hipparchos.*

$\overrightarrow{TC}$ rotates from west to east with a period equal to one tropical month and a constant angular velocity ω_t. It forms an angle $\lambda_m = \Upsilon TC$ with the vernal equinox, called the *mean longitude* of the Moon, and increasing as a linear function of time. The Moon itself revolves on the epicycle from east to west with a constant angular velocity ω_a relative to the line TCA, corresponding to a period of revolution equal to one synodic month. Thus the angle $ACP = a_m$, called the (mean) *anomaly*, is a linear function of time.

The parameters of this model are R, r, and the true longitude $\lambda(t_0)$ of the Moon at a given time t_0. Just as the parameters of the solar theory were found from three different positions of the Sun, Hipparchos was able to find the ratio r/R from three different positions of the Moon, defined by three lunar eclipses (chosen because the eclipsed Moon is diametrically opposed to the Sun so that its longitude can be found from the solar theory).

When the parameters of both models are known it is possible to adapt them to practical computations, that is to calculate the true longitudes of the Sun and Moon from a relation of the form

$$\lambda(t) = \lambda_m(t) \pm p(a_m),$$

87

where $\lambda_m(t)$ is the mean longitude at the time t, and p is a small correction which Ptolemy called the *prostaphairesis* angle (meaning the addition-subtraction angle). It was later called the *equation*. In the solar theory p is the angle under which the eccentricity TD is seen from the Sun, and in the lunar theory it is the angle under which the epicycle radius to the actual position of the Moon is seen from the centre of the Earth. In both cases p is a function of one variable only, namely the mean longitude of the Sun, or the (mean) anomaly of the Moon.

Ptolemy and the Almagest

Practically nothing is known of the development of theoretical astronomy in the time after Hipparchos. A chapter in Pliny's *Natural history* quotes a number of apogeum longitudes λ_a for each of the five planets, and indicates that one tried to describe planetary motion by a system of eccentric circles without epicycles in analogy with the solar theory, thus disregarding the second anomaly of the planets. Only the *Almagest* of Klaudios Ptolemaios (*c*. A.D. 100–*c*. 170) from about A.D. 145 reveals the results of the work of Hellenistic astronomers, with his own formidable achievements as the final contribution to a synthesis destined to last for more than a thousand years.

Ptolemy was the author of a long series of works showing his intention to compile a kind of encyclopedia of applied mathematics. Of his books on statics and mechanics only the titles remain. The *Optics* is partly extant in a Mediæval translation from the Arabic, and his *Planetary hypotheses* is known from a part of the original Greek version and Arabic translations of the rest. Some minor works on the *Analemma* (orthogonal projection for sundials, see Fig. 5.5) and the *Planisphaerium* (stereographic projection of the celestial sphere) are preserved in Greek, and this is the case, too, with the great *Geography*. But Ptolemy's major work is his great exposition on theoretical astronomy which has placed him among the great astronomers of all time. The original title of the work, the *Megalé syntaxis* (the Great Composition) was translated into Arabic in the 9th century as al-Majisti, which was corrupted to *Almagest* by Latin translators of the 12th century. This large work in 13 books summarizes the astronomical theories of the Hellenistic period, and is also important to historians as the only source of Apollonios's and Hipparchos's achievements in this field.

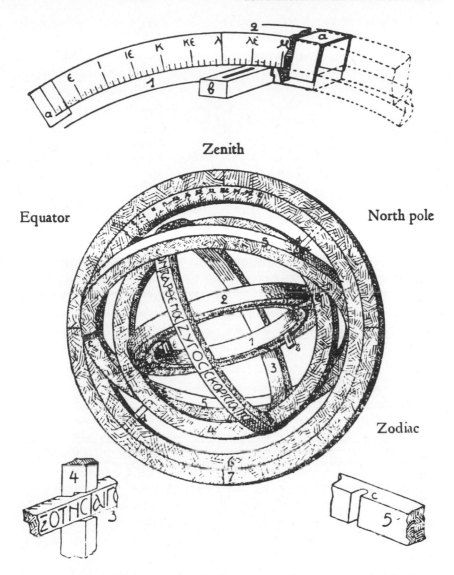

Zenith

Equator

North pole

Zodiac

Fig. 7.14. *The armillary sphere described by Ptolemy in the* Almagest. *The ecliptic and equator are represented by rings provided with diopters. They can be adjusted inside a fixed vertical ring representing the meridian. Ptolemy used it mainly for direct measurements of longitudes and latitudes. No instrument of this type has survived from Antiquity. The figure above is a reconstruction by P. Rome in A. Rome,* Pappus; Commentaire sur l'Almageste', *Città del Vaticano, 1931–33, I, 5.*

In Book I of the *Almagest* Ptolemy states his reasons for adhering to the geocentric conception of the universe. This book reveals that in physics he was an Aristotelian with strong Stoic tendencies. It also contains some mathematical chapters, among which is a table of chords listing the function

$$J = \text{chord } x = 2R \sin \frac{x°}{2} \quad \text{for} \quad R = 60,$$

which in the rest of the work serves instead of a table of sines. Several propositions in plane and spherical trigonometry are derived, mainly by means of Menelaos's theorem. This is the mathematical basis of *Book II* which contains a trigonometrical exposition of spherical astronomy (Fig. 7.14).

Book III introduces theoretical astronomy with an account of Hipparchos's solar theory, which Ptolemy takes over unchanged. Only the parameters are checked from observations, and the theory is made fit for practical use by the addition of tables of the mean motion of the Sun as a function of time, and of the prostaphairesis angle. A beautifully-proved theorem shows that this angle is a maximum at the points where the deferent circle is intersected by a line through the Earth perpendicular to the apogeum line.

The Development of the Lunar Theory

Book IV of the *Almagest* is devoted to a very careful discussion of the concentric lunar theory of Hipparchos. The parameters are newly determined from observations, with the result that the radius of the epicycle is determined as $5^p;15$ for a deferent radius of $60^p;0$. The lunar theory was based, however, on eclipses, that is, on positions of the Moon in conjunction with or in opposition to the Sun. In *Book V* Ptolemy examines the theory in the light of more observations made partly by himself, and he makes the discovery that the theory did not account with sufficient accuracy for the position of the Moon at quadratures, where the computed maximum equation was found to be $5°1'$ whereas the observed maximum was $7°40'$. The computed positions differed from those observed by an amount which proved to be dependent on the elongation of the Moon, that is on the position of the Sun relative to the Moon. In other words, the Moon shows not only the first, but also the second anomaly (see p. 73). Considering that this does not appear visibly in the form of retrograde motions (as in the five planets) this discovery of the *evection* of the Moon was one of Ptolemy's major achievements.

As a result of this discovery Ptolemy was forced to change the lunar model of Hipparchos into another model able to simulate both the first and the second anomalies of the Moon. There is no space for the details of how he succeeded in this, only for the final result. The Moon (Fig. 7.15) still moves in retrograde direction on an epicycle the centre of

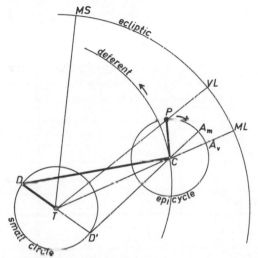

Fig. 7.15. Ptolemy's final lunar theory.

which moves on a deferent circle. But this circle has now become eccentric, and also rotating; its centre D revolves on a small concentric circle in such a way that the angle ATC between the apogeum of the deferent and the epicycle centre is always bisected by the line TM defining the mean position of the Sun according to the solar theory. This produces a coupling between the two luminaries which draws the epicycle nearer to the Earth at quadratures, thus augmenting the maximum equation. Finally, it was found necessary to calculate the position of the Moon on the epicycle not from the true apogee A_v of the epicycle but from a mean apogee A_m, this point being the intersection between the epicycle and a straight line defined by T and the point D' on the small concentric circle opposite to D.

*Fig. 7.16. Ptolemy's parallactic ruler for measuring zenith distances showing
'King' Ptolemy with a crown on his head—in mediaeval times the Alexandrian
astronomer was identified with an earlier Egyptian king of the same name. From
William Cuningham,* The Cosmographicall Glasse, *etc. London, 1559.*

Ptolemy found this model sufficient to account for the longitude of the Moon. The motion in latitude was produced simply by placing the deferent and epicycle in a plane inclined 5° to the plane of the ecliptic. The *Almagest Book V*, concludes with a section on the distances of the Moon and the Sun, containing the first mathematical theory of parallax (Fig. 7.16), after which Ptolemy is able to develop a theory of eclipses comprising the whole of *Book VI*. The only serious flaw in Ptolemy's treatment of the motion of the Moon is that his model led to a variation of the distance of the Moon between the limits of 33 and 64 radii of the Earth. This implies that its apparent diameter varies by a factor of two during each revolution, a glaring contradiction of fact which was accepted by Ptolemy without misgivings, presumably because the theory did account for the longitude and latitude of the Moon which were astronomically much more important.

The Fixed Stars and Precession

In *Book VII* and *Book VIII* of the *Almagest* Ptolemy gives a catalogue of the longitude, latitude, and magnitude of 1022 fixed stars, based on an earlier list (no longer extant) drawn up by Hipparchos. There is also a chapter on precession (see p. 56), which is conceived of as a uniform rotation of the sphere of the fixed stars eastwards relative to the ecliptic at a rate of 1° in 100 years, or 36″ per year.

The Superior Planets

Books IX to XI give the theories of the longitudes of the five planets— Mercury, Venus, Mars, Jupiter, and Saturn. In each case a geometrical model is proposed, its parameters determined from observations, and the model tested against other observations. The problem is to determine the longitude $\lambda(t)$ at a given time t from the mean longitude $\lambda_m(t)$, taking both the first and second anomalies into account. This is done by a model which is essentially the same in the case of the three superior planets Mars, Jupiter, and Saturn.

The first anomaly is accounted for by means of an eccentric circle (as in the solar theory) or deferent, and the second anomaly by an epicycle riding on the deferent (Fig. 7.17). The coupling to the motion

93

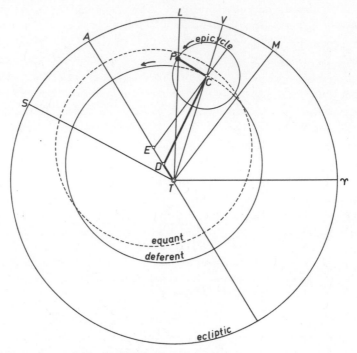

Fig. 7.17. Ptolemy's theory of the superior planets.

of the Sun implied by the existence of a second anomaly is produced by letting the epicycle vector *CP* be parallel to the direction *TM* from the Earth to the mean Sun. This implies that the planet has a direct (eastward) motion on the epicycle (contrary to the Moon), and that the period of revolution on the epicycle is one year for all the three planets.

The principle that uniform circular motion was the only kinematical tool acknowledged by Greek astronomers has been mentioned several times. Ptolemy himself declared it to be the only kind of motion 'in agreement with the nature of the Divine Beings'. Nevertheless, he did not hesitate to abandon it when experience forced him to do so. An example of this is the introduction of the so-called 'equant point' into the theory of the superior planets. Here Ptolemy used three oppositions to determine the radius of the epicycle, on the assumption that the epicycle centre moves upon an 'equant circle' with the centre at the equant point *E* with a motion which is uniform seen from *E* (Fig. 7.18). But without further explanation Ptolemy then states that he

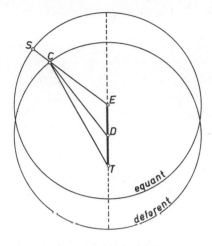

Fig. 7.18. Figure illustrating the transition from the equant to the deferent in the Ptolemaic theory of the superior planets.

found the epicycle to revolve in fact with uniform velocity upon a circle of equal size, but not with the same centre. This other circle is the deferent and its centre D is determined as the mid-point of the line TE. It is obvious that since the motion of the epicycle centre is uniform when seen from E, it will be non-uniform seen from D.

One possible reason for this odd step could be that Ptolemy was experiencing trouble with the apparent size of the epicycle and had to draw it nearer to the Earth at the apogee and push it further away at the perigee. In that case the passage from the equant to the deferent circle can be thought of as the first step in an iterative process which was carried no further, although this is a mere hypothesis. Yet in order to depart from the fundamental dogma of Greek astronomy Ptolemy must have had very strong reasons which only a serious disagreement with observations could provoke.

Venus and Mercury

The theories concerning the inferior planets Venus and Mercury had to account for the fact that these planets are never in opposition, but travel round the heavens in rather close company with the Sun, their maximum elongations being 46° and 28° respectively (see p. 62). This implies that their mean positions are the position of the Sun. Ptolemy obtained this by the simple device of making the period of revolution of the epicycle on the deferent equal to one year. It follows that—compared with the models for the superior planets—the epicycle and deferent have changed roles.

Of all the planetary theories of the *Almagest* the Mercury theory is the most curious. Here, Ptolemy had begun by determining the apogee of Mercury by the line of symmetry of its orbit. As for the other planets, the perigee ought to be placed 180° before (or after) the apogee. But some erroneous observations led Ptolemy to postulate two different perigees placed 120° before and after the apogee, which meant that the epicycle had to be drawn extra near to the Earth in these two directions. From the lunar theory it is known that this can be achieved by a mechanism comprising a moving deferent. Working along similar lines, Ptolemy arrived at the model shown in Fig. 7.19. Here, the planet *P*

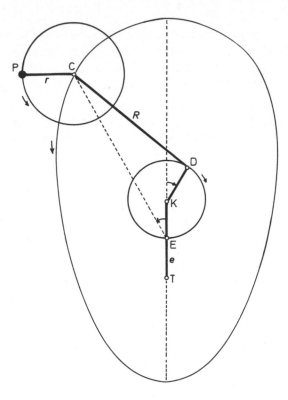

Fig. 7.19.   The Mercury theory on Ptolemy's model showing the oval locus of the epicycle centre which, for small values of the eccentricity, approaches an ellipse.

rotates on an epicycle around the centre C. But C does not revolve on a fixed deferent, but has a constant distance from the point D which rotates on the small circle around K in such a way that KD and EC form equal, but opposite, angles (increasing as linear functions of time) with the apsidal line TD. The point E is an equant point situated in the middle of TD. The locus of C is not a circle as in the other planets, but an oval curve not mentioned in the *Almagest*.

Computation of Longitudes

Although his various planetary theories are somewhat different Ptolemy succeeded in drawing them into a common formula suitable to numerical calculation of the true, ecliptic longitude $\lambda(t)$ of the planet at a given time t. In modern notation and reckoning longitudes from the vernal equinox (Ptolemy usually reckoned from the apogee) this formula can be written

$$\lambda(t) = \lambda_m(t) \pm \varphi(c_m) \pm p(c_m, a_m).$$

Here $\lambda_m(t)$ is the mean longitude of the planet, i.e. the angle ΥTM in Fig. 7.17. It is computed according to the formula

$$\lambda_m(t) = \lambda_m(t_0) + \omega_t (t - t_0),$$

where $\lambda_m(t_0)$ is the mean longitude at a definite epoch t_0 ; ω_t is the tropical angular velocity found from a particular set of tables of mean motion.

The term $\varphi(c_m)$ accounts for the first anomaly. It is called the equation of centre and is a function of one variable, the *mean centrum* $c_m = \lambda_m - \lambda_a$ where λ_a is the longitude of the apogee. The double sign is necessary because negative numbers were unknown in Ancient and Mediaeval mathematics.

The last term $p(c_m, a_m)$ represents the second anomaly and can be interpreted as the angle under which the epicycle radius CP to the planet is seen from the Earth. It depends mainly on the position of CP, characterized by the angle ACP called the *mean argument* (or mean anomaly) and given by $a_m = \lambda_m - \lambda_s$ where λ_s is the mean longitude of the Sun. But it also depends, although to a lesser extent, on the distance of the epicycle from the Earth, which is, again, a function of c_m alone.

97

To make practical computations easier, Ptolemy draws up a special tables of equations. Here $\varphi(c_m)$ is listed as a function of c_m. The more difficult function $p(c_m, a_m)$ is handled by an approximation method. Each table contains two separate columns giving p as a function of the strong variable a_m only, but for two special values of the weak variable c_m. Intermediate values of c_m are found by an interpolation scheme.

Retrograde Motions and Latitudes

Book XII of the Almagest is devoted to retrograde motions. Ptolemy has to show that although the various models are derived mainly from observations of oppositions they will reproduce the retrograde movements and stationary points found by observation (Fig. 7.20). To this

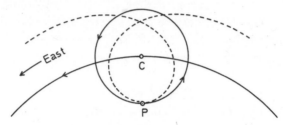

Fig. 7.20. Figure illustrating how the actual orbit of a planet (dotted line) is produced by the combined motion of the epicycle and the deferent (see Fig. 7.1).

purpose he makes use of Apollonios's method of finding stationary points (p. 83). Now this method was based on a model with a concentric deferent and, strictly speaking, was not applicable to the eccentric deferents used by Ptolemy. Nevertheless, it is used as a first approximation because the distance of the planet from the Earth does not vary very much during the rather brief retrograde motion.

The first twelve books of the Almagest deal only with the motion in longitude. In the final Book XIII Ptolemy takes up the problem of latitudes. Here he first places the deferent of the planet in a plane inclined to the plane of the ecliptic and intersecting it along the nodal line. In the superior planets the inclinations are independent of time, whereas in the inferior planets they are oscillating functions. In the lunar theory Ptolemy had placed the epicycle, too, in the inclined plane, but in dealing with the five planets had assumed that the epi-cycles deviate from it which makes the theory of latitude both inelegant and complicated. It is clear that the annual motion on the epi-cycle reflects the motion of the Earth around the Sun in the Coper-nican system (as far as the superior planets are concerned) and it would have been better, therefore if Ptolemy had made the planes of the epicycles parallel to the plane of the ecliptic. It is interesting that Ptolemy made this change in the models underlying the *Handy tables* composed after the *Almagest* had been finished. But it is impossible to learn the reasons which influenced this change. *Book XIII* concludes with a theory of the visibility periods of the planets.

The Impact of Ptolemaic Astronomy

With the *Almagest*, Greek theoretical astronomy had reached its culmination, and for many centuries Ptolemy's Great Synthesis was the obvious foundation both of further progress and of the teaching of astronomy. Firstly, it had achieved the aim of Greek astronomers held since Plato challenged them to create a planetary theory founded on uniform circular motion. The introduction of the equant was, it is true, a departure from this general principle, but it was well hidden and easily overlooked among the more technical details of the various models. Next, it was found that Ptolemy's theories agreed suffic-iently with observations to be acceptable to practical astronomers, although the variation of the apparent diameter of the Moon (p. 81) showed that improvements were still possible. Finally, Ptolemy had set forth his theories in a hypothetical–deductive form, having, at the same time, carefully explained how the parameters of the models were deduced from observations and could be tested against them. This combination of deductive and inductive reasoning makes the *Almagest* a difficult work to study. But it had the advantage of revealing not only the results of Ptolemaic astronomy, but also the way in which astro-nomy is created. It is no wonder that all great astronomers until Kepler's time found inspiration in Ptolemy, even if they abandoned the particular models he had developed.

The most general feature of the *Almagest* is the extremely abstract geometrical character of its theories. There are very few references to the physical properties of the universe. Ptolemy succeeded in furnishing the proof that it is possible to give a complete mathematical description of planetary motions without entering on such questions as the nature of celestial matter or the forces keeping the planets in motion. This abstract presentation contrasted strongly with the picture of a universe composed of ethereal spheres in which Aristotle had tried to embed the astronomy of Eudoxos. Philosophers of Aristotle's kind continued to argue that the universe has a physical structure, and that no description can be complete without taking it into account. Astronomers, and in particular astrologers, saw their main object as the prediction of planetary positions, for which the abstract mathematical models were sufficient.

The Ptolemaic System of the World

Gradually, two different schools of astronomy emerged, one trying to create a physical astronomy, the other concentrating on planetary theory. Though it started in Antiquity, this debate continued throughout the Middle Ages. In general, the 'physical' Aristotelian universe was the basis of more popular expositions, while most professional astronomers held to the mathematical point of view. Because of the abstract character of the *Almagest*, Ptolemy has often been called a theoretician and the same conclusion seemingly emerges from a study of the *Handy tables*, in which Ptolemy reworked the tables of the *Almagest* and presented them in a form better adapted to rapid calculations. The tables were provided with a set of *canones* or instructions on how to use them. But they, too, were completely free of any reference to the physical properties of the universe. However, in his *Planetary Hypotheses* written sometime after the *Almagest* Ptolemy subscribed to a physical universe composed of spheres. Each sphere, he thought, is a spherical shell containing several ethereal bodies representing the deferent and the epicycle, with some additional bodies to keep the latter in the correct place. The picture (Fig. 7.21) from a late Mediæval treatise is very similar to the illustrations found in Ptolemy.

An interesting feature of the *Hypotheses* is Ptolemy's attempt to compute the size of the universe from the assumption that the celestial spheres fit closely together so that, for example, the sphere of Jupiter starts where the sphere of Mars ends. Absolute measures are derived from the maximum and minimum distances of the Moon found in the

Theorica trium fuperiorum: & Veneris.

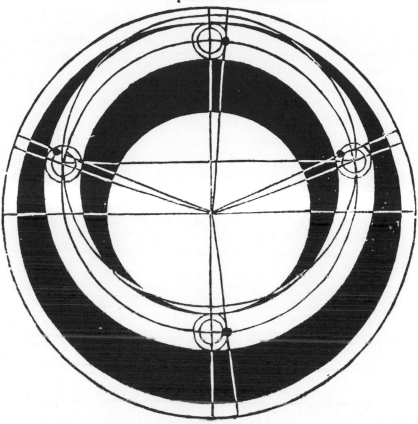

Fig. 7. 21. Ptolemy's theory of the superior planets and Venus in a late Mediaeval 'physical' representation with ethereal spheres. From Peurbach, Novae theoricae planetarum *(1460), Augsburg, 1482.*

Almagest (p. 93), and show that the spherical shell containing the whole lunar mechanism must have an inner radius of 33 and an outer radius of 64 radii of the Earth. As a result, the inner sphere of Mercury has an inner radius of 64 Earth-radii. Its outer radius can then be found from the ratio of its maximum and minimum distances implied in the Mercury theory. In this way Ptolemy arrived at an outer radius of the sphere of Saturn of 19,865 radii of the Earth, or about 120,000,000 km. (less than the actual value of the radius of the orbit of the Earth).

Ptolemy's solar system was rather small in dimensions and surrounded by the eighth sphere containing the fixed stars. But Ptolemy does not indicate the thickness of this sphere or the total radius of his universe. It is quite impossible that he regarded it as infinite considering his adherence to the general principles of Aristotelian cosmology, and the various proofs that a rotating universe cannot be infinite. It is worth noticing that outside this finite spherical universe nothing exists, not even an empty space. To Aristotle, space is that which contains something, so that the space of the universe is the surface of the outermost sphere containing everything which exists. The question, what is outside the universe? is to Aristotle and presumably to Ptolemy, a pseudo-problem based on a misconception of the word 'space'.

Chapter Eight

The Foundation of Statics

S TATICS is the study of the equilibrium of a body under the action of given forces and arose as the art of using the balance and the lever. Since it was closely connected with the practical problems of life, this art developed very early in the history of civilization. A small balance and the weights belonging to it were found in Egypt and dated to a period

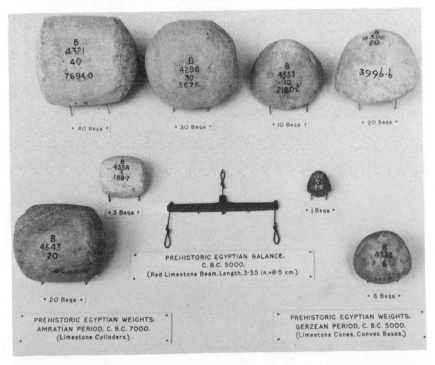

Fig. 8.1. *Balance and weights of stone from Naqada in Egypt, probably dating from the fifth millenium* B.C. (By courtesy of *Science Museum, South Kensington, London.*)

about 3500 B.C., that is before the beginning of Egypt's dynastic history (Fig. 8.1). It is also known that the Egyptians were familiar with the use of levers as tools, in building, for example where they applied the principle that a heavy load on a short lever arm can be raised by a small force on a long lever arm. It can be safely assumed that the Greeks, too, were familiar with balance and lever at an early date.

The Greek who first attempted to develop a true science from the practical art of statics remains anonymous. Archytas of Tarentum (early fourth century B.C.) is often quoted as the founder of 'mechanics', which can be taken to mean statics. If this is correct, then his writings on the subject have since been lost, as have any other works on statics which may have existed before Archimedes of Syracuse (c. 287–212 B.C.) wrote the first extant treatises on scientific statics. Archimedes' accomplishment was so exact and so full that almost overnight this new science became the most complete branch of ancient mathematical physics which here reached a high standard never achieved in other fields.

Archimedes and the Foundation of Statics

For most of his life, Archimedes worked in his home town Syracuse, having studied at Alexandria in his youth. By the third century Alexandria had superseded Athens as the centre of Greek learning, and here Archimedes made the acquaintance of Eratosthenes, the mathematician Conon and others. Most of Archimedes' works deal with problems in pure mathematics, but, two of his treatises are concerned with mathematical physics. The first of these works is *On the equilibrium of planes*, which discusses the statics of rigid bodies. The second called *On floating bodies*, is the foundation of hydrostatics.

The treatise *On the equilibrium of planes* begins, without any form of introduction, with the words 'I postulate the following', succeeded by seven 'postulates', or 'axioms' and then by fifteen 'theorems' (propositions) which are proved from the axioms. There is nothing more. Later the work was supplemented by a second part containing ten further propositions, and the whole is very reminiscent of a geometry textbook. What the modern physicist finds particularly surprising is that the axioms are chosen, as it were, out of thin air, even though they deal with physical phenomena and are by no means self-evident.

Today, an equivalent publication would begin with a statement of experimental evidence and then explain how the proposed assumptions for the theoretical discussion were selected. Archimedes does neither of these things. Naturally, he relies on his familiarity with instruments such as balances but he studiously avoids any reference to them. Perhaps he has understood Plato's desire to take geometry as the paradigm for the description of nature to mean that physics should give its results as abstractions, without mentioning the preceding experimental work. Certainly this would be in keeping with Plato's call to astronomers to divert their attention from the visible phenomena, but, there is evidence to suggest that Archimedes considered it beneath his dignity to discuss such things. His biographer Plutarch (c. A.D. 46–120), says of him:

> Archimedes possessed such a lofty spirit, so profound a soul, and such a wealth of scientific theory, that although his inventions had won for him a name and fame for superhuman sagacity, he would not consent to leave behind him any treatise on this subject, but regarding the work of an engineer and every art that ministers to the needs of life as ignoble and vulgar, he devoted his earnest efforts only to those studies the subtlety and charm of which are not affected by the claims of necessity.

It may well be that such words tell more of the idealistic bias of their author than of Archimedes; but it is certain that Archimedes can, as the many stories about him maintain, be credited with mechanical inventions. Through these inventions he acquired such a great insight into the practical aspects of the study of equilibrium that he knew beforehand what the results of theoretical statics should be. He then endeavoured to find a system of axioms from which these results could be derived by deduction. In the final exposition both the strength and the weakness of Greek physics are conspicuous: the strength is the clear and logical establishment of theory, the weakness is the lack of a presentation of the experimental evidence.

Doctrine of the Centre of Gravity

The seven postulates beginning Archimedes' statics read as follows:

1. Equal weights at equal distances are in equilibrium, and equal weights at unequal distances are not in equilibrium, but incline towards the weight which is at the greater distance.

2. If, when weights at certain distances are in equilibrium, something be added to one of the weights, they are not in equilibrium, but incline towards that weight to which the addition was made.

3. Similarly, if anything be taken away from one of the weights, they are not in equilibrium, but incline towards the weight from which nothing was taken.

4. When equal and similar plane figures coincide if applied to one another, their centres of gravity similarly coincide.

5. In figures which are unequal but similar the centres of gravity will be similarly situated.

6. If magnitudes at certain distances be in equilibrium, (other) magnitudes equal to them will also be in equilibrium at the same distances.

7. In any figure whose perimeter is concave in (one and) the same direction the centre of gravity must be within the figure.

The first observation which can be made about this set of axioms is that, in order to prove the results which follow, it is unnecessary to use precisely these axioms. It can easily be shown that the following five axioms could have replaced the seven and simplify the logic involved:

I. Any heavy body has one, and only one, centre of gravity.

II. Similar bodies have centres of gravity in similar points.

III. A body's centre of gravity remains unchanged in so far as two equally heavy parts of it are brought together at the mid-point of the straight line which previously joined their centres of gravity.

IV. If a body is composed of two bodies which have a common centre of gravity, this will also be the composite body's centre of gravity.

V. A rigid body suspended at its centre of gravity will be in equilibrium.

Formulated in this way the first four axioms imply a definition of a geometrical point called the centre of gravity, while the physical content of this system of axioms is arrived at only in Axiom V.

Among Archimedes' seven postulates, the sixth calls for special consideration, since at first glance it seems to be tautological: if two weights A_1 and A_2 keep one another in equilibrium at distances l_1 and l_2 from the point of suspension it seems obvious that weights B_1 and B_2

will also be in equilibrium at the same distances, providing that $A_1 = B_1$ and $A_2 = B_2$. Yet it is hardly conceivable that a philosopher of Archimedes' calibre would have introduced a tautology into an otherwise carefully selected set of axioms.

Postulate 6 can be shown to have its rightful place in the set. If one considers how the weights A_1 and A_2 could possibly differ from B_1 and B_2 the answer is clear: they could be different in form. From a physical point of view, this axiom contains therefore an assumption that an alteration in the shape of the weights will not affect the equilibrium when the weights and the centres of gravity are the same as before. This is the same as an assumption that a body can only influence the lever by its own weight and the position of its centre of gravity. In the proof of the law of the lever, argued below, the assumption is in fact used, and the proof does not, as is sometimes maintained, depend on assumptions not contained in the stated axioms.

The most curious thing about the set of axioms is that the concept of centre of gravity is never defined. It is introduced in Postulate 4 as something well known and obviously familiar to the reader. As earlier authors do not mention the concept, and because Archimedes did not define it elsewhere in his known works, it has been conjectured that the concept of a body's centre of gravity was formed by Archimedes himself, and presented in a work which has since been lost. This treatise, *On the centre of gravity*, should then have preceded his work *On the equilibrium of planes*. This is of course only a hypothesis, but it does explain the apparent discrepancy.

Centre of Gravity Determination

On the basis of his postulates, Archimedes next determined the position of the centre of gravity for a series of plane figures. Here, (with a few unimportant changes) his proof that the centre of gravity of a parallelogram lies on the straight lines joining the mid-points of opposite sides (Proposition 9) is considered. In a parallelogram $ABCD$, let the mid-point of AB be E and that of DC be F. The aim is to show that the centre of gravity lies on EF.

First, assume that the centre of gravity T does not lie on EF. Through T a straight line can be drawn parallel to AB to cut EF in G. EB is then divided into equal parts, each of which is shorter than TG; this is always possible. *AE* is divided into parts of equal size and lines are drawn parallel to *AD* through all the points thus marked out. *ABCD* is consequently divided into an even number of small congruent parallelograms (see Fig. 8.2) which are numbered in the diagram from 1 to 10.

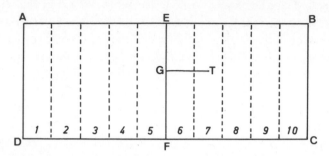

Fig. 8.2. *Archimedes' determination of the centre of gravity of a parallelogram.*

The centres of gravity of the parallelograms must lie on a line parallel to *AB* (Postulate 4 or 5). If numbers 1 and 10 are now imagined as moving to cover parallelograms 5 and 6 respectively, this cannot affect the position of the centre of gravity *T*, since they are the same size and have been moved through the same distance. Similarly parallelograms 2 and 9, 3 and 8, or 4 and 7, could be moved to cover 5 and 6, showing that the centre of gravity must lie inside 5 and 6—contrary to the assumption that it lies at *T*, which is found in parallelogram 7 (Postulate 7). Clearly, such a situation occurs whenever *T* lies off the line *EF*. It follows that *T* must lie on *EF*.

A similar argument shows that *T* must lie on the line joining the midpoints of *AD* and *BC*. The centre of gravity lies, therefore, at the centre of the parallelogram (point of intersection of the diagonals).

The Law of the Lever

Another example of the way in which Archimedes proves his theorems on the basis of these axioms is his derivation of the law of the lever, stated as Proposition 6, which can now be studied. The physics contained in the theorem is not that of Archimedes, but was known long before and referred to by Aristotle.

Let A and B (Fig. 8.3A) be the centres of gravity of two bodies whose

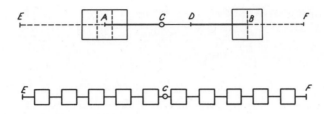

Figs. 8.3A. and B.   Archimedes' derivation of the law of the lever.

weights are taken to be commensurable in the ratio $p:q$, where p and q are whole numbers taken here, for simplicity, to be 3 and 2. C is the point which divides AB internally in the ratio $q:p$ and D is chosen so that $AC = DB$, that is, D divides AB in the ratio $p:q$. On the line through A and B lines are drawn to the left from A and to the right from B equal, respectively, to AD and BD. Thus C becomes the mid-point of the extended line EF. In Fig. 8.3B the two bodies are distributed evenly over ED and DF respectively in $2(p + q)$ equal small parts; such a distribution is always possible, since the lines on which the weights are spread symmetrically about their original positions, A and B, are proportional to the weights, and consequently to the number of small parts. From Postulates 3 and 4 it follows then that Fig. 8.3B has its centre of gravity at the mid-point C, since the two outermost parts can first be brought into C, then the next two, and so on, without changing the centre of gravity of the whole system. But from the same postulates it follows similarly that this centre of gravity lies at the centre of gravity of the original system, as without altering the centre of gravity C the situation of Fig. 8.3A can be reproduced. It has been shown, therefore, that for the commensurable case the centre of gravity of two bodies lies on the line joining their individual centres of gravity and divides it in a ratio inversely proportional to their weights. Moreover, from this proof it can be said, further, that if these two bodies constitute parts of a more extensive system, its centre of gravity will not change if the two bodies are concentrated together at their common centre of gravity C, since replacing the two bodies by the system shown in Fig. 8.3B causes no change in the centre of gravity of the system, nor does a subsequent recombination of the small parts in C.

The extension to two incommensurable weights is investigated by Archimedes in Proposition 7, and the law of the lever is shown to hold in this more general case.

The 'Method' of Archimedes

Archimedes made a very great impression through his vigorous mathematical methods, particularly because his writings presented the final results in a very elegant form, although without revealing how they were first found. Plutarch says:

> It is not possible to find in geometry more profound and difficult questions treated in simpler and purer terms. Some attribute this success to his natural endowments; others think it due to excessive labour that everything he did seemed to have been performed without labour and with ease. For no one could by his own efforts discover the proof, and yet as soon as he learns it from him, he thinks he might have discovered it himself, so smooth and rapid is the path by which he leads one to the desired conclusion.

Secrecy about methods was not unusual in ancient times, when new theorems were often published without proof so that others could test their abilities on them; and sometimes Archimedes knowingly published incorrect results to force other mathematicians to think critically. But even if results were often presented synthetically, there must have been a foregoing analytical treatment. It was therefore immensely important to the whole knowledge of ancient science when, at the beginning of the 20th century, Professor J. L. Heiberg of Copenhagen found a manuscript in Constantinople which contained, among other things, a book, *The method*, by Archimedes, a treatise Archimedes was known to have written, but believed to be lost. The manuscript was a so-called 'palimpsest'. It originally contained Archimedes' text, copied in the tenth century, but in the thirteenth century an attempt was made to erase this and to write over it with a different text. Luckily, however, most of the original text could be restored.

The work *The method* is dedicated to Eratosthenes. The preface reads:

> Seeing moreover in you, as I say, an earnest student, a man of considerable eminence in philosophy, and an admirer [of mathematical inquiry], I thought fit to write out for you and explain in detail in the same book

the peculiarity of a certain method, by which it will be possible for you
to get a start to enable you to investigate some of the problems in mathematics by means of mechanics . . . for certain things first became clear
to me by a mechanical method although they had to be demonstrated
by geometry afterwards because their investigation by the said method
did not furnish an actual demonstration. But it is of course easier, when
we have previously acquired, by the method, some knowledge of the
questions, to supply the proof than it is to find it without any previous
knowledge.

Initially, Archimedes states that he will use as an example for Eratosthenes, the expression for the area of a segment of a parabola, followed
by the axioms on which his methods are built. These are simply the
statics theorems which had already been proved in the book *On the
equilibrium of planes*. As an example of the use of the mechanical method
we shall examine the derivation of the expression for the volume of a
sphere.

The volume of a sphere is four times that of a cone which has its base
equal to the greatest circle of the sphere and its height equal to the radius
of the sphere.

Figure 8.4 shows a section through a sphere, a cone and a cylinder,

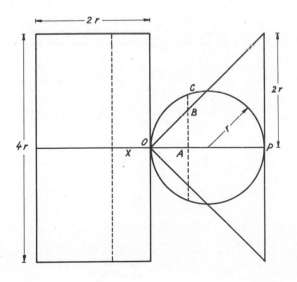

Fig. 8.4. Archimedes' derivation of the volume of a
sphere.

with a common axis and dimensions as given in the diagram. Cutting the sphere and cone in a plane, perpendicular to the axis at the point A, distance x from o, two circular surfaces are found. The sum of their areas is

$$\pi[(AB)^2 + (AC)^2] = \pi[x^2 + (2r - x)x] + 2\pi\, rx.$$

As the figures are considered to be homogeneous, these two circles, placed perpendicular to the axis at its end-point P, will, together with a circle from the cylinder, distance x from O and perpendicular to the axis, have their centres of gravity at O, since $2\pi rx \cdot 2r = \pi(2r)^2 \cdot x$. By regarding each of the three figures as a 'sum' of such circles Archimedes transports all the parts of the sphere and cone to P and arrives at the result that the sphere and cone, brought together with common centre of gravity at P, together with the undisturbed cylinder, have their centre of gravity at O. (If the axis is regarded as the beam of a balance, it could therefore be said that the sphere and cone placed together at P balance the cylinder on the opposite arm.) It follows that

(volume of cone + volume of sphere) $\cdot 2r$ = volume of cylinder $\cdot r$,

or volume of sphere = $\frac{1}{2}$ volume of cylinder − volume of cone

$$\tfrac{1}{2}[2r \cdot \pi(2r)^2] - \tfrac{1}{3}2r \cdot \pi(2r)^2 = \tfrac{4}{3}\pi r^3.$$

The weakness in this kind of reasoning which seems quite logical is, of course, that a three-dimensional object cannot be considered as a 'sum' of the parallel planes in it. To make this argument correct one could divide the object into a finite number of thin slices and afterwards proceed to the limit. Archimedes was quite aware that the above method was not a rigorous mathematical proof. This is clear from his comments to Eratosthenes and from the final form which he gives to his proofs.

Another theorem investigated by this method is the proposition 'A cylinder which circumscribes a sphere has a volume one and a half times that of the sphere' (cf. the corresponding theorem on the surface areas of the two bodies), a theorem of which Archimedes was so proud that he wanted its diagram engraved on his tombstone.

In this way Archimedes' theorems were based on results which he had previously obtained from studying the mechanics of the centre of gravity and equilibrium. But they were directed towards the formulation of purely mathematical relationships which have nothing whatever to do with mechanics. One could say that here mechanics was repaying something of the debt it had owed to mathematics ever since physics took over the use of mathematical methods. More and more, such examples of interaction between mathematics and physics mark the development of the two sciences.

Hydrostatics

Archimedes' treatise *On floating bodies* plays the same role in hydrostatics as *On the equilibrium of planes* does in the statics of rigid bodies. This work is also built up axiomatically, but with the peculiarity that at the beginning one single axiom is stated as the basis for the series of fundamental theorems that follows. This postulate states first, that a fluid is a continuous medium; next, that the parts of the fluid on which the pressure is least are pushed away by those parts on which the pressure is greatest; and finally, that all parts of the fluid are subjected to a pressure from the parts of the fluid situated above them.

Following this introduction Archimedes gets down to proving the theorems, of which the second, for example, states that:

> If a quantity of fluid is at rest, its surface is the surface of a sphere whose centre is at the centre of the earth.

Proposition 3 deals with bodies which volume for volume, have the 'same weight as a fluid' which is, in ancient physics the closest approach to the concept of specific gravity. About such bodies it is said that they will neither sink to the bottom nor rise to the surface, but will be suspended in the fluid. Proposition 5 says that:

> If a body, lighter than the fluid, is placed in the fluid, it will sink to such a depth that the body's weight will be equal to the weight of displaced fluid

and Proposition 6:

> If a body, lighter than the fluid, is forced down in the fluid, it will experience an upward force equal to the difference between the weight of displaced fluid and the body's weight.

113

Proposition 7 is the famous 'Archimedes' Principle' which is formulated:

> A solid heavier than a fluid will, if placed in it, descend to the bottom of
> the fluid, and the solid will, when weighed in the fluid, be lighter than
> its true weight by the weight of the fluid displaced.

The proof of the principle runs as follows: Archimedes first considers a body A, heavier than the fluid to the extent that the volume of fluid corresponding to the volume of A weighs g, while A itself weighs $g + h$. When A is placed in the fluid it will sink to the bottom. Next a body B is considered, which is lighter than the fluid, weighing g when the corresponding volume of fluid weighs $g + h$. When B is placed in the fluid it floats with a part of its volume above the surface. If A and B are now brought together to form one continuous body, this body will have the weight $(g+h)+g$, while the corresponding fluid volume weighs $g + (g + h)$, which is the same. The collective body will then be able to remain stationary in the fluid. The force which would act downwards on A alone must therefore be equal to the force which acts upwards on B alone, providing that B is completely submerged in the fluid. But this force is, from the previous result of Proposition 6, equal to the difference between the weights of the body and of the displaced fluid, or $(g + h) - g = h$. When submerged in the fluid, A therefore experiences a downward force h, but when out of the fluid it experiences a downward force of $(g + h)$. It has thus lost the weight g which, from the original statement, is equal to the weight of displaced fluid. Assuming that gravity acts on the submerged object with undiminished strength, Archimedes' principle can also be expressed in such a way that the immersed body is acted upon by an upward force—upthrust—whose magnitude is equal to the weight of fluid displaced.

The method of reasoning applied in the proof suggests a reconstruction of the method by which Archimedes, according to the well-known story, found the value of King Hieron's crown. If the weight of the crown is called W, and the weights of gold and silver contained in it W_1 and W_2, respectively, $W = W_1 + W_2$. By weighing the crown first in air and then in water, it would be possible to determine the loss in weight t suffered by its original weight W on immersion. Similarly, by weighing pure gold of weight W a loss in weight of t_1, would be obtained, and pure silver of weight W would give a loss in weight ot t_2.

The losses in weight of the quantities of gold and silver contained in the crown would then be, respectively,

$$\frac{W_1}{W} \cdot t_1 \quad \text{and} \quad \frac{W_2}{W} \cdot t_2$$

Thus we have the condition,

$$t = \frac{W_1}{W} \cdot t_1 + \frac{W_2}{W} \cdot t_2$$

and obtain the equation

$$\frac{W_1}{W_2} = \frac{t_2 - t}{t - t_1}$$

for the relative weights of gold and silver in the crown. Although this method closely follows the proof of Archimedes' principle, Vitruvius (first century B.C.) writes that Archimedes made use of the volumes of liquid displaced by the crown and by equal weights of pure gold and silver. The solution of this problem therefore awaits further evidence.

After Proposition 7 Archimedes introduces a new axiom:

> Let it be granted that any body which is thrust upwards in the fluid is thrust upwards along the vertical drawn through its centre of gravity.

From this assumption he solved a long series of problems concerning the equilibrium of floating bodies. There is no space here to expand on these questions which demonstrate in particular Archimedes' acute perception, except to mention that in determining the equilibrium position for a floating paraboloid of revolution he not only found the two obvious positions, with the axis vertical, but also a position of equilibrium with the axis inclined to the vertical, since the two portions into which the liquid surface divides the body may have centres of gravity lying on the same vertical.

Archimedes was thus among the first, and also one of the greatest mathematical physicists of all time. His achievement is unique in Antiquity, and knowledge of his works later became the starting-point for further advances.

Later Statics—the Problem of the Inclined Plane

Authors after Archimedes often dwell on problems in statics, frequently in connection with technical mechanics or the study of machines, building and so on. One of the best known of these writers was Heron of Alexandria (*c.* A.D. 60), who has bequeathed a series of texts on surveying (Fig. 8.5), optics, pneumatics (machines worked by

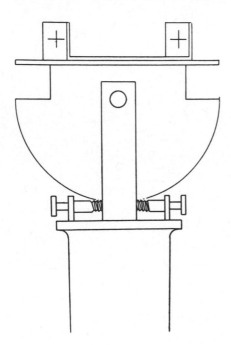

Fig. 8.5. Hero's Dioptra in A. G. Drachmann's reconstruction. Above is a diopter which can turn on a graduated circle. The inclination of the latter can be adjusted by the endless screw. A. G. Drachmann, Centaurus *vol. 1, Copenhagen, (1950).*

air or steam pressure), on automata, and a textbook on mechanics for engineers. This textbook, preserved only in an Arabic translation, contains the theory for the five so-called *simple powers*: the lever, windlass, screw, wedge and pulley, and for combinations of these machine elements (Fig. 8.6). Particularly interesting is Hero's correct solution of the problem of the bent lever: there is equilibrium when the weights are inversely proportional to the perpendicular distances from their points of suspension to the vertical through the fulcrum of the lever.

Hero was, however, less fortunate with another problem which was to become, for many generations of scientists, the great stumbling

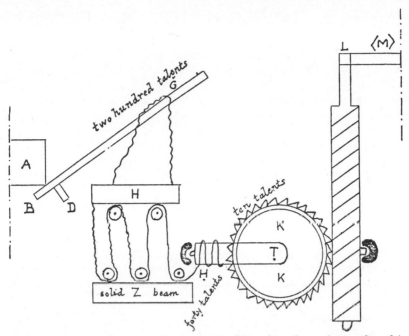

Fig. 8.6. Illustration from an Arabic manuscript of Heron's mechanics showing four of the five simple machines. A is a weight which can be lifted by a lever BG with the fulcrum D. HZ is a kind of a block of pulleys. T is a windlass connected with a tooth wheel KK turned by the endless screw L with the handle M. From A. G. Drachmann, The Mechanical Technology of Greek and Roman Antiquity, *Copenhagen, 1963.*

block of statics. It concerns a body's equilibrium on an inclined plane. Hero considers a cylinder on such a plane (see Fig. 8.7) and assumes the

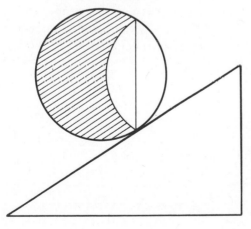

Fig. 8.7. Heron's attempt to solve the problem of the inclined plane. From A. G. Drachmann, The Mechanical Technology of Greek and Roman Antiquity, *Copenhagen, 1963.*

117

relationship that the force required to prevent motion down the slope is to the cylinder's weight as

$$\frac{\text{force}}{\text{weight}} = \frac{\text{shaded area}}{\text{area of circle}}.$$

The commentator Pappos (fourth century B.C.) also investigates this problem and gives a solution which is even more erroneous than that of Heron; it is based on the relation

$$\frac{F}{K} = \frac{OA}{BA}$$

where F is the force required and K the force which would draw the cylinder along a horizontal plane (see Fig. 8.8), since ancient dynamics,

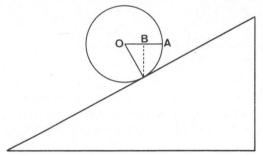

Fig. 8.8. *Pappos's theory of the inclined plane.*

as discussed in the next chapter, supposed that a force was necessary for maintaining uniform motion in a horizontal direction.

These fruitless attempts to solve a simple problem of such theoretical and technical importance show the limited scope of ancient mechanics, and emphasize in particular the difficulties which arose because the theorem of the parallelogram of forces was not yet known (see p. 212).

Chapter Nine

Dynamics in Antiquity

The Problem of Motion

COMPARED with the results achieved by Archimedes in statics, the work on dynamical problems, whose aim was to give speculations on dynamics a rational, mathematical structure, did not turn out so well. While Archimedes' work is thus of lasting importance, evolution caught up with the dynamics of Antiquity long ago. However primitive it appears today, it must be considered because of the great influence it came to exert on physical views for almost 2,000 years. Aristotle's dynamics is the key to the understanding of Mediæval physics, and without it one cannot understand the problems with which modern physics had initially to grapple.

Aristotle defined physics simply as the theory of every form of motion. There were several varieties of motion, as also qualitative changes, such as those of colour, were considered as motion. Here, however, only motion of material bodies from place to place, will be considered, that is, *local* motion.

The Concept of Velocity

The physics of Antiquity differs distinctly from that of later times in its conception of velocity as a quality of the moving body. A body is fast in the same way as it is hot or red. Velocity is the intensity, or degree, of a local motion, and it is taken for granted that this intensity has a numerical measure, of which the definition is not specified. In the same way, slowness is looked upon as a quality similar, but opposite to velocity. A motion is defined as fast when the body traverses a long distance in short time, and as slow when the opposite takes place.

Aristotelian kinematics considers only uniform velocities, as it does not possess mathematical ideas which allow a definition of non-uniform velocities. The most important result at this first stage was stated as the theorem of the parallelogram of velocities, which occurs for the first time in a treatise under the name of Aristotle's *Mechanica*, presumably written by one of Aristotle's pupils. The theorem states that when a body moves in a certain ratio, that is when it has two linear velocities of a constant ratio, 'it must move on a right line, which is the diagonal in the figure formed by the two right lines which have the given relation'. The theorem is proved by means of similar figures.

Three Types of Motion

The classification of different types of motion plays an important role in dynamics. Aristotle distinguishes between three classes, of which the two first are significant to physics.

The first type is *natural motion*. Natural motion is performed by bodies under the influence of forces seated in the body proper. There are three forms of natural motion in the world. First, there is the downward fall of heavy bodies; this is due to a force existing in the body called gravity. In addition, there is the rise of light bodies like smoke and fire. This is due to a similar inner force, levity. A body is always either heavy or light. Gravity and levity are two mutually exclusive qualities. The conception of levity as only a small degree of gravity is quite alien to this point of view. When there are no external obstacles to the motion, the natural motion will cause the bodies to seek their 'natural place'. For the heavy bodies this is the centre of the world, while to the fire, for example, it is a spherical shell just inside the lunar sphere and adjacent to it. The third type of natural motion is the rotation of the heavenly bodies. This, too, is due to inner forces in the spheres, but differs from the terrestrial motions in being eternal and unchangeable.

The second category of motion is *forced motion*. This is caused by external forces and hinders the natural motion. Examples are projectile movement, lifting of loads, and so on. The third and final type is *voluntary motion*, found only as the intentional movements of living beings, which will not be dealt with here.

The Moving Force

The need for a moving force is common to all types of motion, and modern recognition of inertial motion as the uniform and straight motion of a body without any action of force is alien to the physics of Antiquity. All motion, according to ancient theory, requires an active force, and the physicist's task is to explore its nature and way of action. Natural motion on the Earth was considered to be due to the inner forces of gravity and levity, but it was more difficult to explain the natural motion of celestial bodies. Aristotle resorted to explaining their eternal rotation as caused by a rational (intelligent) soul in the spheres, an explanation springing from much older animistic conceptions of the nature of the celestial bodies. According to Aristotelian psychology this meant that the celestial bodies had to be conceived of as quasi-living beings.

The underlying cause of forced motion gave rise to great difficulties. When, for instance, one lifts a weight or pulls a carriage there is no problem in finding the origin of the force. But when a stone is thrown, it becomes more difficult, at least from the moment when the stone has left the hand. The force cannot be seated in the stone, as in this case it could perform only its natural downward motion. Nor can it come directly from the hand, as the physics of Antiquity does not accept the possibility of action or force at a distance. Aristotle then advances the theory that moving force is transmitted from the hand to its adjacent layers of air, from these layers to the neighbouring ones, and so on, until the stone is reached, which is driven forward by a force seated in the air immediately behind it.

In this case, the so-called *antiperistasis* theory is applied to the motion. The word comes from a Greek verb which means to 'turn to the opposite side' and is a direct allusion to the fact that during the movement the air in front of the stone moves behind it. From the text it appears, however, that the stone is kept moving not by the circulation of air round the stone, but by the transmitted moving force. The relationship is obscure, and many interpretations have been advanced to explain it.

Force and Velocity

Force and velocity are connected. In every mechanical theory some sort of relationship must occur connecting kinematic concepts such as velocity or acceleration with dynamic concepts like force, as for example, in Newton's second law. In Aristotle the connection has the form:

> If the moving force A can move the body B a distance C during the time D, then the same force will be able to move the body $1/2B$ the distance $2C$ during the same time D, as it will be able to move $1/2B$ the whole distance C during the time $1/2D$.

This may be more simply expressed: the velocity is proportional to the force. If we call the force F and the velocity v, then

$$v = \text{const.} \times F$$

Aristotle adds that from this it cannot be concluded that, for instance, half of the force can move the whole body half the distance in the given time. Half the force may not be able to move the body at all, as the resistance might be too great. It is characteristic of Aristotelian physicists that they always take the resistance of motion into account, unlike Galileo, for instance, who solved the problem of the fall neglecting the resistance of air. If the resistance to motion is designated R, the force-velocity relationship above can be generalized to

$$v = \text{const.} \times \frac{F}{R}.$$

But if movement is to start at all, F must be greater than R.

An important consequence of this fundamental mechanical relation between force and velocity is the Aristotelian argument of the non-existence of a vacuum. Imagine a body moving in water under the influence of a given force. The body will achieve a certain velocity, determined by the resistance of the water. If, thereafter, the same body is imagined as moved by the same force, but in a thinner medium, exerting only half the resistance, the velocity would be doubled. In a still thinner medium, the velocity would be further increased, and if the medium disappeared completely so that the body would be moving in empty space, its velocity would be infinite but would be the effect of a finite force. This must be impossible, and it was therefore concluded than an empty space cannot exist.

The Fall of Heavy Bodies

Aristotle's treatment of free fall was influenced by these considerations. He nowhere ignores the resistance which is defined partly by the density of the medium (looked upon as a qualitative concept, not as specific weight in an exact sense), partly by the form and size of the body. The Aristotelian equation of motion thus leads to the theory: when in the same medium equally heavy bodies of similar form and size will fall equally fast, i.e., they will take the same time to fall through a given distance. If the two bodies are of similar form and size, but of different weight, the heaviest body will take the shortest time. When falling in different media, the same body will have the shortest time of fall where the density is smallest. If we call the time of fall t, the density of the medium d and the weight of the body p, the formula

$$t = \text{const.} \times \frac{d}{p},$$

covers all possibilities.

In Antiquity the acceleration during the fall was explained in many ways. Aristotle argued that as the heavy body approaches the Earth (which is its natural place), it acquires its heavy 'form' in a more perfect way, that is it becomes even heavier and therefore falls faster.

The Mechanics of Johannes Philoponos

Even in Antiquity the many discrepancies in Aristotelian dynamics gave rise to doubts. The criticism levelled against the Aristotelian theory of fall by Johannes Philoponos (c. A.D. 500) is particularly interesting. It is contained in a treatise which although it appeared as a commentary to Aristotle's *Physics* contained several new ideas, and showed the progress made in dynamics during the previous 800 years. Johannes concentrates on a crucial point:

123

> Aristotle assumes erroneously that the times of fall [for the same body] through different media correspond to the density of the latter [. . .] but it is not easy to discover the error because it is impossible to find out how many times water is denser than air.

The last sentence refers to the idea of specific weight which had not yet been elaborated and disposes, too, of the grounds for the impossibility of a vacuum. According to Philoponos, the time of fall does not at all tend towards zero as the density of the medium decreases. On the contrary he assumed that there will be a finite time of fall t_0, even with zero density. If the fall takes place in a medium of density d, this only implies that the 'natural' time t_0 is increased. By means of this argument Philoponos replaced the Aristotelian expression for the time of fall by

$$t = t_0 + T(d),$$

where $T(d)$ is an unknown but not necessarily a linear function of the density. Concerning the fall through the same medium, Philoponos maintains that the times of fall of different bodies are not simply proportional to their weights:

> For if two weights, the one many times heavier than the other, fall from the same height, you will see that the ratio of the times is not the same as the ratio of the weights, but that the difference in the times is very small.

Nothing indicates that Philoponos himself conducted this experiment, which later, in the hands of Stevin and Galileo became famous in the history of mechanics. It is more likely that he obtained his insight from everyday observations of falling bodies, but formulated it as a thought experiment.

Philoponos also had new ideas about forced motion. The Aristotelian antiperistasis theory maintained that a projectile moves because it receives a moving force from the surrounding air. This idea is refuted by means of another thought experiment: a stone could be suspended by a rope and the air behind it set in motion by 'innumerable machines'. But this would not impart any considerable motion to the stone. Philoponos concluded, further, that since the air is unable to start the motion, it is also unable to maintain it, and therefore put forward a completely different explanation. When a stone is thrown it will continue its motion because a certain immaterial moving force was impressed upon it during the short time when the stone was in contact with the hand. This force is supposed to exist after the immediate contact has ceased, whereas the air has nothing to do with the continuation of motion, and acts only as a resistance.

The postulation of an impressed force means that the cause of a projectile's motion resides inside the body itself. Philoponos therefore upholds the Aristotelian axiom that every motion needs a moving force, and his description of the forced motion is not at all similar to the 17th century concept of inertial motion, requiring no outer or inner force. Philoponos' explanation turned up later under the name of the *impetus* theory (p. 237), and it will be discussed in more detail in connection with Mediæval dynamics. The only remark relevant here is that motion in a vacuum would be possible since it can be sustained by the inner impetus without any interaction with a medium.

Chapter Ten

The Conception of Light and Sound

The Nature of Light and Colour

IN Antiquity, prevalent ideas on the nature of light were very different from those of today. It is true that in the pre-Socratic period there were philosophers like Democritos of Abdera (c. 460–370 B.C.) who, through his general atomic theory, regarded light as small particles emitted from a source and moving with a finite velocity. But this theory was never widely accepted. Both Empedocles of Agrigentum (483 c. 425 B.C.) and Plato considered light to be a kind of fire, or something related to the fiery element.

The predominant theory was developed by Aristotle and was closely connected with his philosophical ideas of actual and potential qualities. Aristotle began from the fact that there exist transparent bodies, like air or water, whose transparency is not due to the elements of air and water, but must be ascribed to a certain attribute found in pure form in the heavenly bodies, which are both material and completely transparent. According to Aristotle, an attribute or quality may be present either actually or potentially in the body containing it. The quality exists actually if it is clearly discernible and maintained by a cause already in existence. It has potential existence if it is hidden and needs an effective cause to 'actualize' it. Aristotle's theory of light rests, therefore, on the assumption that transparency of air or water is a potential quality which has to be actualized before such substances can be seen as transparent. The efficient cause changing transparency from potential to actual existence is fire. Without fire, the transparency remains potential and the bodies seem to be filled with darkness. But through the action of fire (the pure, invisible element, see p. 146) the transparent substance is imbued with a certain colour. This colour is what Aristotle understands by light.

So philosophical a definition is obviously directed against the older conceptions of light as a fiery emanation from a source. Aristotle wanted to underline the idea that light in transparent bodies does not consist of particles separated from the original source, but is a certain state of the medium. This implies that a transparent medium is essential to the transmission of light to the eye, and that we would see nothing if there were no such medium.

The problem of the velocity of light was frequently mentioned in Antiquity. Since lightning comes before the thunder it was concluded that light travels faster than sound. According to Aristotle, light is transmitted instantaneously (that is with infinite velocity), and this idea was almost universally accepted.

Colours were explained as mixtures of light and darkness: much light and little darkness gave a red colour, a more even mixture gave green, and violet or purple contained much darkness but little light. This theory was used, among other things, to explain the changing colours of clouds at sunset, and the colours of the rainbow (in which only three colours were distinguished). Aristotle developed a theory in which the rainbow appears when light from the Sun is reflected from a dark cloud. The outer rim was supposed to be nearest to the light and therefore red, and the inner rim violet. The secondary rainbow, with the colours in reverse order, was explained as a reflection of the primary bow. Seneca (c. 4 B.C.–A.D. 65) mentioned the colours produced by the refraction of light in rods of glass, but such colours were never investigated by the physicists of Antiquity.

Euclid's Optics

While the work on physical optics led to no achievements of lasting value, a very different situation existed in geometrical optics, a subject considered as a separate science by Greek mathematicians and another testimony to their efforts to give the description of nature an axiomatic form in the language of mathematics. The pioneer work may well have been performed by Euclid, whose *Optics* is the first extant treatise on the subject. It has exactly the same logical structure as the first book of his *Elements*, and begins with seven definitions, or, more correctly, axioms, the first of which is

> Let us assume that the rectilinear rays emanating from the eye diverge in infinity.

Euclid therefore supported the opinion that the eye functions by means of rays emanating from itself. Today, this seems very strange, and even in Antiquity was much discussed although without any final solution being found. Although Aristotle held light to be an activity caused by the source of light, in his theory of the rainbow he clearly speaks of visual rays coming from the eye. Within geometrical optics, however, the question has no real importance, since the path of light from one point to another is reversible. But it is important to notice that geometrical optics is not concerned with light as such, but with the rectilinear rays mentioned in Euclid's definition. Like geometry, geometrical optics was created through a process of abstraction. The real properties of the physical world were described through abstract concepts, and the visual rays of the new optics had nothing more to do with material reality than the straight lines of pure geometry.

The theorems proved in Euclid's *Optics* are mostly concerned with visual angles. As an example, the following propositions can be quoted:

> 5. Equal magnitudes at different distances from the eye are seen as unequal, greater the nearer they are.
> 6. When seen from a certain distance, parallel lines do not seem equidistant.
> 8. Equal and parallel magnitudes at different distances from the eye do not seem to be [inversely] proportional to their distances from the eye.
> 45. There is a common point from which unequal magnitudes seem to be equal.

As time went on, this section of optics developed into the new theory of perspective. In the Middle Ages the terms *optica* and *perspectiva* were frequently used in the same sense.

The Reflection of Light

The science of the reflection of light was called *catoptric*. Aristotle knew the fundamental fact that the angle of incidence and the angle of reflection are equal, and made use of it in his theory of the rainbow. From Antiquity some interesting attempts at proving this proposition are known. There are some indications that Archimedes was the first to venture on a proof along the following lines.

In Fig. 10.1 AT is a plane mirror. B is the eye from whence the visual

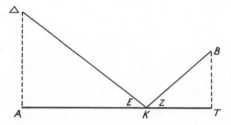

Fig. *10.1. Figure illustrating Archimedes' proof of the law of reflection.*

rays proceed, and Δ is an object seen by means of the ray BK. Z is the angle of incidence, and E the angle of reflection. There are three possible relations: $Z = E$, $Z > E$ and $Z < E$. In the case $Z > E$ the eye and the object could be interchanged, making E the angle of incidence and Z the angle of reflection, whence $E < Z$. A similar contradiction would result from the third hypothesis $Z < E$, and the only possible relation remaining is $Z = E$.

It is difficult to believe, however that a 'proof' of this kind was formulated by a mathematician as ingenious and acute as Archimedes. The most obvious defect is that it is not known whether the reflection still takes place at the point K, after B and Δ have been interchanged. If K moves to another position the whole argument is invalidated. But perhaps Archimedes was aware that K has the same position in the two cases. Heron of Alexandria described an experiment in which the point K is covered in wax, with the result that the object becomes invisible, both before and after the interchange. As a result, only a single small area around K is active at the reflection. If this experiment was known to Archimedes, the proof holds good, its only incompleteness being that all-important experimental presupposition was left unmentioned.

Ptolemy's *Optics* contains the first direct, experimental proof of the law of reflection. His apparatus was a round copper disc with the two diameters *GE* and *BD* engraved perpendicularly to each other, and the rim divided in degrees (Fig. 10.2). A diopter is placed along the radius

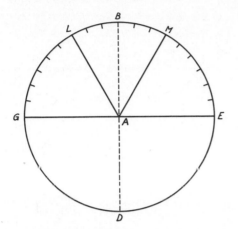

Fig. 10.2. *Ptolemy's experiment on the reflection of light.*

LA, and a small mirror of polished steel put on the disc along *GEA*. A small coloured object *M* is moved along the rim of the disc until it is seen through the diopter. The numbers of degree at *L* and *M* are read off the graduated circle, and the arcs *LM* and *BM* are seen to be equal.

In his *Catoptric*, Heron of Alexandria was interested in explaining optical illusions produced by mirrors. This led to some important investigations of a theoretical nature. Hero started from the axiom that

> what moves with constant velocity follows a straight line. An example is an arrow which we see shot from the bow. For, because of the forward moving force the moving body strives to follow the shortest path since it cannot afford the time for a slower motion, that is a longer path. The moving force does not allow such a delay. Thus the body tries to follow the shortest path because of its speed, but between the same endpoints the shortest of all lines is the straight line.

Although the first sentence of this passage is wrong, the rest is extremely interesting as it forms one of the first attempts to formulate a general minimum principle in mechanics. Hero applies it to the motion of light, assuming an infinite velocity as the explanation of the rectilinear path of the rays. But the same axiom also enables him to deduce the law of reflection.

In Fig. 10.3 *AB* represents a plain mirror, *G* is the eye, and *D* an

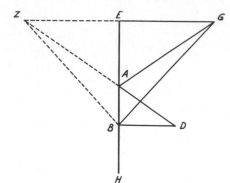

Fig. 10.3. *Heron's proof that a ray of light follows the shortest path when reflected from a plane mirror.*

object seen by means of the reflecting ray *GAD* for which the law of reflection is satisfied. Let another ray be *GBD*. Now Heron proves that *GA* + *AD* < *GB* + *BD*. That proof (which may be marked in detail by the reader) reveals that the principle of the shortest path is valid with regard to the reflection of light. Whatever Heron's conceptions of the physical nature of light the fact is that he deduces a fundamental law in optics by means of a mechanical analogy. It was unfortunate that his mechanical axiom, together with the old dynamics in general, should turn out to be incorrect, but that is of minor importance compared with the fact that analogies between optics and mechanics were destined to play a very considerable role in the development of physics.

Heron was also interested in spherical mirrors and proved that the rays will converge after reflection. Fig. 10.4 shows a spherical mirror with

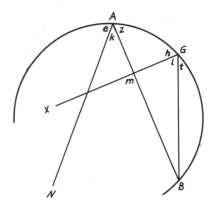

Fig. 10.4. *Heron's theory of reflection in a spherical mirror.*

the eye (that is the source of the visual rays) placed at *B*. The two rays *BG* and *BA* are reflected as *GX* and *AN*, respectively. Since arc *AB* > arc *GB* it follows that angle *z* > angle *t*, and, by the law of reflection, *e* > *h*. Then for the two remaining angles 1 > *k*. From *m* > 1, *m* > *k* follows, so that the two rays will intersect after reflection. Heron does not examine how a complete bunch of rays is reflected and fails to discover the caustic curve of the mirror (compare Fig. 19.1).

The Refraction of Light

Burning mirrors of polished metal were well known in Antiquity, and as early as the fifth century B.C. the Greeks had some practical knowledge of the refraction of light, and knew how to make burning glasses. But the phenomenon was not thoroughly investigated before Ptolemy, who gave a masterly description of it in his *Optics*. First he mentions a kind of lecture experiment with a *baptistir* (Fig. 10.5), that is

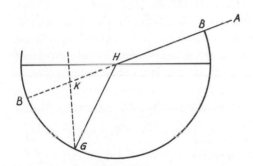

Fig. 10.5.   Ptolemy's baptistir.

a kind of bowl. If, he says, one looks from *A* just over the rim at *B*, a coin at *G* will be invisible. But if the bowl is filled with water the coin seems to be lifted to a higher position, from which it can be seen. Ptolemy postulates that the apparent place of the coin is the point *K* where the vertical through *G* and the prolonged ray *AB* intersect.

After this introductory demonstration Ptolemy proceeded with a quantitative experiment, using the graduated copper disc or goniometer from the reflecting experiment. The disc is now placed in a vertical position inside a jar filled with water up to the level of the diameter *DB* (Fig. 10.6). The centre *E* and a point *Z* are indicated by small

133

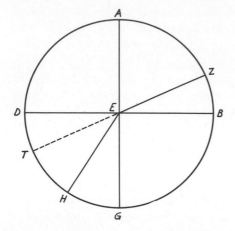

Fig. 10.6. Ptolemy's experiment on the refraction of light.

coloured marks. Beneath the surface a pointer is moved along the circumference until it is seen at H, on the continuation of the incident ray ZE. The angle of incidence $AEZ = i$ and the angle of refraction $HEG = r$ can then be read off on the instrument. Similar measurements are made with rays passing from glass to air (still according to the assumption of the rays coming from the eye). In this case a half-cylinder of glass is fixed to the lower half of the copper disc, and the positions of the eye and the mark Z are interchanged. Finally, Ptolemy investigated refractions from glass to water. His results were given in the form of tables frequently quoted in optical literature until the discovery of the precise law of refraction in the seventeenth century.

Ptolemy's Table
of Refraction

i	r
$10°$	$8°$
$20°$	$15\frac{1}{2}°$
$30°$	$22\frac{1}{2}°$
$40°$	$29°$
$50°$	$35°$
$60°$	$40\frac{1}{2}°$
$70°$	$45\frac{1}{2}°$
$80°$	$50°$

Neither in Antiquity nor in the Middle Ages did scientists succeed in formulating the results in a mathematical relation of satisfactory exactitude. Sometimes the formula i/r = const. was used which does, however, give only a very approximate expression for the values found. It seems that Ptolemy himself preferred to apply the formula

$$r = ai - bi^2$$

which, with adequate values for a and b, gives a somewhat better approximation.

Atmospheric Refraction

The Greeks were also the people who discovered the refraction of light in the atmosphere. In *On the circular motion of the heavenly bodies* Cleomedes (first century A.D.) relates that he has observed a lunar eclipse before sunset. At first, he could not explain this for such an eclipse presupposes the Sun, Earth and Moon to be on the same straight line, so that the Sun and Moon, seen from the Earth, must be diametrically opposite in the sky, and one ought not to be able to see the Sun at the same time as the eclipsed Moon. Cleomedes therefore advanced the hypothesis that the Sun's rays do not pass through the atmosphere of the Earth in a straight line (Fig. 10.7). If it is assumed that they bend

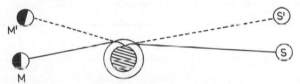

Fig. 10.7. *Cleomedes' discovery of atmospheric refraction. The Sun S and the eclipsed full Moon M are diametrically opposite with respect to the Earth. In fact, because of the refraction, the Sun is seen at S', and the Moon at M'.*

towards the Earth, the discs of both Sun and Moon will apparently be raised over the horizon, and the phenomenon explained. Cleomedes shows that such an explanation is reasonable by means of the experiment mentioned above: if a coin is put in a jar, so that it is hidden below the rim when the eye is in a certain position, it becomes visible when water is poured into the jar. That is, the picture of the coin is raised by the water in the same way as the Sun and Moon are apparently raised by the atmosphere.

Acoustics

The *Optics* of Ptolemy differs from many other physical works of Antiquity in that the experimental results on which the mathematical theory is founded are always carefully described. This characteristic feature is also found in the extant treatises on sound.

The Greeks' interest in acoustic problems undoubtedly arose from their preoccupation with music. After Terpander of Lesbos organized the first school of music in Sparta, in about 700 B.C., music became one of the most beloved arts. Thus, as in so many fields, practice preceded theory. The theory was chiefly concerned with two sorts of problems: the physical nature of sound, and the mathematical description of the system of tones in music.

Even in Aristotle's works the opinion is voiced that sound is transmitted like vibrations in the air. There were no means available at the time to decide how fast it spreads, although Plinius writes that the velocity of sound is less than that of light. In his (partly extant) treatise *Sectio canonis*, Euclid goes through the doctrine of sound in an axiomatic-deductive way which reminds one of his *Optics*. He assumed that sound is a movement and that the height of a tone depends on the 'number of motions' in the air in a given time, or, in modern terminology, on the number of vibrations per second. The important idea of frequency was thus introduced into physics. The last part of the treatise concerns the theory of musical instruments, and here it is assumed that the frequency of the tone produced by a string is inversely proportional to the length of the string. The Stoic Chrysippos (c. 280–207 B.C.) clearly conceived of sound as a wave motion in the air:

> We hear because the air between the speaker and the listener is touched and expands in shell-formed waves, just as the waves in a pond expand when you throw a stone into it.

Of later authors, Vitruvius dealt with acoustics along Aristotelian-Euclidian lines and at the same time himself laid the foundations of room acoustics through his reflections on the transmission of sound in buildings. Later still Ptolemy's *Harmonics* contrasts with Euclid's brief treatise as an attempt to collect everything known about sound and notes in a unified exposition, similar to what Ptolemy had already achieved in astronomy and geography; thus his book contains not only physical acoustics but also physiological considerations of the sense of hearing and the Pythagorean doctrine on the harmony of spheres.

Theory of Music

According to a unanimous tradition the mathematical part of ancient acoustics was derived from the Pythagoreans who were the first to perform systematic experiments with the monochord (page 19) which revealed that musical intervals can be characterized by numerical ratios. Theon of Smyrna (c. A.D. 130) says that

> if the single string in the monochord is divided into four equal parts, the sound produced by the whole length of the string forms with the sound produced by three quarters of the string (the ratio being 4:3) the consonance of a fourth.

Similarly, the ratio 2:1 corresponds to the octave, and the ratio 3:2 to the fifth. In this way the surprisingly numerous scales used in Greek music could be investigated and characterized by ratios found by empirical methods. The table below shows, as an example, the ratios of the diatonic, Lydian scale which was very similar to our modern major modes.

Diatonic scales

		Lydian	Pythagorean
1	doh	1:1	1:1
2	re	9:8	9:8
3	mi	5:4	81:64
4	fa	4:3	4:3
5	so	3:2	3:2
6	la	5:3	27:16
7	si	15:8	243:128
8	doh¹	2:1	2:1

Among these intervals the fifth is a consonance characterized by a ratio 3:2 composed of small integers, while the seventh is a dissonance with a ratio 15:8 composed of larger numbers. This means that the numerical ratio not only defines the size of the interval, but also its harmonical quality.

These investigations also revealed the important law that two intervals can be added together if their corresponding ratios are multiplied. For example, the relation

$$\text{fourth} + \text{fifth} = \text{octave}$$

corresponds to

$$\frac{4}{3} \times \frac{3}{2} = \frac{2}{1}$$

Similarly, two intervals can be subtracted if their corresponding ratios are divided. This insight led to an attempt to construct musical scales from purely mathematical principles. But since the Pythagoreans were unwilling to accept any ratio not based on the integers 1, 2, 3, and 4 which to them had a particular philosophical importance, they were unable to arrive at the empirical values for the third, the sixth, and the seventh given in the table above. Their alternative proposal was the slightly different ratios of the so-called Pythagorean scale.

This situation split the Greek theorists of music into two hostile camps. The *canonists* followed Pythagoras who, according to Boethius (*c.* A.D. 480–524),

> abandoned the auditory sense as a criterion of judgement and had recourse to the divisions of a measuring rod. He had no faith in the human ear, which suffers change not only naturally, but by reason of external accidents and varies with age [. . .]. Instead he sought long and ardently for a method by which he might learn the fixed and unalterable measurement of consonances.

Considering his general mistrust of sense experience it is no wonder that Plato adhered to this school in his discussion of music in the *Republic*. The opposite point of view was taken by the *harmonists* who refused to yield to mathematical speculations and wished to maintain an empirical basis for the study of music, without discarding the new mathematical insight. Their first important spokesman was Aristotle's pupil Aristoxenes of Tarentum (*c.* 320 B.C.) who said that

> the voice follows a natural law in its motion, and does not place the intervals at random. And of our answers we endeavour to supply proofs that will be in agreement with the phenomena—in this unlike

our predecessors [. . .]. Our method rests in the last resort on an appeal to the two faculties of hearing and intellect. By the former we judge the magnitudes of the intervals, by the latter we contemplate the function of the notes.

Aristoxenes' extant *Elements of Harmonics* was later followed by the *Enchiridion harmonies* by Nicomachos of Gerasa (*c.* 100 A.D.) which in a Latin version by Boethius became the principal source for the Mediæval science of *musica*.

Chapter Eleven

The Structure of Material Substances

The Problem of Material Change

Iᴛ could easily be conjectured that it was the highly developed chemical technology of Antiquity which gave rise to the first philosophers' many speculations about the structure of matter. The early tradition shows however, that the Ionian thinkers approached the problem of matter and changes of matter from a more universal angle. They do not seem to have taken special interest in chemical problems, but were preoccupied with explaining the multitude of things in the world, ever generating, changing, perishing, on the assumption of a fundamental unity in nature below the surface of changing phenomena. They put forward various solutions to this question (see p. 14). To Thales, the world had emerged from water, and water was therefore considered as the fundamental element. To Anaximenes, it was the air, to Anaximander the 'undetermined', and to Heraclitos it was fire.

All these first theories of matter were monistic: things were believed to originate from one single, fundamental substance. They develop in all sorts of ways because this original substance, either of itself or on account of particular inherent forces, is able to modify its outward appearance. Consequently, in the early philosophical tradition no real distinction can be found between matter and force. The strongest expression of this monistic attitude was by Parmenides and the Eleatic philosophers, who considered all changes to be unreal and who emphasized the fundamental unity and immutability of the universe. Other philosophers stressed the existence of a multitude of phenomena,

and adopted an infinite number of 'principles' or fundamental substances, an extremist outcome of a pluralistic conception of nature. Both
monism and the following pluralistic conception of nature seem to have
given rise to much intellectual perplexity even at the start, for it was
extremely difficult to explain the plurality of phenomena through the
assumption of a single element. How could a single substance transform
itself? But also the assumption of many elements offended a deep-
rooted tendency in Greek thought to reduce the number of principles
in an attempt to achieve unity. The solution, therefore, became a
compromise which during the fifth century B.C. developed into the
theory of the four elements.

The Origin of the Doctrine of Elements

The ancient tradition unanimously mentions Empedocles of Agrigentum (see p. 127) as the author of the Greek theory of elements. In
the large existing fragment of his didactic poem *De natura*, are found
the outlines of a theory by which all changes in the world can be traced
to the combination and separation of a few, eternal and unperishable,
aboriginal substances, which Empedocles called 'roots'. Originally,
the theory involved four substances, later known as the four elements: fire, air, water, and earth. But to these were added two more,
equally material substances, called 'friendship' and 'strife'. The former
Empedocles stated, acts as a force which unifies the elements; the latter
separates them.

Using these six principles Empedocles sketched a theory for the
development of the universe as a whole. Its conception is clearly
pluralistic, in contrast to the monism of the Ionians. He does comply
with the Eleates however, in as far as something unchangeable (that is
the elements), is the basis of the universe. In a certain sense, he continues, nothing is generated nor perishes—apart from the many different
combinations of elements replacing each others during the course of
the world.

> Nothing is originated from what perishes, nor is there destruction in the
> annihilating death, but only a mixture [of elements] and a separation of
> what has been mixed. Origin is only a name given by man.

The original sources of the theory of elements are inextricably lost, but later, the Greeks often regarded the theory as obvious. When a piece of fresh wood is thrown into a fire it is split up into four constituents: the fiery element appears in the flames, the air element in the smoke, the watery element as tar and the earthy element in the ashes. However, popular explanations like this do seem to have been worked out to illustrate an already existing theory, rather than initiating the theory. A more plausible possibility is that Empedocles may have been inspired by oriental (Persian) theories and that he was influenced by the role which the four elements seem to have played in certain initiation rites. It may be significant, too, that Empedocles seems to have been an adherent of the Orphic religion. The theory of the elements soon became the dominating doctrine of matter in classical Antiquity. After Empedocles' time, philosophers strived both to give it a rational foundation, and also to apply it in many different fields.

Plato and the Four Elements

The best-known attempt to give the doctrine of elements a logical form is that of Plato. It was Plato who replaced the Empedoclean term '*roots*' with '*elements*' (stoicheia) of which the original meaning was 'letters'. The combination of elements into compound substances was apparently conceived to be analogous to the combination of letters into words. Plato restricted the number of material elements to the four well-known ones, and regarded the uniting and separating forces as immaterial. Finally, Plato tried to produce a philosophical and mathematical explanation of why the elements must necessarily be four in number.

ˈ The first of Plato's proofs starts from the fact that material bodies can be seen and touched. But nothing is visible except fire, and nothing can be touched without a certain hardness. Therefore the *Timaios* maintains that 'in the beginning God created the body of the universe from fire and earth'. But according to Plato the two substances cannot be kept together without intermediates.

> Therefore God placed water and air between fire and earth in the same mutual proportion (as far as possible) so that air is related to water as fire to air, and water to earth as air to water.

143

This fundamental relation,

$$\frac{\text{fire}}{\text{air}} = \frac{\text{air}}{\text{water}} = \frac{\text{water}}{\text{earth}} \text{ ,}$$

is admittedly one of the most obscure details of Platonic philosophy, and from it an innumerable series of comments has arisen. One hypothesis is that the equation defines the ratios between the quantities of the four elements in the universe; another hypothesis is that Plato had in mind the thicknesses of the four elementary spheres. No commentator has yet been able to solve the question completely, and the text remains an enigmatic attempt to apply the Platonic scheme of a mathematical description of nature to the theory of matter.

The second Platonic proof is found in another part of the *Timaios*. Here, Plato identifies the geometrical forms of the four elementary substances with the regular polyhedra of geometry. Presumably this idea stems from Philolaos the Pythagorean. As the most inert and immovable element, the earth is designated as the cube. Water is identified with the icosahedron, air with the octahedron, and fire with the regular tetrahedron, with its sharp and stinging corners. Plato was aware of the existence of a fifth regular body, the dodecahedron. Accordingly, he argued, there must be a fifth substance different from the other four elements. This fifth elementary substance he identified with the 'ether' introduced by Plato in his later years as the substance of the heavenly bodies (p. 65). This geometrical deduction seems even more obscure than the original, even if the same attempt to give the theory of the elements a mathematical foundation is manifest. Chalcidius's Latin translation of the *Timaios* stopped just before this point and the relationships between elements and polyhedra did not become an integral part of the doctrine of elements in the early Middle Ages.

Aristotle and the Elements

In the works of Aristotle, the doctrine of elements assumes quite a different character. It is obvious that he attaches little importance to the Platonic attempts at a mathematical reduction of the number of elements, but instead proceeds in a more physical way. First, he defines an element as 'one of the substances into which the other substances can be split up, but which cannot itself be split up into others , a definition which can be applied to the later concept of elements. Next, Aristotle tries to connect the doctrine of elements with his hylemorphism, according to which the world contains only one aboriginal substance: the undetermined and undifferentiated *materia prima*, which can appear as a substance only after combination with some sort of form. This theory does not destroy the conception of elements, which are merely regarded as particularly stable 'elementary forms' of matter.

The Structure of the Elements According to Aristotle

Dominating Quality	Secondary Quality	Element
dryness	coldness	earth
coldness	moistness	water
moistness	heat	air
heat	dryness	fire

Aristotle deduces the number of elements on the basis of an idea which presumably stems from early Greek (Hippocratic) medicine, in which it is assumed that four fundamental 'qualities' or properties of material bodies, hotness, dryness, coldness, and wetness, are found in all bodies to a greater or lesser degree (Fig. 11.1). These qualities are also found in the elements, which are, however, presumed to be combined in the simplest possible way.

Now Aristotle argues, one cannot be content with assigning to a given element only one of these properties, as a single principle to explain their mutual changes. On the other hand, an element cannot contain three fundamental properties, as two of them will be opposite (as, for example, hotness and coldness) and thus incompatible in the same element. Each element must therefore contain two, and only two, non-opposite elementary qualities; this gives the four possible combinations enumerated in the table above. With this, Aristotle considers he has given a physical deduction of the number of elements.

The table above indicates that one of the basic qualities in each element is thought to dominate the other one. This belief forms the cornerstone of the theory of elementary transmutations: under the influence of adequate causes one element can be transformed into another. If, for example, water is heated by fire, the hotness will

145

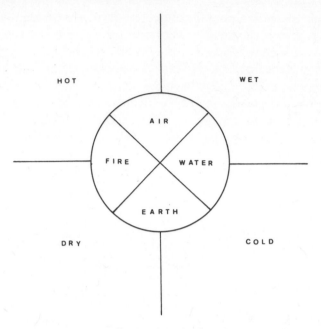

Fig. 11.1. Diagram showing how the four elements partake in the four elementary qualities.

destroy the coldness of water and replace it, while at the same time the secondary property—moistness—of water will be promoted to a dominant position. A new element is now created, with moistness as the dominating and hotness as the secondary property, that is air, as happens when water is transformed into steam. Other changes of state are explained in a similar way.

By the combination of elements, according to Aristotle, all the known substances of nature are produced, each in such a way that it is quite impossible to distinguish the pure elements. Two elements unite when their mutual attraction, due to their common qualities, is stronger than the repulsion of their opposite qualities. Thus fire and earth may combine to form embers because both are dry, and their dryness can, when circumstances are favourable keep them together, overcoming the repulsion between the coldness of the earth and the hotness of the fire. But because there are opposite qualities in all compound substances, these substances are inevitably less stable and must, eventually, dissolve again into their constituent elements, which then recombine in another way. In this way, all substances of the material world are created and pass away in a never-ending succession. Each compound substance is characterized by its *complexion*, that is, the ratio of its constituent elements. Only this complexion is changed during chemical processes.

Aristotle also postulates the existence of a fifth element, but for reasons different from Plato's. Aristotle is convinced that the heavenly bodies remain unchanged for all time. It is therefore impossible for them to contain opposite qualities which sooner or later would destroy them. They cannot be mixtures of the four elements, nor can they consist of any element in its pure form because the natural motion of the elements is rectilinear and not circular. Therefore, the heavens are made from a fifth kind of matter, unchanged, eternal, and unknown in the lower parts of the universe. Aristotle, too, calls this fifth element the *ether*.

Physics of the Stoics

The doctrine of elements was modified once more by the Stoic philosophers in the third century B.C. Contrary to Aristotle they defined the elements by one quality only, according to the scheme below. At the same time they stressed the Aristotelian distinction between fire and air as active, and water and earth as passive elements. This means that to the Stoics heat and cold were primary forces in nature, and can be explained by their interest in the processes taking place in living organisms: life presupposes breath (air) and develops heat (fire). But as the same scheme is applied to inanimate objects, the distinction between active and passive elements can be regarded as a first germ of the conception of the whole universe as a thermodynamic system.

The structure of the elements according to the Stoics

Quality	Element	
heat	fire	active
coldness	air	
dryness	earth	passive
moistness	water	

The Stoics later came to regard a combination of fire and air as a definite substance called *pneuma* which, they thought, permeates the whole universe, acting as a kind of glue which keeps its parts together, and confers characteristic properties on a multitude of substances. This universal pneuma is in a state of tension or stress and has elastic properties. It can therefore perform oscillations in many different ways. Each particular substance is distinguished from others by the nature of the oscillations of its inherent pneuma (cf. the Stoic doctrine of sound, page 136). Using a backwards projection of modern theoretical concepts, it would be tempting to regard the Stoic conception of matter as an example of a 'field theory' in which the material substances with all their various qualities appear as modifications of a field, namely the Stoic pneuma.

Application of the Doctrine of Elements

The later and less important changes in the theory of elements will not be studied here, although throughout the Middle Ages and up to the breakthrough of the new chemistry in the seventeenth and eighteenth centuries the theory was the accepted foundation of all speculations concerning the properties of material bodies.

The Four Elements in Physiology and Psychology

Humour	Dominating Element	Corresponding Temperament
blood	fire	sanguine
yellow bile	air	choleric
phlegm	water	phlegmatic
black bile	earth	melancholic

From its very inception, through the work of Empedocles himself and the medical school of Hippocrates, the doctrine of elements permeated Greek medicine. One single, anonymous author of a treatise *On the ancient medicine* tried to resist this movement, pointing out that not everything known of food or digestion can be explained by the four qualities. But this was ignored, and Empedocles outlined a medical theory asserting that the health of the body depends on the proper equilibrium between the four elements. Later Hippocratics turned this into a theory of health based on an equilibrium between four bodily *humours*, each dominated by one particular element. Finally, the foremost medical authority in Antiquity, Galen of Pergamon (*c.* A.D. 130–199) created a psychological theory of the four 'temperaments', a word originally meaning the proportion of elements, each temperament depending on an excess of one of the four humours. This theory became an integral part of Mediæval psychology.

Physics and astronomy also made use of the doctrine of elements, as in Plato's notion of a universe of spheres, each consisting of one particular element (p. 65). Aristotle incorporated the doctrine in a much more organic way in the very foundation of his mechanics. Earth and water were supposed to be provided with a special quality called *gravity*, while air and fire possess an opposite quality called *levity*. This enabled Aristotle to link the theory of elements with the mechanical concepts of natural place and natural motion (cf. p. 120). Gravity was defined as an inherent tendency in heavy bodies to move towards the centre of the world, which is the natural place for the heavy bodies, while levity was considered an opposite tendency for movement away from the centre. If all hindrances are removed, it was argued, a body dominated by the element earth must move downwards along a straight line towards its *natural place* at the centre. If a substance is dominated by the element of fire, it is light by nature and performs a natural motion upwards towards the natural place of light bodies, that is, the inner surface of the sphere of the Moon. In between are the natural places of water outside the earth, and air inside the fire. The elements air and water were considered to be *media*, because they are placed in the middle between the extremes of earth and fire, which explains the subdivision of the elementary world into four sublunary spheres between the centre of the world and the sphere of the Moon.

The assumption that heavenly bodies consisted of a fifth element, the ether, was founded on their immutable nature. Aristotle supported this idea by pointing to the uniform, perfect, self-perpetuating circular motion of the spheres. As this motion is totally different from the upward or downward natural and finite motion of the elements, the heavenly bodies cannot, for this reason consist of elements. The doctrine of elements thus created that connection between Aristotelian physics and astronomy which was, in the long term, to be one of the greatest stumbling-blocks in the creation of a new picture of the universe.

The Atomic Theory

At about the same time as the doctrine of elements was formulated a quite different conception of the structure of matter was developed by Leucippos of Miletos (*c.* 440 B.C.) and his pupil Democritos of Abdera (*c.* 460–*c.* 370 B.C.). These philosophers are usually mentioned together, and it is difficult to distinguish their individual contributions to the new theory as, apart from a few fragments their works have been lost. In the broadest sense their way of presenting the problems is the same as that of Empedocles, but their basic principle of all being are the *atoms* and the *empty space*.

To Leucippos and Democritos, any substance is a gathering of atoms, that is of countless small bodies of varying size and form. They are eternal, unchangeable, and indivisible (hence the name), in contrast to the elements, which were perceived as continuous (infinitely divisible). The atoms of Democritos move permanently hither and thither in the empty space, they collide and enter into a multitude of combinations which are later decomposed by new collisions. But in addition they perform, as a whole, a spontaneous, vortex motion during which innumerable worlds are created.

Leucippos's and Democritos's theory did not only differ from the theory of elements in that it gave another solution to the problem of unity and plurality. It also implied an essential distinction between primary or objective, and secondary or subjective properties (sense qualities). Only the size, form, hardness, and spatial configurations of the atoms are primary qualities on this theory. Properties like colour, smell, taste, warmth, cold, and so on, are subjective impressions, which in one way or another are provoked by the primary qualities and only attributed to the subjects of the outside world by convention. Even gravity is a secondary quality caused by the original, vortex motion of the atoms, a conception which suggests a primitive understanding of the centripetal force of circular motion.

Democritos maintains that the soul, too, consists of atoms. As all atomic movements are regular—even if their regularity cannot be detected—also mental processes are causal. For Democritos the atomic theory is deterministic and leaves no place for the freedom of the human will. Later, Epicuros of Athens (341–271 B.C.) tried to change this as part of an attempt to give the atomic theory a new framework. First, he discards Democritos' conception of an infinity of atomic forms: there is a finite number of different atomic types, each represented by an infinite number of individua. In saying this, Epicuros approximates the ancient atomic theory to the fundamental idea of the modern one. Next, he has a different understanding of the motion of atoms which he regards as an infinite 'fall' through space. Collisions between atoms become necessary if new combinations are to be forged. As impact cannot take place during the parallel motion of fall of the atoms, Epicuros maintains that the fall is, from time to time, disturbed by small, spontaneous aberrations not provoked by any external cause. Through these aberrations some of the atoms approach each others and can unite by means of the small hooks with which they are provided. These spontaneous movements make the theory non-deterministic, and Epicuros succeeds in 'saving' the concept of free will.

Atomism Refuted

The doctrine of the elements was a typical continuum or field theory, particularly in the way in which it was developed by the Stoics, who considered all material qualities to be modifications of the all-pervading pneuma. The atomic theory, on the other hand, derived all qualitative differences from purely quantitative structures, namely the various geometrical arrangements of unchangeable atoms in empty space. The more recent development of physics has shown that it may be necessary to use both these descriptions, even if they seem incompatible from a purely logical point of view. But in Antiquity, all philosophers took sides. Most of them adopted the doctrine of the elements, and eventually the atomic theory was almost completely discarded, although it was not always refuted with scientific arguments. It was often maintained that the doctrine of the atomists implied a materialistic philosophy and left no place for the freedom of either man or gods. Epicuros tried to forestall this objection which was, perhaps, not as serious as it might seem, for Stoic philosophy—intimately connected as it was with the doctrine of elements—achieved widespread support, although in many ways, it was just as materialistic as that of Democritos.

Undoubtedly, the atomic theory was abandoned mainly for physical reasons. It was unacceptable to most philosophers simply because it presupposed the existence of empty space. Greek philosophy was, for several reasons, always reluctant to acknowledge the idea of the void. First, for most pre-Socratic philosophers it was an unspoken assumption that any existing being must have a material character; an actually existing vacuum was a concept without meaning. Secondly, Aristotelian physics left no place for the void, as its existence seemed to lead to insurmountable difficulties in mechanics. The fundamental relation of Aristotelian dynamics (p. 122) implies that a finite force would move a body with infinite velocity through empty space, which is absurd. Moreover, Aristotle maintained that if a body is supposed to move through the void:

> nobody will be able to tell why it should stop anywhere, for why rather at this place than at another? Therefore, it will either remain at rest or move on in infinitum until something stronger hinders its motion.

A statement of this sort is reminiscent of Newton's first law of inertial motion but has, of course, a completely different background and cannot be thought of as a precursor of the law of inertia. To Aristotle, any idea of infinite motion (apart from that of the heavens) was incompatible with the doctrine that, because of their own nature, bodies will come to rest at their natural places where they are deprived of any intrinsic tendency to further natural motion.

The arguments against the atomic theory are all more or less speculative or theoretical, and have to be weighed against the reasons given in favour of the doctrine of elements. It must be remembered that at no essential point was this doctrine at variance with any well-established facts about the properties of matter. The final outcome of the battle is therefore not unexpected, and the atomic theory was so heavily defeated that not even the primary literary sources have been conserved. When atomistic ideas appear in Ancient or Mediæval works, they can usually be traced back to a secondary source, namely the long didactic poem *De rerum natura* by the Roman author Lucretius (*c.* 98–55 B.C.) who remains the chief authority on the opinions of Leucippos and Democritos.

Chapter Twelve

Properties of Matter and Chemical Change

Early Experiments on Air

Although theoretical speculations dominate the Greeks' efforts to describe the properties of matter the more practical and experimental investigations of the same problem must not be forgotten. The Greeks knew of an experimental approach; it was often ignored by the professional philosophers but was closely connected with technology and handicraft, particularly in the late Alexandrian period.

The nature of optical and acoustical experiments has already been studied. Here another aspect of Greek experimental science in two different fields will be examined; first, a rather limited series of experiments on the properties of air and the effects of heat, and secondly the development of Hellenistic alchemy.

In the second book of Empedocles' tract *On nature* we find the first authentic report of an important physical experiment, performed, for physiological purposes to support a theory concerning the motion of blood inside the body. Empedocles thought that blood rises and falls in the vessels. When the blood is falling, he maintained, air is sucked in through the pores of the skin, to be blown out again when the blood is rising. The experiment is described in Homeric style, in the form of a parable, but the text proves that a real experiment was actually performed.

> Just as when a girl plays with a clepsydra of shining brass, placing the mouth of the tube on her well-shaped hand, and dipping the clepsydra in the yielding mass of silvery water—the stream does not flow into the

vessel, but is kept out by the air within pressing against the close-set perforations, until she opens for the compressed stream. Then the air escapes and an equal volume of water runs in.

The experiment was then varied, the vessel now being filled with water which is unable to escape until the mouth is uncovered. This is explained by the assumption that the air outside is pressing upon the water through the perforations, preventing it from running out.

Which instrument Empedocles refers to is not clear. Usually the term *clepsydra* meant a water clock as mentioned on p. 156, but here it seems to be a kind of siphon. The description suggests a longish vessel of brass with an orifice at one end and several narrow perforations at the other. Aristotle mentions that Anaxagoras also made similar experiments with a clepsydra, and that he used inflated bladders to show that the air inside resisted his efforts to compress them.

The results of such experiments had a far-reaching importance. They proved that the air inside the empty clepsydra and the inflated bladder acts as a material body resisting influences from the outside, and keeping out other bodies (water) from the space it occupies. This experimental proof that air is a material body was, presumably, an essential condition for its being reckoned among the four elements by Empedocles himself.

Aristotle's account of these experiments shows that they were immediately used by the opponents of the atomic theory as a proof that no vacuum exists in nature. But in fairness, he points out the fallacy of this conclusion; to show that an apparently empty vessel contains a material substance, namely air, is not, he says, the same as proving that 'an extensive domain not occupied by any substance perceptible to the senses may exist anywhere'. Since he, too, opposed the atomic theory Aristotle had to find other arguments to put against the existence of a vacuum in nature (p. 122).

Pneumatics

In the third century B.C. the properties of air were experimentally investigated once more, this time by Strato of Lampsakos who from 286 to 268 B.C. was head of the Peripatetic school in Athens, succeeding Theophrastos (*c.* 372–*c.* 288 B.C.). Although he belonged to the

Aristotelian tradition, Strato criticizes Aristotle on a number of very essential points. Strato discarded the idea of levity, and ascribed a definite gravity to all elements, implying that air and fire also have weight. In doing so, Strato implied a criticism of the doctrine of natural places. He maintained the Aristotelian conception of matter as infinitely divisible, but went along with the atomists in assuming an actual vacuum inside material substances, but not outside the world. This attempt at uniting the atomic theory with the doctrine of elements came too late to exert any considerable influence on the development of physics, but Strato's many simple experiments on the behaviour of air reveal his efforts to construct his physics from empirical facts.

The school at Alexandria soon took the lead in the study of air and established the science of its properties as a new, independent discipline, usually called *pneumatics*. This science was a kind of technical physics,

Fig. 12.1.   *Philo's double-acting lift and force pump.* (*Usher*, A History of Mechanical Inventions, *New York 1927*).

as physical principles were studied with a view to their practical use. As far as is known such studies were begun by Ktesibios of Alexandria (presumably *c.* 250 B.C.) who was one of the greatest inventors in Antiquity. He dealt with pneumatics in a treatise which is now lost but whose subject matter can be reconstructed from information in the works of Heron, Vitruvius, and others.

Ktesibios constructed a pressure pump (Fig. 12.1) with cylinder, piston and valves. This was later developed into the perfect technical device seen in the Roman fire engines with two reciprocating pumps (Fig. 12.2). Ktesibios also invented the so-called water organ which,

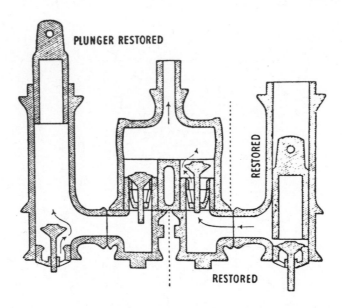

Fig. 12.2.   A Roman lift and force pump, partly restored from a specimen found at Bolzena, Italy. From A History of Technology, *II, 376, Oxford, 1954.*

with its pipes, windchest and keyboard was the prototype of all later organs. The air did not come from bellows, but from piston pumps, its pressure being stabilized by an air vessel which contained a rising and falling volume of water. Finally, Ktesibios developed the unsophisticated clepsydra into a much more reliable water clock (Fig. 12.3). It had two vessels of which the upper one was provided with an overflow pipe to keep the water level constant. The water therefore flowed

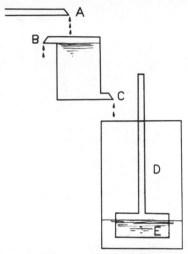

*Fig. 12.3. The water clock of Ktesibios recon-
structed by A. G. Drachmann. Because of the
overflow B there will be a constant water level in
the upper vessel from which the water flows
through C with constant velocity into the lower
vessel. Here the float E will rise in proportion to
the time read off the scale D.*

through the orifice below with a constant velocity and was collected
in the lower vessel where its level rose proportionally to time.

Sometime after Ktesibios, the engineer Philo of Byzantium worked
in Alexandria and on the island of Rhodes. He wrote a long treatise on
mechanics containing many sections on construction of ports, military
engines, and so on. One section of the work was a *Pneumatica*, extant
in Arabic translation, and mainly derived from Ktesibios. Several new
hydraulic and pneumatic inventions must, however, be attributed to
Philo himself.

Another sphere of Philo's experimental work was concerned with
the interaction of air and heat. In Fig. 12.4 the lead flask *a* is provided
with a close-fitting tube *b*, one end of which is close to the bottom,
while the other end is placed in the vessel *c* which is filled with water.
When the Sun shines the air inside *a* expands and escapes through the
tube. When, later, *a* is cooling, the water is sucked into the tube. The
apparatus could be used as a primitive thermoscope indicating the
temperature of the air in *a*, but it seems that Galileo was the first person

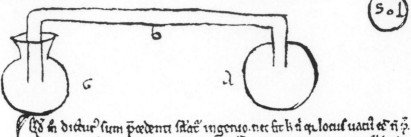

Fig. 12.4. *Philo's experiment on the expansion of air by heat. (From a Latin manuscript of c.A.D. 1200).*

to use it in this way. Neither Antiquity nor the Middle Ages had any clear understanding of the different notions of temperature and heat concepts which were distinguished only in the eighteenth century.

In another experiment Philo burnt a candle inside a jar whose mouth is placed under water. He describes how the water rises in the jar in order to prevent the creation of a vacuum when the flame destroys the air. Nature's presumed horror of the void (*horror vacui*) had, at this point in time, become a natural starting point for the explanation of pneumatic observations, although other principles were eventually applied. Philo explains how water follows air into a pipette by saying that the two bodies seem to be glued together. Yet it should be remembered that Galen (A.D. 129–199) clearly realized the necessity of air in combustion, a process which he compared with the production of heat in animals.

The pneumatics of Antiquity reached its climax at the time of Heron of Alexandria whose contributions to mechanics were mentioned on p. 116. His extant *Pneumatica* has a theoretical introduction in which the problem of the void is viewed in a new light. An extensive vacuum, he maintains, although it is not found in nature, might be produced in some artificial way. Heron assumes, however, that space exists between small parts of bodies, an assumption founded on the fact that air can be compressed because the particles of the airy element approach each other in the surrounding vacuum. Such ideas obviously stem from

158

Strato of Lampsakos (p.154). Following his introduction, Heron describes a number of pneumatic devices of many different kinds, such as various types of siphons, an improved water organ (Fig. 12.5), drinking

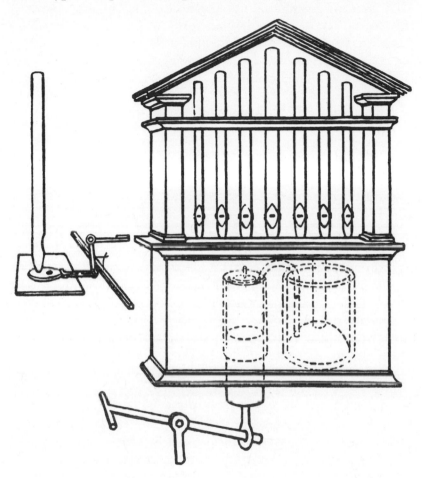

Fig. 12.5. A reconstruction of Heron's water organ. From The Pneumatics, *ed. B. Woodcroft, London, 1851.*

animal figures, a machine opening the doors of a temple when the fire is kindled on the altar, an improved thermoscope, and a small reaction turbine run by steam and many more. Most of these inventions look like toys or gadgets and were probably not devised for any serious scientific purpose. This does not in itself reduce their importance as testimonies to the technical abilities of ancient science, but it certainly does reveal that Hellenistic society had not reached a stage where it could profitably use advanced technology.

The pneumatics and mechanical technology of Antiquity show the existence of a most valuable practical tradition which—outside the narrow circles of the natural philosophers—contributed to the knowledge of the laws of nature and their possible use for practical purposes. This tradition continued through the Latin Middle Ages, thanks mainly to the great technological manual *De architectura* written by the Roman engineer Vitruvius (first century B.C.).

The Beginning of Chemical Theory

While the pneumatics of Antiquity was an experimentally-based doctrine, the first attempts at the creation of a proper chemical theory are quite different. They found in the fourth book of Aristotle's *Meteorologica* containing the germ of the later Stoic distinction between active and passive qualities, a distinction which is the starting point for explanations of how heat and cold act upon bodies through processes like maturation, boiling, calcination, desiccation, solidification, dissolution, and so on.

Following this introduction, a series of substances are classified in different categories such as fusible and non-fusible, combustible and non-combustible, plastic and non-plastic. Each property is described in detail. Later, an attempt is made to state the proportion between the elementary qualities in various substances. Although it borrows a number of facts from chemical technology the treatise contains very few references to chemical experiments. But despite this it can be considered as the first known work on theoretical chemistry because it represents an effort to arrange and to classify substantial properties and chemical (as well as physical) processes by means of the doctrine of elements.

Alchemy

A far more intimate connection between theory and practice was established by the ancient alchemists. Alchemy is known through a series of Greek treatises, all of which came originally from Alexandria, but most of these writings are incorrectly attributed to either legendary or historical persons who had not the slightest connection with alchemy. Examples are the philosopher Democritos (p. 150), the neo-Platonist Iamblicos (c. A.D. 300), the prophet Moses, Queen Cleopatra, Ostanes who was a legendary Persian sage, Hermes who is identical with the Egyptian god Thoth, and Agathodaemon who was a Phoenician snake-god. Obviously alchemy was not a purely Greek doctrine, but had strong attachments with the religion and mythology of the Orient.

The oldest texts on alchemy are a now lost tract, called *Physica et mystica* and ascribed to an Egyptian, Bolos Democritos of Mendes, plus several extant treatises by Isis, Moses and Ostanes which cannot be dated with certainty, but which most likely stem from the 1st century A.D., although Bolos may have lived in about 200 B.C. The tracts deal with subjects such as the production of gold, silver, precious stones, colours, etc. The techniques described are rather simple, and generally involve only surface treatment or alloying of metals.

A later group of alchemical fragments is associated with the name of the Jewess Maria, who was probably a real person living in Alexandria some time before A.D. 300. These are interesting because they describe highly developed apparatus for distillation and sublimation.

A third group of writings can be traced back to one Zosimos who in about A.D. 300 wrote a long series of tracts which, as regards technology, built on Maria's writings. There is, here, a definite influence from Christianity, and the tracts reveal the alchemists as forming an esoteric society which reveals only reluctantly its ideas or methods to the uninitiated.

From classical Antiquity, no alchemical literature exists in Latin. But through the Arabs, Hellenistic alchemy was brought to Europe, where from the 12th to the 18th centuries it gave rise to a torrent of Latin tracts, from which chemical literature, in the modern sense, emerged only slowly. After this, the later history of alchemy contains few new features.

Because alchemy was veiled in mystery from its inception, interpretation of texts is often difficult. The alchemists' use of a secret and symbolic language which did not persist often makes it impossible to identify the substance or process referred to. Often this symbolism is curiously anthropomorphic; the solution of a substance may be described as its 'death' and the following precipitation as a 'resurrection'.

In the same vein distillation is conceived of as the separation of a soul (pneuma, spirit) from a body.

In the Middle Ages this symbolism was pursued even more vigorously. Sometimes it is fairly easy to decipher. The much-used picture of a green lion devouring the Sun, for example, simply means that *aqua regia*, represented by the lion (the king of the animal world) dissolves gold, represented by the Sun. At other times the symbols are so obscure that identification with any known chemical process appears impossible. It is most likely that some of these texts are not concerned with chemistry at all, but are mystical or ascetical treatises on the progress of the human soul through the various stages of birth, life, death, and resurrection. Quite probably, many alchemical authors were unable to see any fundamental difference between chemical processes occurring in nature and spiritual processes of human life. A unified conception of this kind was made possible through the pneuma-philosophy mentioned below.

Apart from its possible philosophical interpretations, the study of alchemy also had materialistic aims, of which the most conspicuous was the transformation of base metals into gold. Such attempts are known as early as about A.D. 300 from Egyptian papyri containing recipes of substances having a superficial resemblance to gold or silver (see p. 161). Several methods were employed. Sometimes, the surface of the base metal was covered with coloured substances, such as a lacquer or sulphide; at others, the metal was alloyed with a small amount of gold and so treated that the base metal was dissolved and removed from a thin layer of the surface, which then remained covered with pure gold.

The dream of making gold was not as naive as might be thought today. Before the development of modern analytical chemistry there were no reliable methods of identification of specific substances, and the concept of a chemical element was unknown. A heavy, reddish and chemically resistant alloy could thus be taken for gold without any conscious deception. Archimedes' exact determination of the amount of gold in King Hieron's crown (p. 114) was a unique achievement in ancient science. It might have led to a method of identification based on the notion of specific weight, but the latter was also unknown, and methods of this kind did not appear until much later (p. 187).

Alchemical Theories

The early Egyptian recipes belong more to the domain of chemical technology than to alchemy in the true sense because they are completely devoid of any theoretical explanation of the processes involved, whereas alchemical texts include, according to Greek tradition, an attempt to find a theoretical foundation for chemical phenomena both in nature and in the laboratory. Apart from the desire to transform metals into gold, alchemy thus had an ulterior motive, namely to put the methods employed on a rational basis according to what was known, or believed to be known, about the structure of matter.

Two of the starting points of alchemical theory were the doctrine of the four elements (p. 145) and the Aristotelian hylemorphism (p. 32) from which a general theory of transmutation could be developed. On this theoretical basis it was thought possible that copper for example, could be transformed into gold, first by bereaving the metal of the 'form' of copper, and secondly by replacing this form with that of gold. Such processes presupposed the existence of a number of efficient causes, and their determination became one of the main objectives of alchemy.

In the organic world, new living forms are continually emerging while others disappear. Plants, for instance, grow under the combined influences of air, moisture, and heat, then they wither leaving behind fruits or seeds of a completely different form. But under the influence of the same forces, the fruit is transformed once more into the form of the original plant. One of the earliest texts on alchemy, *The Gold-Making of Cleopatra*, uses this analogy:

> Then Cleopatra said to the philosophers: Look at the nature of the plants from where they come. For some come down from the mountains and grow up from the earth, and some grow up from the valleys, and some come in the plains. But look how they develop, for you must collect them in certain seasons and on certain days [. . .] and look at the air which serves them, and the nourishment surrounding them that they do not perish and die. Look at the divine water providing them with drink, and at the air reigning over them after they have acquired a body in a singular being.

The similarity between changing forms of life in nature and those of metals is explained on the basis of the Aristotelian idea that metals are formed by condensation of volatile gases or 'exhalations' from the earth. This is connected with the Stoic doctrine of the ubiquitous 'pneuma' (p. 148) as the real uniting and form-giving force in nature. According to Aristotle, metals are condensed from the 'breath' of the earth, later regarded as composed of sulphur and mercury. Both these substances thus became the true 'principles' of all metals, but according to the pneuma-theory all substances whatever are formed through the action of the pneuma upon the passive elements.

Notions such as these were the basis of the belief that, from all metals except gold, a certain, subtle, pneumatic substance can be separated which is able to exert a specific influence upon other substances and change their 'formal' properties, that is, their essential nature. This process was thought to be supported by fire, and the whole idea was founded on the results of distilling volatile, etheric components from various natural substances. Alcohol is still called 'spirits', or *spiritus vini*, a reminder of the conception of alcohol in Mediæval alchemy as the volatile 'spirit' or 'pneuma' separated from wine by distillation. In a similar way, Mediæval alchemists regarded nitric acid as the liberated 'spirits' of salt petre. The conspicuous action of these 'spirits' on metals was cited in support of the theory.

The general belief that real transmutation of substances is impossible without the influence of celestial forces—just as the growth of plants can only be brought about by the rays of the Sun combined with the surrounding heat and air—was decisive in the evolution of alchemy, for it created a connection between alchemy and astrology. Perhaps this belief that the co-operation of the heavenly bodies is necessary for any chemical process is the feature which most radically distinguishes alchemy from chemistry. The marriage between alchemy and astrology was clearly a result of Babylonian and Persian influences, but it also fitted in with the idea of the microcosm and the macrocosm (Fig. 12.6) of Plato's *Timaios* (p. 29). The historian Diodoros Siculos (*c.* 50 B.C.) tells of the Egyptians that:

> They say that these gods [Isis and Osiris] contribute much to the generation of all things because of their natures. For the one has a hot and active nature, and the other a moist and cold, but both have something of the air [. . .] so that every single substance in the world is completed and perfected by the Sun and Moon the properties of which are five in number as mentioned before: a spirit or pneuma with vitalizing force

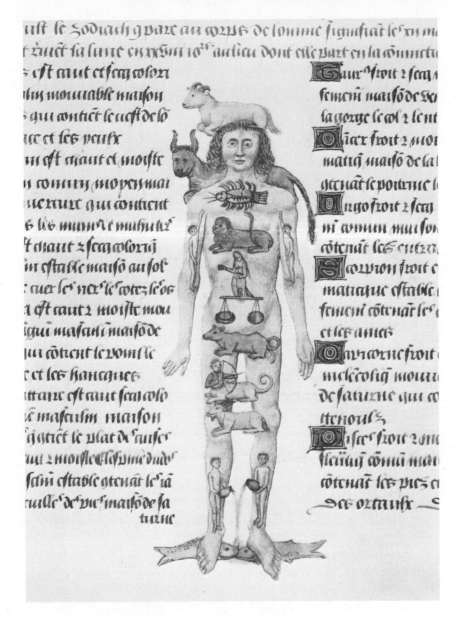

Fig. 12.6. Microcosm with the signs of the zodiac placed at the respective parts of the body. From a fifteenth-century French manuscript in the Royal Library, Copenhagen.

heat or fire, dryness or earth, moisture or water, and air of which the world consists, just as man consists of head, hands feet and other limbs. . . .

The doctrine of the elements, the pneuma-philosophy and astrology were thus combined never to be separated as long as alchemy existed. An innocuous result of the union was a particular terminology, naming each of the known metals after one of the planets, the properties of which dominated the metal. More annoying to the work of the alchemists was the firm belief that chemical operations could only be carried out at certain definite times when a favourable constellation of the planets made possible the processes involved.

The Achievement of the Alchemists

Although theoretical alchemy seems to be a curious mixture of the Greek doctrine of elements, Babylonian astrology and oriental mysticism, it must not be forgotten that in the experimental field the achievements of the alchemists had a far-reaching importance. While their theoretical speculations remained barren, we are still enjoying the fruits of their practical skill. It was the alchemists who created the first research laboratories equipped with specially adapted instruments, and many of their methods are still in use. Most important was the art of distillation, invented as a means of driving off the volatile pneuma or spirit from a substance. The Sumerians (p. 2) certainly carried out distillations, but in classical Antiquity the first descriptions and pictures of stills are found in the writings ascribed to the Jewess Maria. Figure 12.7 shows a bottle or flask with a long, narrow neck, on the top of which is placed a cap (the alembic) with three side tubes. The vapours were condensed in the alembic and the distillate ran out through the tubes into three different receptacles, so that the distillate was completely separated from the remaining part of the liquid. Compared with the primitive Babylonian apparatus of 3,000 years before (Fig. 1.1) Maria's still represents an almost perfect solution of the problem of distillation. The alchemists also developed particular stills for distillation of mercury and for sublimation, that is, the distillation of solid substances which evaporate without melting.

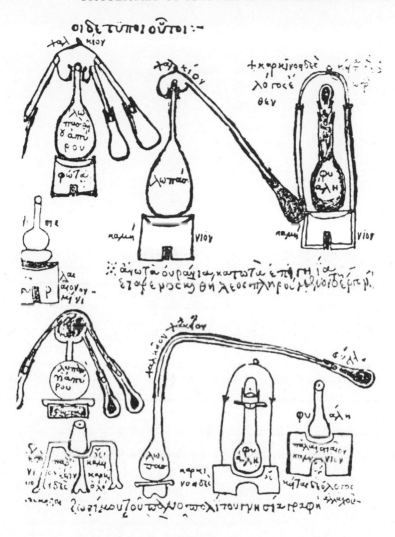

Fig. 12.7. Hellenistic chemical apparatus from a Byzantine manuscript dating from the ninth or tenth century A.D. *From Berthelot,* Collection des anciens alchimistes grecs, *Paris, 1887.*

The *kerotakis*, shown in Fig. 12.8 is particularly interesting. It was used for the surface treatment of metallic objects with vapours and makes use of the principle of reflux cooling, the vapours continually condensing at the top of the apparatus and returning to the solvent.

167

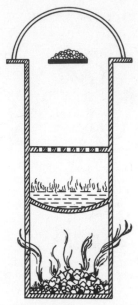

Fig. 12.8.   Greek kerotakis *using the reflux principle for the surface treatment of metallic objects with vapours.*

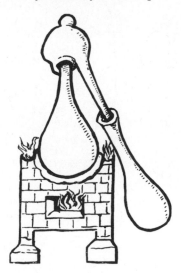

Fig. 12.9.   *Distillation apparatus or alembic of a type used from Antiquity to the nineteenth century. From F. Sherwood Taylor:* An Illustrated History of Science, *London 1956.*

As time went by, several new apparatus were developed, some of which are direct precursors of modern laboratory equipment (see Fig. 12.9). The skill of the alchemists in gradually mastering processes like distillation, sublimation, calcination (the conversion of minerals to powders by means of heat, sometimes an oxidation in the modern sense), circulation (reflux distillation) and crystallization was their lasting contribution to the development of chemistry, as were the many new and useful substances they discovered and produced in more or less pure form. The mineral acids are a good example. Even if the alchemists failed in their search for the 'Philosophers' Stone' or the 'Tincture'— the substance supposed to bring about the final transmutation of base metals into gold—their achievement has, nevertheless, borne fruit, by contributing to our knowledge of chemical substances and their properties.

Fig. 12.10.   Illustration from a Byzantine tenth-century manuscript showing a naval battle using the Greek Fire, presumably a mixture of naphtha and other inflammable liquids. Reproduced from A History of Technology, *II, p. 376, Oxford, 1954.*

Chapter Thirteen

The Transmission of Greek Science

The Impact of Greek Science

THERE are many reasons in support of the view that the achievements of the Greeks were the factors most decisive in the origin of modern science. It is true that pre-Greek civilizations had made important findings both in mathematics and technology, and that the Babylonians, in particular, had succeeded in giving astronomy a concise mathematical form contemporary with the unfolding of the geometrical astronomy of Eudoxos, Apollonios, and Hipparchos. Nevertheless, the briefest of comparisons between Babylonian and Greek science shows the extent to which the Greeks succeeded in reshaping and developing scientific methods by a conscious effort reflected in their philosophy of science. The elegance and perspicacity of the Babylonians are undeniable and admirable. But to a modern scientist it is equally impossible not to feel more at home in the worlds of Eudoxos or Archimedes. All later science is indebted to the Greeks because of their adherence to the rational concept of natural law. Their discovery of how suitable hypothetical-deductive theories are to the description of nature, their introduction of the concept of mathematical models, and their first tentative uses of the experimental method are features of such lasting importance that the development of science would have taken a completely different course without them.

The results of the Greeks are all the more remarkable because the period in which their science matured and flourished was a brief one. Only 400 years lie between Thales and Hipparchos, by which time the creativity of the Alexandrian school began its long decline. An impressive number of mathematicians, astronomers, physicists, and engineers did carry on, but their efforts were less comprehensive, and their methods less original, than those of their predecessors.

Backed by their enormous library, Alexandrian scholars turned more and more to the systematization of past results. They were brilliant compilers, editors, and encyclopedists. Only Ptolemy's eminent and creative work in astronomy is an exception. When with the advent of Plotinos (3rd century A.D.) neo-Platonic philosophy captured the minds of creative thinkers, the tide had turned. Later commentators like Proclos or Johannes Philoponos might occasionally venture new ideas of lasting value, and the writings of Simplikios must be regarded as an original contribution, as well as useful to a better understanding of traditional doctrines; but, in general, Hellenistic science had become stagnant. The reason is still a puzzle to historians, who can only register the decline and try to follow the various channels through which Greek science penetrated into other civilizations and eventually into the modern world.

Byzantium

When the Roman Empire disintegrated into eastern and western sectors, the Greek tradition was kept alive inside the Byzantine Empire around Constantinople, for here the Greek language maintained its position, and the old authors were still read. However, the history of Byzantine science is still obscure, and only a few characters emerge from a general background of commentators and scholiasts. In the early Byzantine period, Anthemius of Tralles (who died c. A.D. 534) wrote on technology, optics, and mathematics, and also arrived at the so-called gardener's construction of an ellipse, based on the constant sum of the two radii vectors. Later, Heron of Byzantium (tenth century A.D.) wrote on machines and surveying, thus continuing the tradition of his Alexandrian namesake, but no major discovery can yet, it seems, be ascribed to Byzantine scientists. The importance of keeping alive a great tradition which, in the years before the fall of Constantinople in A.D. 1453, spread to the Latin world as a new stimulus to both philosophy and scientific research, should not, however, be underrated.

Greek Science in the East

Meanwhile, Greek science had begun its long march towards the east. Even in the first centuries of our era Greek centres of importance were established in Gaza, Antioch, and other towns on the eastern shores of the Mediterranean. Eventually, these schools became Christian like the School of Alexandria, and survived, unlike the old Platonic Academy of Athens, the last stronghold of a non-Christian, philosophical tradition, which was closed in A.D. 529 by the Emperor Justinian. Dogmatic controversies between the Christians in Alexandria and elsewhere started a long wave of emigration of religious minorities. In particular, the Nestorians travelled as far as Nisibi and Edessa in Mesopotamia and even reached Jundishapur in Persia. In these, and in other places, Greek schools were established, and Greek culture brought into closer contact with oriental doctrines of many different kinds. Especially important were the many translations of Greek texts into other languages, particularly Syriac, a semitic language related to Aramaic and Arabic. Syriac translations became singularly important after the expansion of Islam, and were the most obvious factors in the speed with which the Muslims adopted Greek science.

Hindu Astronomy

Even before the Nestorians carried Greek science into Persia a strong western influence existed in India which began as early as the Dark Age of Hindu civilization, between 500 B.C. and A.D. 500, from which very few texts are conserved. The division of a synodic month into 30 equal parts called *tithis*, is Babylonian in origin, as is the accepted ratio 3 :2 between the longest and the shortest days; this relationship is found all over the east regardless of latitude. The odd speculative Indian chronology shows traces of Babylonian and Persian influence. Thus some very long periods were used in Hindu cosmology to describe the cyclic life of the universe, the present period being the *kaliyuga* of 432,000 years beginning in 3102 B.C. But this is the round number 2,0,0,0 in the sexagesimal system, and well known by Babylonian authors as the age of the Babylonian kingdom before the Flood. Even Babylonian planetary theory with its linear algebraic methods penetrated as far as southern India, and served as a basis for Tamil astronomy down to the nineteenth century.

The way in which Greek astronomy reached India is still somewhat obscure. According to legend it was King Rudradāman I (c. A.D. 130—c. A.D. 160) who introduced Greek horoscopy and astronomy. His capital Ujjain defined the standard meridian of Mediæval Indian astronomy and became known as Mount Arim to Latin astronomers in the west. It seems that initially, Hindu science was influenced by Alexandrian astrology, and that some at least of the Babylonian elements also came via Alexandria. Round about A.D. 500 Hindu astronomers mastered the advanced geometrical methods of the *Almagest* without having any direct translation of it.

From this date, for several centuries, a series of astronomical textbooks were produced in Sanskrit called *Siddhāntas* (meaning established conclusions) and containing epicyclic theories of planetary motion, tables and so on. Some of these are anonymous, like the *Romaka-Siddhānta* (the Roman Established Conclusions) whose title clearly points to a western source. Other *Siddhāntas* and supplementary works were written by several astronomers who made India the scientific centre of the world between the eclipse of Alexandria and the rise of Islam. The first of these astronomers were Āryabhata and his disciple Lātadeva who worked in about A.D. 500, Varāha Mihira (c. 550) and Brahmagupta who published the important *Brāhmasphuta-Siddhānta* in A.D. 628. All these astronomers used advanced trigonometrical methods, and while Ptolemy had satisfied himself with a table of chords, the Indians introduced the sine function. A peculiar feature of their planetary theories is their computation of the corrections to the mean longitude (the prostaphairesis angles of Ptolemy) by means of models with concentric deferents, similar to the epicycle models of Apollonios (see p. 81).

The Rise of Muslim Science

In the years following Muhammed's death in A.D. 632, the Arab conquest of the Middle East began, and by the end of the century Islam had established itself over an area ranging from Spain in the west to India in the east. Nothing is heard of any special Arab science in the first century after the conquest. It was only after the foundation of Baghdad in A.D. 762 that the 'Abbāsid kaliphs began to take an active interest in science and learning. They established the House of Wisdom, a kind of academy provided with a large library and an observatory, and similar institutions were later created in Cairo, Cordoba, and many other cities. Placed as it was between Europe and India, Muslim science became influenced from both west and east.

The first record of the translation of a scientific work into Arabic is in A.D. 772, when a Hindu scholar brought a work called *Sindhind* to the court of al-Manṣūr, who ordered a young astronomer to translate it. There is no doubt that the *Sindhind* was one of the *Siddhāntas*, so that Muslim astronomers were from the start introduced to the special methods of the Hindus. In the ninth century Greek scientific works were translated in large numbers, at first from earlier Syriac versions, but later directly from Greek manuscripts found in the existing schools now under Arabic rule, or acquired in Byzantium. An excellent translation of the *Almagest* was made by Isḥaq ibn Ḥunain who died in A.D. 910. An immense scientific activity thus began within Islam

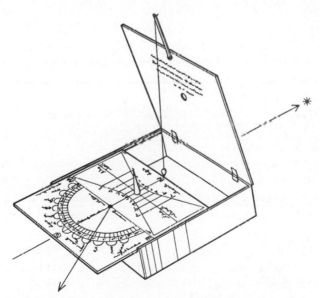

Fig. 13.1. Syrian sundial from the fourteenth-century showing both the hour of the day and the direction to Mecca (the Qibla). From A History of Technology, *III, p. 599, Oxford, 1954, after S. Reich and G. Weit.*

integrating many different elements: Greek mathematics and science, Hindu trigonometry and astronomy, astrological doctrines from Mesopotamia and Persia, plus independent Muslim contributions. Original Muslim work was often provoked by new practical needs such as navigation in the Indian Ocean. In particular, the determination of geographical co-ordinates was of obvious importance for finding the *qibla*, the direction of Mecca, necessary for the correct attitude in prayer (Fig. 13.1). It should be remembered, however, that although this new civilization is called 'Arab' only very few of its protagonists were Arabian. The rest came from the whole of the vast area dominated by Islam, and many of them were non-Muslim, Jews, Christians, and pagans. The unifying principle of Islam as a civilization exists solely in the Arabic language, which, in the standard form used in the *Koran*, played a role similar to that of Sanskrit in India or Mediæval Latin in western Europe.

In the field of mathematics, Arab science differed from the Greek in attaching much greater weight to both algebra and trigonometry. Indian influence is very conspicuous in the work of the Persian al-Khwārizmī (*c.* A.D. 850), the first important Arab mathematician who also composed a *zīj*, or collection of astronomical tables, with instructions for their use. Many of them are derived from the *Handy tables* of Ptolemy, but the Indian methods are used for computing planetary equations. This work—known only in a version by Maslama al-Majrīti (*c.* 950–*c.* 1007)—was later commented on by Ibn al-Muthannā (possibly tenth century) and also translated into Latin (p. 190). This *zīj* is usually the source of the Indian methods which emerged from time to time in Latin astronomy even after the Ptolemaic tradition had become universal. Astronomical tables were composed in great numbers by Arabic observers using several methods and often new parameters deduced from observations. Notable are the so-called *Hakemitic tables* compiled by the great astronomer ibn Yūnus who worked in Cairo about A.D. 1000, and the *Toledo tables* derived by the Moorish astronomer al-Zarqālī (*c.* 1029–1087), the Azarquiel of the Latins). The *Toledo tables* were adapted by Latin astronomers of the eleventh and twelfth centuries to many localities in Europe, and were also the paradigm of the *Alphonsine tables* computed by a team of Moorish, Jewish, and Spanish scholars at the command of King Alphonsus X (the Wise) of Castile. Until Kepler's time they dominated European astronomy.

Fig. 13.2.   Astrolabe constructed in A.D. 1220–21 by Muhammed Ibn Abu Bakr, Isphahan.
Now in the Museum of the History of Science, Oxford.

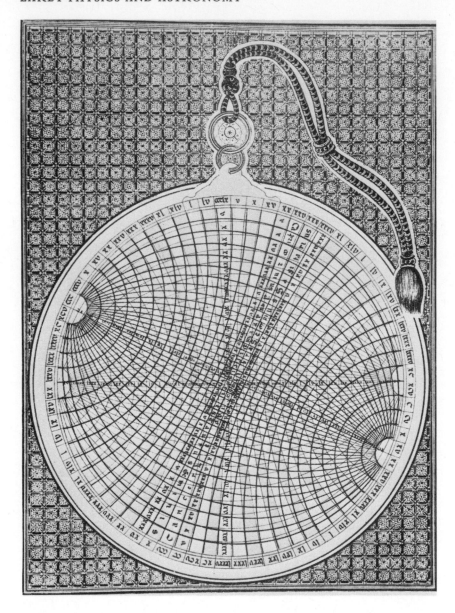

Fig. 13.3. Constructional drawing of the mater of al-Zarqāli's ṣafīka. From the Alphonsine Libros del saber, III, Madrid, 1864.

In the practical field, the astronomers of the Muslim world were very active, constantly developing new or greatly improved instruments and making careful observations enabling them to re-determine astronomical parameters with greater accuracy. The Jewish astronomer Māshāllāh, who did the surveying when the foundations of Baghdad were laid, wrote a treatise on the astrolabe, an instrument dating from

Fig. 13.4A. Modern reconstruction of a Gothic astrolabe calculated for the latitude of Aarhus, showing (A) the frontside on facies and (B, p. 180) the backside or dorsum. Institute for the History of Science, Aarhus.

Antiquity and whose theory was worked out by Ptolemy in his *Planisphaerium* (Fig. 13.2). The Arabs developed the astrolabe to a degree of great perfection and Māshāllāh's treatise became the standard book on the subject, and remained so in the Latin Middle Ages when it was known as the *De compositione Astrolabii*, by the author Messahalla. The instrument was developed further by al-Zarqāli, whose *azafea*, or *safika*, was an astrolabe provided with many extra engraved circles

Fig. 13.4B.

(Fig. 13.3) so that more problems in spherical astronomy could be worked out in a mechanical way (see p. 248f).

Small and handy astronomical devices like the astrolabe were supplemented by large observatory instruments for making precision measurements. A detailed description by the historian al-Maqrīsī (who died A.D. 1442) tells of how in A.D. 1119 someone made a great meridian instrument for the observatory of Cairo. This instrument had a graduated copper circle with a diameter of 10 cubits (about 5.80 metres); its casting pressed the ability of the technicians to its limit, and there were great difficulties in preventing it from bending. Obviously an instrument of such bulk represented the upper limit of Mediaeval technology.

Muslim Geodesy

A particularly impressive achievement of Muslim astronomers was their effort to improve the measurements of the size of the Earth. The work started, with public support, under al-Ma'mūn, who entrusted two groups of scholars with the operation which was carried out on the plains of the Arabian desert. One group travelled north, the other south and each measured the distance traversed until the altitude of the celestial North Pole had changed one degree. In the final analysis, 1° on the surface of the Earth was represented by $56\frac{2}{3}$ Arabian miles of 4,000 cubits; one cubit was about 58 cm, so that the circumference was worked out at about 47,325 km. This is no better than the values arrived at by Eratosthenes and Poseidonios (see p. 53). One interesting new method among several other Earth measurements was proposed and tried in India by the great and immensely learned Persian scholar and indologist al-Bīrūnī (A.D. 973–1048). He first measured the height of a mountain dominating the ocean, and next the depression of the horizon as seen from the top of the mountain, with the result that an arc of 1° represented 58 miles.

Theoretical Astronomy

The first phase of Muslim science is best remembered in history for its preoccupation with theoretical astronomy and planetary theory. To begin with there was some rivalry between the Hindu methods introduced with the *Sindhind* and adopted by many of the makers of tables, and purer Ptolemaic models known from the *Almagest* translations. After a period of indecision, the Ptolemaic system was victorious but not until one had felt how difficult it was to master the *Almagest*. Against this background a number of new manuals of theoretical astronomy appeared. A brief exposition in 30 chapters was composed in Baghdad by the Persian astronomer al-Farghānī (in Latin, Alfraganus, *c.* A.D. 800–*c.* 870). It gives a systematic, but very introductory, sketch of Ptolemaic astronomy with a few Oriental additions, and became very popular as a textbook in Muslim schools. In a Latin translation by John of Spain, entitled *Rudimenta Astronomica* (or *Liber 30 differentiarum*) it spread through the universities of Europe and was finally published in print by Melancthon as late as 1537.

Al-Farghānī included an elementary description of the geometrical models of the *Almagest* in his work, but omitted all the observations and calculations on which they were founded. This defect was soon remedied by the Mesopotamian astronomer al-Battānī (*c.* 850–929) who had his observatory at al-Raqqa on the Euphrates. His *Opus astronomicum*, or, in the Latin version of Plato of Tivoli—*De scientia stellarum*,—is the work of a highly professional scientist who gives careful accounts of his observations and the astronomical methods he employed in trying to supplement the results of his predecessors. In modern times a critical edition of this work (by C. A. Nallino) is the starting point of all scholarly research into the history of Arab astronomy.

All the theoretical astronomers of the early Muslim period made use of the same geometrical treatment of planetary theory as that of Ptolemy's *Almagest*, supplementing it with new trigonometrical methods. Little attention was paid to the physical structure of the universe, and in this field Aristotelian views held sway, as can be seen from a brief paper *On the concentric structure of the universe* by the Arab philosopher al-Kindī (*c.* A.D. 805–870). An attempt to unite the different conceptions was made by the great Cairo scientist Ibn al-Haitham (Latin, Alhazen, *c.* A.D. 965–1039). His astronomical manual is founded upon the *Planetary hypotheses* of Ptolemy, and regards all parts of the celestial machinery as composed of material bodies, or 'spheres'. These alternative mathematical and physical interpretations of astronomical theories were the starting point of the great debate on the nature of scientific hypotheses which, at a later date, arose in the west (p. 267) and which for a while threatened to destroy the very foundations of Ptolemaic astronomy.

The combination of interests in theory and observation led to important new results. While Ptolemy could compare his observations only with those made by Hipparchos some 300 years earlier (although in some cases he had older Babylonian data to go by) Muslim astronomers were in a more fortunate position. Working in the ninth and tenth centuries, 700 years after Ptolemy and 1,000 after Hipparchos, they stood a much better chance of discovering secular variations only discernible over long periods of time. Al-Battānī was able to prove the secular motion of the solar apogee (supposed by Ptolemy to be a fixed point relative to the fixed stars and taking part only in the common motion of precession). Making observations in A.D. 871 he found the longitude of the solar apogee to be $\lambda = 82°; 17$; this is $16°; 47$ more than the Hipparchian value of $65°; 30$ quoted in the *Almagest*, and more than precession can account for.

At about the same time as al-Battānī was making his measurements the theory of precession was investigated by Thābit ibn Qurra (c. A.D. 826–901), who had revised Ishāq ibn Hunain's translation of the *Almagest*, in a brief treatise *On the solar year*. Using a number of observations from A.D. 830–832 he redetermined the parameters of the solar theory, finding the apogee to be $\lambda_a = 82°; 45$ and the eccentricity to be $e = 2^p; 2, 6$, the radius of the deferent being 60^p. This led to new tables of mean motion and anomaly of the Sun by means of which Thābit was able to compare Hipparchos's equinoctial observations with those of his own time. Observations of Regulus (a Leonis) made by both Hipparchos and the Arabs proved that in 958 years of 365 days the fixed stars had moved $13°; 23$ farther to the east than the equinoxes. This gives the mean rate of precession as

$$13°; 23 \div 958 = 0°; 49, 39 \text{ per year}$$

Comparing his own solar observations with some of Hipparchos's determinations of the autumnal equinox, Thābit found that during 976 revolutions relative to the stars the Sun had to move $13°; 23$ further beyond the equinox in order to return to the same star. This took $13 \frac{3}{5}$ days, and had to be added to the $976^a 236^d 19$ separating the two observations. The result is $976^a 1^d 9\frac{1}{4}^h 8^m$ which, changed into days and divided by 976 gives $365^d; 15, 23$ as the length of the *sidereal year* and represents another improvement of a fundamental astronomical constant. Thābit also draws attention to the period of return of the Sun to the same point on its eccentric orbit. Being independent of the equinox, and also of the motion of the apogee relative to the stars, this *anomalistic year* must, he said, be of a more fundamental nature than either the tropical or the sidereal year.

The correction of fundamental constants by means of observation generally implied the determination of small quantities. In this measurement, observational errors were unavoidable and led, in some cases, to curious results. A historically important example is Thābit's attempt to improve the theory of precession. Ptolemy regarded precession as a linear function of time, increasing the longitudes of the fixed stars at a constant rate of 36″ per year. Compared with this value the rate of 49″ 39‴ found by Thābit was much better (the modern value is 50″.4). But other astronomers had arrived at different results and Thābit suspected that precession was not a linear function of time, and his result not a constant but only a mean value of a periodically variable, annual rate, a mistake due to nothing more than observational error. Thābit also supposed the obliquity, ε, of the ecliptic to be a decreasing function of time, which is correct, although the rate of change is smaller than Thābit believed. In fact, in Hindu astronomy we have $\varepsilon = 24°$, and in Ptolemy 23°; 51, while Muslim astronomers at Thābit's times found $\varepsilon = 23°; 33$.

In his treatise *On the motion of the eighth sphere* Thābit tried to account for both phenomena in a single theory. The ecliptic EQ in Fig. 13.5 is supposed to perform an oscillatory motion around its highest point Q (the Tropic of Cancer) which lies on a great circle PQB through the pole P of the equator so that the arc BQ is $\varepsilon_0 = 23°; 33, 30$, equal to the mean obliquity of the ecliptic. The oscillations, he argued, are governed by a mechanism comprising a small circle with centre Υ_0 (the mean equinox) on which a point D revolves with uniform angular velocity, the argument of obliquity $i = B\Upsilon_0Q$ being a linear function of time. The line QD intersects the equator at Υ which is the true equinox. When D revolves on the small circle, Υ will oscillate to and fro about Υ_0; thus a single mechanism produces both a change in the obliquity and—to use the Latin terminology—a 'trepidation' of the equinoxes (Fig. 13.5).

The curious theory of the trepidation of the equinoxes gave rise to much discussion. In his *Opus astronomicum* al-Battānī devoted a special

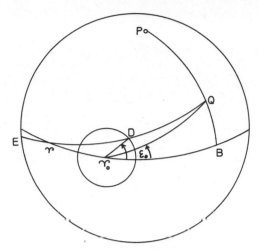

Fig. 13.5. Figure illustrating the
theory of trepidation according to
Thābit ibn Qurra.

chapter to refuting it, adhering to a linear theory of precession with the
improved rate of 1° in 66 years, or 54″ per year. But, according to
Thābit, al-Battānī later came to support the theory. Among the other
eastern astronomers, Ibn Yūnus rejected it, giving the still better value
of 1° in 70 years, or 51″ per year, of the constant of linear precession.
In the west it found eager supporters in al-Zarqālī and many later
astronomers.

Muslim Physics and Technology

In the field of physics, the Arabs made a lasting contribution to optics,
mainly through the experimental and theoretical research of al-Haitham.
He investigated the reflection of light from concave mirrors made of
steel, turned and polished on a machine specially constructed for
the purpose. Al-Haitham outstripped the Greeks (p. 132) in discovering spherical aberration, and showed that parabolic mirrors can
concentrate a parallel beam of rays at a single point. The formation of
images was also investigated, and the position of the image of a given
object constructed geometrically.

185

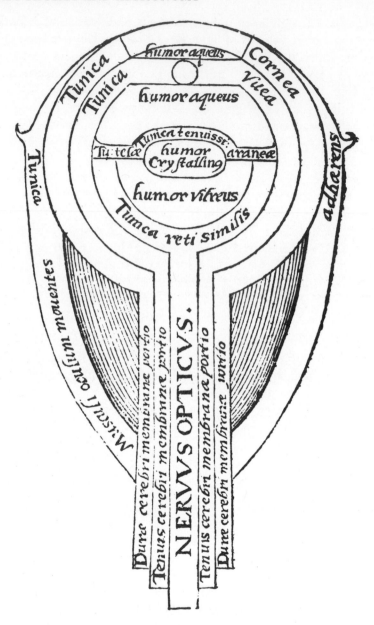

Fig. 13.6. Ibn al-Haitham's description of the eye. From the printed version of his Opticae thesaurus, *ed. Risner, Basle, 1572.*

Al-Haitham is significant among Ancient and Mediæval scientists in assuming a finite velocity of light. He tried to explain refraction by a theory on which the velocity is split into one component parallel to the surface between the two media, and another perpendicular to it. When light is passing from a less dense to a more dense medium the parallel component, he maintained, is diminished so that the angle of refraction becomes smaller than the angle of incidence. Yet al-Haitham did not succeed in discovering the exact law of refraction although his theory does lead to a reduced velocity of light in a denser medium (contrary to Newton's theory, but in agreement with modern experiments). Al-Haitham's assumption that light always follows the 'easiest way' is the same as that of Heron (p. 131).

The Arabs did not surpass the scientists of Antiquity in theoretical mechanics. They generally followed Archimedes in statics, and Aristotle and his commentators in dynamics. Al-Bīrūnī stressed the concept of specific weight and used Archimedes' principle to determining specific weights of a large number of metals and gemstones as a means of identification. This was an important step forwards, since a simple quantitative method of identification would make it possible to distinguish gold from silver and thus, in the long term, undermine some of the basic tenets of alchemy, a subject eagerly studied by the Arabs, particularly Geber (Jābir ibn Hajjan, c. A.D. 770).

In the realm of technical mechanics, the Arabs were much more progressive than in the theoretical field. Numerous authors described machinery such as watermills, water clocks, automata, and so on. A prime-mover of such economic importance as the windmill seems to have originated somewhere in the eastern part of the Muslim world, from where it spread to northern Africa and finally, in the late twelfth century, to Europe.

The Arabs paid particular attention to magnetism. It was known in classical Antiquity that natural lodestones attract iron and this was explicitly mentioned as early as the sixth century B.C. by Thales. In China, a work from the eleventh century A.D. mentions that a freely movable magnetic needle will point towards the north. But there are other indications that the magnetic compass and its application to navigation was first introduced by the Arab seamen who had to navigate in the great Indian Ocean. The question of priority is still a matter of discussion; there is, however, no doubt that the Arabs used the primitive floating compass before it became known in Europe, where it was used in the Mediterranean and later by Scandinavian sailors in the Atlantic.

Science in Latin Europe

While Arab science flourished in the east there was little progress in the west. After the disintegration of the Roman Empire, knowledge of the Greek language disappeared, and the few early Mediæval scholars were in general unable to study the principal works of Greek science which had never been translated into Latin. Roman authors had left a number of compilations of secondary importance, such as the *Natural history* of Pliny (A.D. 23–79) and some encyclopedic works by Macrobius (c. A.D. 400) and Martianus Capella (fifth century A.D.) later commented on by Remy d'Auxerre (?–A.D. 903) and others, containing brief summaries of the various results of ancient science and scholarship. A few of Aristotle's logical works were available in a translation by Boëthius (c. A.D. 480–524). The first part of Plato's *Timaios* had been translated into Latin by Chalcidius (fourth century A.D.) and was virtually the only available source of Greek cosmology. But the *Timaios* represented a very primitive stage in the evolution of science which, however, fitted in well with the neo-Platonic philosophy dominating late Antiquity. It is a pity that a translation of the *Almagest* by Boëthius was lost almost immediately. Had it been preserved, Latin astronomy would not have been compelled to start with a delay of more than 700 years.

The few scientific writings produced in Europe in the seventh and eighth centuries, including for example a widely used encyclopedia *On the origin of words* by Isidore of Seville (c. A.D. 560–636) and the various computistical treatises of the Venerable Bede (c. A.D. 673–735), were derived from these Latin works. The teaching at the many monastic and cathedral schools of the ninth century was also based on them with a curriculum centred around the so-called *liberal arts*. This concept developed in Antiquity, where the liberal arts were the occupation of free men, in contrast to the servile or mechanical arts and crafts performed by slaves. In the Middle Ages the meaning gradually changed, so that the liberal arts became those theoretical activities which liberate the soul from its native darkness and ignorance. The seven liberal arts were generally classified in two groups, namely, *trivium:* grammar, dialectics, rhetoric, and *quadrivium:* arithmetic, geometry, astronomy and music, to which some authors added medicine and architecture, but as practical arts they had no permanent place in the system.

On this foundation, the work of the early Mediæval schools could build no new scientific results, for although the Latin sources transmitted some of the knowledge of Antiquity, they were silent about the methods by which it was acquired. It is to the credit of the schools that without the stimulus of experiment the old Pythagorean *quadrivium* was retained in the curriculum on a level with the humanistic disciplines of the *trivium*. Thus the liberal arts kept alive the conviction that a proper education must comprise knowledge both of nature and of human affairs. In addition, despite the scientific poverty of the schools—or perhaps because of it—Antiquity was looked back on as a golden age that had to be re-established in Europe. Throughout the Middle Ages the dream of a cultural or scientific renaissance was kept alive as an efficient inspiration for conscious attempts at reconquering the past and 'transferring the studies from Athens to Paris', to quote an oft repeated slogan of the time.

The Translations

In about the year A.D. 1000 the situation clearly began to change. A number of previously unknown treatises started to circulate in Latin translations often made by wandering scholars who had visited Spanish monasteries, and there acquired some knowledge of Arabic science. Most prominent among them was Gerbert of Aurillac (died A.D. 1003), who in A.D. 999 became Pope Sylvester II.

During the twelfth century the wave of translation grew, and its quality improved. Not least, the disciplines of the *quadrivium* were now enriched with new material of a scientific standard previously unknown in the Latin world. The three most prolific translators of the twelfth century, Adelard of Bath, Gerard of Cremona, and Domenicus Gundissalinus, made Latin versions of at least 116 treatises, of which no less than 84 were on mathematics and natural science. The value of this new material can be seen from the following table of some of the more important translations (cf. the much more detailed list in A. C. Crombie, *Augustine to Galileo*, London 1952).

Author	Work	Translation from	Date	Place	Translator
al-Khwārizmī	Arithmetic	Arabic	1126?		Adelard of Bath
al-Khwārizmī	Astronomical tables	Arabic	—		Adelard of Bath
al-Khwārizmī	Algebra	Arabic	1145	Segovia	Robert of Chester
Ibn al-Haitham	Optics	Arabic	12th cen.	Toledo	
al-Biṭrūjī	Astronomy	Arabic	1217	Toledo	Michael Scot
Aristotle	Meteorology	Greek	1156?	Sicily	Henricus Aristippus
Aristotle	Physics	Greek	12th cen.		
Aristotle	Physics	Arabic	12th cen.	Toledo	Gerard of Cremona
Aristotle	Physics	Greek	1260?		William of Moerbeke
Aristotle	Metaphysics	Greek	12th cen.		
Aristotle	Metaphysics	Greek	1270?		William of Moerbeke
Euclid	Elements	Arabic	1126?		Adelard of Bath
Euclid	Optics	Greek		Sicily?	
Apollonios	Conics	Arabic	12th cen.		Gerard of Cremona
Archimedes	On the Circle	Arabic	12th cen.	Toledo	
Archimedes	Many Works	Greek	1269		William of Moerbeke
Heron	Pneumatics	Greek	12th cen.	Sicily	
Heron	Catoptrics	Greek	after 1260		William of Moerbeke
Ptolemy	Almagest	Greek	1160?	Sicily	
Ptolemy	Almagest	Arabic	1175	Toledo	Gerard of Cremona
Ptolemy	Optics	Arabic	1154?	Sicily	Eugenius of Palermo

The University Curriculum

Within the traditional schools it was impossible to assimilate this vast amount of new knowledge. Some tried to cope with the situation by specializing in particular subjects. In the late twelfth and thirteenth centuries, these special schools merged into universities offering an integrated system of higher education, the schools serving as faculties. In the universities, teaching was based on the system of liberal arts taught in the faculty of arts, where all students had to take a degree before passing on to a more specialized education in medicine, law, or theology. In this way, at least a rudimentary knowledge of the exact sciences was incorporated in the basis of any kind of university study. Within the *quadrivium* itself a series of new disciplines like mechanics and planetary theory were treated with a previously unrivalled thoroughness and penetration.

190

A number of new disciplines, of great importance to thirteenth century science, were difficult to place in the Aristotelian classification. Aristotle had distinguished physics from mathematics in such a way that physics became the study of natural phenomena in their material form, whereas mathematics became the study of their abstract, immaterial, quantitative properties. Strictly speaking this distinction excluded sciences like statics, optics, theoretical astronomy, and several others which made the material world the subject of a mathematical description, so that strictly they belonged neither to physics nor to mathematics. In the thirteenth century they were often classified as intermediate sciences (*scientiae mediae*), examples of which are the statics of Archimedes and the astronomy of Ptolemy. The *scientiae mediae* mark the beginnings of mathematical physics.

Aristotelianism

Very important in this period was the change in the philosophy of science following in the wake of Aristotelian philosophy which now gradually displaced neo-Platonic trends as the common intellectual framework of Mediæval scholars. The Aristotelians classified the various sciences in a rational way according to their degree of abstraction (p. 34) and in so doing gained a greater insight into the connections between the sciences than could be reached from the previous, somewhat arbitrary classification. The principle of experience as the basis of any kind of knowledge was once more maintained and strengthened through the teaching of such philosophers as Albertus Magnus (A.D. 1194–1280), Thomas Aquinas (A.D. 1225–1274), and Roger Bacon (c. A.D. 1210–after 1292). From this followed a deeper understanding of scientific method, the role of physical theories, and the hypothetical status of general statements about natural phenomena. Thomas Aquinas and his followers distinguished between philosophy and theology in a much more radical way than the Augustinians who previously dominated philosophy. Scientific investigation was thus made more independent of theological presuppositions, in the fields of methodology and epistemology.

Fundamental changes such as these could and did not take place unopposed. Aristotelian philosophy was smuggled into university teaching long before the authorities gave it their official permission.

As late as 1277 Archbishop Etienne Tempier of Paris published a long list of doctrines which were condemned as heretical or disallowed. A similar condemnation, though of a much shorter list of propositions, was directed at the English universities by Archbishop Robert Kilwardby. Interventions like these meant temporary troubles for progressive teachers, but did not stop the march of Aristotelianism; at most they did prevent the 'Philosopher' (Aristotle) from being regarded as an absolute authority. Aristotle's philosophy won the adherence of Mediæval scholars through its own qualities. His strict logic and rational common sense, combined with his wide knowledge was an irresistible contrast to the former vague mixture of Platonism and theology, although his teaching was not regarded as inviolable. In the words of Albertus Magnus: 'Had Aristotle been God he would have been infallible, but since he was a human being like the rest of us, he could, of course, be in error'.

Chapter Fourteen

Experimental Physics in the Middle Ages

Grosseteste and Optics

A NEW understanding of scientific methods emerged in the latter period of the Middle Ages. The way in which this happened is illustrated by the 'renaissance' of optics in the thirteenth century. The pioneer figure in the revival was the English scholar Robert Grosseteste (c. A.D. 1175–1253), author of a long series of philosophical writings among which are a number of small treatises on optical problems. His sources are the optical works of Euclid and Ptolemy, the sections on atmospheric optics in Aristotle's *Meteorologica*, and the work of Ibn Sīnā (Avicenna). There is no evidence that he knew Ibn al-Haitham's optics.

In his tract *On the rainbow* Grosseteste divided optics into three parts: the science of vision and the eye; the science of mirrors and reflection; and the science of lenses and refraction. On the theory of lenses and refraction he added that it had been left 'unmentioned and unknown among us until our days' although it is an important part of optics; this fact motivated his own choice for research. In another treatise *De lineis* Grosseteste shows how a ray of light passing from a less dense into a denser medium will approach the normal to the refracting surface, whereas passing into a less dense medium it will recede from it. And in *On the rainbow*, Grosseteste postulated that a ray of light refracted from air to water bisects the angle between the normal to the surface and the prolongation of the incident ray, that is he gives the law of refraction the form $r = \frac{1}{2}i$, where i and r are the angles of incidence and refraction respectively.

Grosseteste also maintained that the phenomenon of refraction obeys the general principle of natural philosophy, that is, that any natural action takes place in the most complete, regular, brief, and best possible way (see p. 131).

It is doubtful whether Grosseteste himself conducted optical experiments of this kind, but he certainly investigated the way in which rays pass through a lens. In a treatise *De natura locorum* he says that

> if one takes a filled jar of a spherical shape—for example an urinary flask—and places it in the strong rays of the Sun, the rays passing through the round part of the flask will converge from beyond it to a point between it and the person holding it because of the twofold refraction mentioned before. At this point and around it is the burning place and if anything combustible is brought there it will be kindled.

Following this is a more qualitative description of the path of the rays, including their points of entry to and exit from the flask. Grosseteste most probably observed the formation of images in such a spherical lens for in the treatise *On the rainbow* he says that

> if it [refraction] be properly understood it will show us how we might so arrange it that small and distant objects appear quite close, and large, near objects seem to be far away; and how we can get small objects placed at a certain distance to appear with whatever size we like, so that it would be possible to read the smallest letters at incredible distances, or to count sand, or grain, or seed.

Such prophecies about the power of optical instruments reveal the scope of Grosseteste's ideas, even if they were far from fulfilment and, for example, had to await the art of making real lenses which could replace spherical bottles.

Grosseteste tried to apply his optical insight to a new theory of the rainbow which at this time was still explained along Aristotelian lines, as a reflection of the rays from the surface of a cloud. Instead, Grosseteste assumed the rainbow to be caused by refraction of the rays in a whole cloud, so that it acts as a kind of lens, and it is likely that he had observed coloured rings of light behind his flask in previous experiments. Even if the explanation was wrong, introducing refraction into the theory was an important step in the right direction.

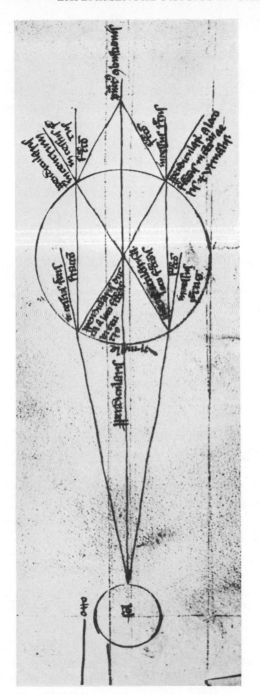

Fig. 14.1. Robert Grosseteste's theory of the refraction of light in a spherical lens. The Sun is on the left, and the various rays converge to the right at a point called punctus combustionis. From a manuscript in the British Museum of Roger Bacon's Opus Maius.

It is interesting that Grosseteste was so deeply impressed by optics that, as a philosopher, he tried to develop a complete metaphysical doctrine based on the laws of the propagation of light. This 'meta-physics of light' concurred with Augustine's epistemological ideas of 'the eternal light' in which the soul perceives the essence of things, and was wide-spread among philosophers of the thirteenth century.

Clearly, Grosseteste was attempting to 'mathematize' the description of nature as a whole by means of the most developed branch of mathematical physics, that is geometrical optics. This may have been a vehicle for the general idea of nature as a mathematical structure, and of natural philosophy as an essentially mathematical discipline. Grosse-teste himself expressed this quite clearly:

> The usefulness of considering lines, angles and figures is very great since it is impossible to grasp natural philosophy without them. They are absolutely important both in the universe as a whole, and in its individual parts.

Roger Bacon

Grosseteste's interests in optics were reflected in his Oxford disciple Roger Bacon (c. A.D. 1214–1292), one of the most brilliant natural philosophers of the thirteenth century. Bacon's dream was the creation of a completely new and comprehensive philosophical system by which Christianity could be defended against Islam. In it, theology, philosophy, mathematics, and natural science were to be united into a harmonious whole with sufficient persuasive force to defeat both rival philosophies and other religions.

In the field of optics, Bacon has often been credited with some of the ideas of his master, such as the prophecy of optical instruments. Bacon also learned the properties of lenses from Grosseteste, but unlike him he worked with real lenses (Fig. 14.2), which is apparent from his main work, *Opus maius*, of A.D. 1267, and from the fact that he sent a lens to Pope Clement IV, on whose request the work had been written.

Bacon knew of Ibn al-Haitham's optical investigations and followed his Arab predecessor in assuming a finite velocity of light, but the most interesting part of his optical work is his account of the eye as an

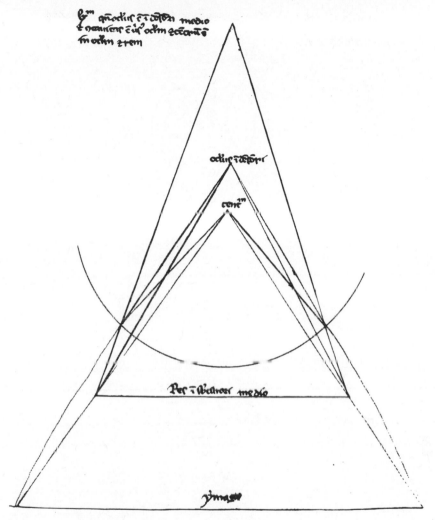

Fig. 14.2. The formation of images in a spherical lens from Roger Bacon's Opus Maius. *From a manuscript in the British Museum.*

optical instrument (Fig. 14.3). Following al-Haitham and Ibn Sīnā he clearly distinguished the psychological impression of vision from the physical processes in the eye from which they arise. He realized that the physical function of the eye has to be explained from existing knowledge of lenses and his only serious mistake was his conception

197

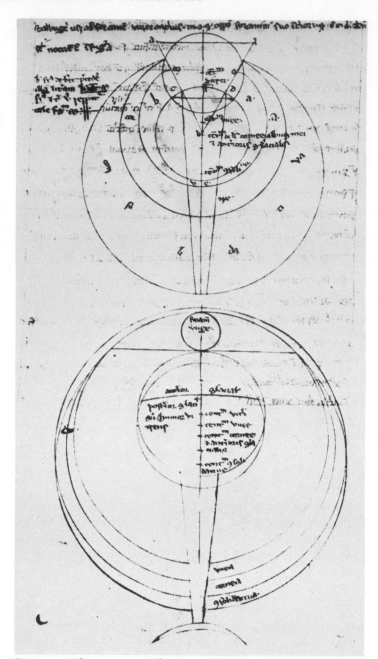

Fig. 14.3.   The eye as an optical instrument from Roger Bacon's Opus Maius. *From a manuscript in the British Museum.*

of the eye lens as the sensitive part of the system. Yet he was aware that vision is completed not in the lens but in the optic nerve, whose anatomy and function he investigated.

The Problem of the Rainbow

One of Roger Bacon's greatest achievements was to introduce the physiological optics of the Arabs into Europe. But equally interesting, from a methodological point of view, are his attempts at giving the theory of the rainbow a solid foundation. First, he searched for the conditions necessary for a rainbow to appear, and tried to explain them in an inductive way. The rainbow he considered to be a complex phenomenon of colour, figure and size, and for each of these components Bacon tried to find a number of analogous phenomena in nature.

The colours of the rainbow, Bacon argued, can be observed and demonstrated objectively and subjectively when light is directed through crystals. Subjectively, he went on, they can be seen in the spray from oars or watermills, in dewdrops in the meadows on a summer morning, around a narrow opening when it is looked at with the eyes screwed up, along the rim of a hat held in the rays of the sun, in the shadow of the eyebrows, when the rays of the Sun pass through a jar of water or an oil lamp, and, finally, when one blows a bubble out of one's mouth and looks at it from the side.

For Bacon, the shape of the rainbow was a more difficult problem, for in crystals the coloured fringes are straight. Yet it is possible to observe coloured bands along the eyebrows, the eyelashes, and around small holes in rags; in a cluster of dewdrops or round a burning candle a complete circle may be seen, just as haloes and mock suns are seen round the Sun.

Bacon found the size of the rainbow in a more direct way—he simply measured it with an astrolabe (see p. 248) and found that the Sun, the eye of the observer, and the centre of the bow are on a straight line, and that the maximum altitude of the bow (with the Sun at the horizon) is 42°.

It seems impossible that Bacon's final explanation of the rainbow, based on these facts should be wrong. But it was wrong, and for an obvious reason: he was aware of the subjective character of the pheno-menon—each observer sees his own rainbow, and it cannot therefore be located at any definite place—but the colours produced by refraction in a crystal can be demonstrated objectively. Bacon thus concluded that the rainbow has nothing to do with refraction, and in the end he fell back on the Aristotelian theory. Nevertheless, his investigation is of great importance as a testimony to his understanding of both empirical, qualitative and quantitative methods, and of inductive reasoning.

Vitello and Dietrich of Freiberg

The most complete optical exposition surviving from the Middle Ages is the great *Perspectiva* of 1270 written by the Silesian Vitello (born *c.* A.D. 1230). The work shows the influence of Grosseteste and Bacon, and al-Haitham. Vitello realized that the most urgent problem in optics was to find the exact law of refraction and tried to deduce it by experiments with a goniometer derived from the similar instrument of al-Haitham. Like his predecessors, Vitello formulated his results as a table of corresponding values of the angles of incidence and refraction. He also tried to fit the results into a general formula, but could only deduce a number of inequalities of no great practical use. Vitello also described other experiments such as the production of artificial rainbows through refraction in crystals or water bottles. His description of how to make parabolic mirrors was new in Europe, but it, too, was derived from al-Haitham. Vitello's *Optics* became a standard work of reference and as late as 1604 was commented on by Kepler.

From a theoretical viewpoint the most important result of the renaissance in optics was achieved by the Dominican Dietrich of Freiberg, who in his book also entitled *On the rainbow* finally succeeded in giving the correct explanation of the phenomenon which, since Aristotle's time had been a perennial challenge to philosophers. Dietrich described, first a series of experiments with water bottles and quartz crystals, and continued with a successful combination of earlier theories. Aristotle had explained the rainbow by reflection in the surface of a cloud, Grosseteste by refraction in the cloud as a whole. Dietrich now showed how all the known phenomena can be explained from the assumptions that the rays are refracted and reflected in the individual drops, and that the refracted ray is split into a bundle of neighbouring coloured rays, each of which is refracted in a particular way. This was the first time that dispersion (differential refraction) of light had been described and incorporated in a scientific theory. According to Dietrich, the primary rainbow (Fig. 14.4) is produced by rays which

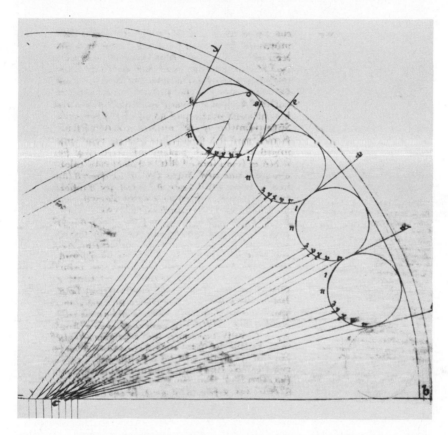

Fig. 14.4. The explanation of the primary rainbow from the model in Dietrich of Freiberg's De iride. *From a manuscript in the University Library, Basle.*

are refracted when they enter the waterdrop, reflected from its inner surface, and refracted once more upon leaving it. The explanation of why the secondary bow shows the colours in the inverse order and has a greater radius (Fig. 14.5) was achieved by assuming that the rays are reflected twice inside the drop. The only part of the phenomenon left unexplained is the size of the rainbow. Dietrich was unable to calculate it because he had insufficient knowledge of the law of refraction and,

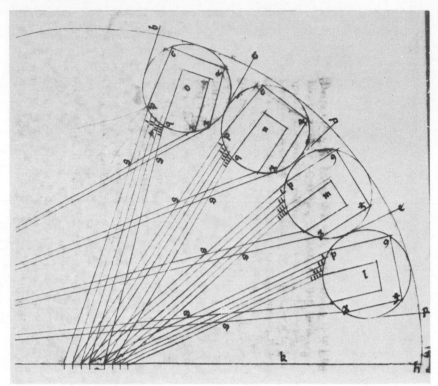

Fig. 14.5. Dietrich of Freiberg's theory of the secondary rainbow. From a manuscript in the University Library, Basle.

curiously enough, he quotes a wrong empirical value despite Roger Bacon's previously correct measurement.

The rainbow theory of Dietrich of Freiberg is thus qualitatively correct and could be improved only in the 17th century after the discovery of the exact law of refraction. This was done by Descartes who, without mentioning Dietrich, transformed his theory into quantitative terms from which he could deduce the radii of both the primary and secondary bows.

Optical Inventions

A more practical result of the new optics was the invention of spectacles. As early as 1284 Venetian guild records mention them; in a contemporary chronicle the Dominican Alessandro della Spina (died 1313) is reported as the first to make them, and at about the same time spectacles appeared in France and in the Netherlands. The first lenses were made either of glass or from the semi-precious stone beryl, from which the German *Brillen* is derived. During the fourteenth century spectacle-making spread all over Europe (Fig. 14.6). This

Fig. 14.6. Spectacles, as depicted in an evangelistary from about A.D. 1380. From a manuscript in the Bibliothèque Nationale, Paris.

invention has, perhaps, contributed more than most others to the advancement of learning.

The Investigation of Magnetism

The experimental attitude to optics had its counterpart in the thirteenth century investigation of magnetism. The phenomenon had been known since Antiquity, although not clearly differentiated from electricity, but in 1269 it was established as a separate branch of physics by one of the most revolutionary of single treatises in Mediæval science, entitled *De magnete* and written by Pierre de Maricourt (Petrus Peregrinus), whom Roger Bacon mentions in most laudatory terms.

The slim volume is divided into a physical and a more technical part. The physical section is subdivided into ten chapters describing the main physical properties of the loadstone. Pierre had a specimen of a spherical shape on which a small compass needle could be placed and its direction marked by a line drawn on the loadstone. When the needle was moved around the loadstone a system of lines appeared inter-secting each other at two opposite points and resembling the meridian circles of the heavenly sphere. Next, the loadstone was placed in a wooden tray floating on water like the usual sailors' compass. One of the points defined before then turned to the north and the other to the south, which were called respectively the north and south poles of the magnet. Then another loadstone with known poles was brought near to the floating magnet, and the mutual interaction between similar and different poles investigated.

In the chapters that follow the author describes induced magnetism in an iron rod which has been touched with the loadstone, and also shows how the loadstone can be broken into smaller magnets whose poles are determined on the basis of definite rules. Finally, Peregrinus puts forward the hypothesis that the action of the loadstone is caused by the heavens and not, as previously assumed, by large deposits of magnetic ore at the north pole of the Earth. He suggests an experiment to verify this hypothesis using a spherical loadstone turning around an axis through its poles parallel to the axis of the heavenly sphere and believed that the stone would follow the heavens in the diurnal rotation. The actual experiment was not carried out, presumably because of technical difficulties.

In the second part of his treatise Pierre proposes three new technical inventions based on magnetism. The first is a combination of a floating compass and a diopter by which the azimuth of a star can be found. This means that the instrument can be used as a sundial. The second invention is an improved form of the compass (Fig. 14.7) in which the needle to longer floats on water but turns about a vertical axis suspended on pivot bearings; another new feature is a graduated circle making the compass a real measuring instrument. Finally, Pierre de Maricourt describes a project for a magnetical *perpetuum mobile* (Fig. 14.8) which

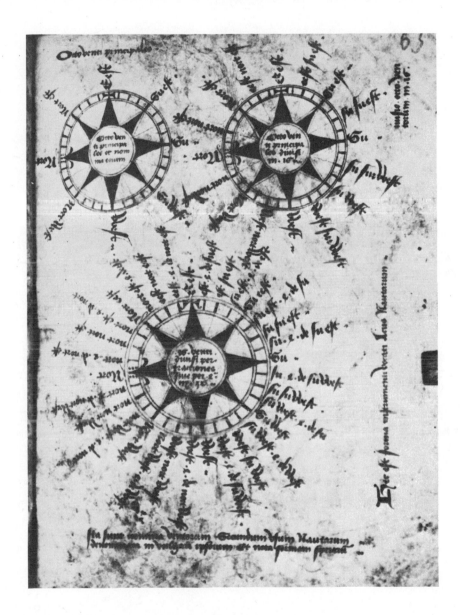

Fig. 14.7. Compass cards from the fourteenth-century. From a manuscript in the Biblioteca Bongarsiana, Berne.

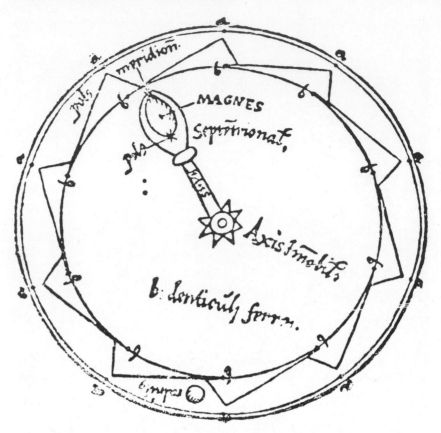

Fig. 14.8. Magnetic perpetuum mobile. From Petrus Peregrinus, Epistola de magnete. Reproduced from S. P. Thompson's edition, London, 1902.

implies that he believed such a machine to be feasible despite the objections raised by many philosophers, including Thomas Aquinas. This does not reduce the physical and methodological value of the treatise, which is a gem in experimental physics, and remained the best work on magnetism until Gilbert's famous *De magnete* of 1600.

Physics as an Experimental Science

The optical and magnetic investigations of the thirteenth century prove that by that time some sections of physics had acquired an experimental status. But was true experimental method achieved? The

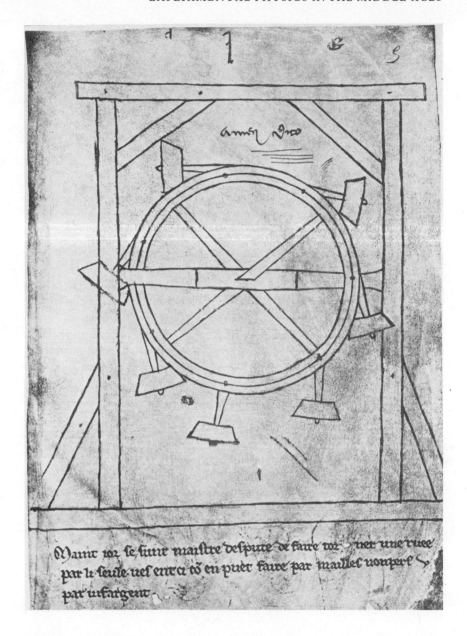

Fig. 14.9. Mechanical perpetuum mobile, *devised by Villard de Honnecourt, a French architect and engineer who was the author of one of the rare technical sketch books from the thirteenth century. From a manuscript in the Bibliothèque Nationale, Paris.*

answer, of course, depends on how the experimental method is defined. In the sixteenth century Galileo performed quantitative experiments, with the aim of obtaining numerical data to test his hypotheses. Compared with Galileo, the physicists of the thirteenth century did not get very far. Nearly all their experiments were qualitative, although precise measurements were made occasionally—the astrolabe was sometimes used for physical purposes, and the goniometer employed in optical experiments.

Despite their qualitative approach scholars like Roger Bacon, Vitello, Dietrich and Pierre de Maricourt clearly understood the empirical basis of physical science. All started their work with personal observations of phenomena, either in the field or in the laboratory, and used inductive methods quite naturally when analysing their empirical results. The great weakness of these scientists was that they were not able to carry the mathematical description of nature very far, although in principle they realized its importance. Undoubtedly, one reason for this was that thirteenth century mathematics could not use the rich heritage of the Greeks to the full. Another reason was that purely qualitative experiments did not provide numerical data from which relationships could be computed. Thirteenth century physics was only a first step. Further progress presupposed a greater knowledge of mathematics on the one hand, and new techniques of quantitative measurements on the other. In the century that followed, great advance took place in the field of mathematics, while the art of making quantitative experiments did not really develop until the days of the great physicists of the sixteenth and seventeenth centuries.

Chapter Fifteen

The Development of Statics

Statics as a Scientia Media

THIRTEENTH-century scholars and experimentalists realized in principle that optics must be a mathematical science, although they were unable to produce any quantitative mathematical formulation of the laws of light. In sharp contrast, thirteenth-century statics acquired a consistently quantitative character, and much higher standards were thus attained. One significant fact was that Archimedes' treatise *On floating bodies* had been translated in the previous century, and most of his other works were made accessible through a new translation in 1269. Archimedes' statics and hydrostatics were an outstanding example of the way in which natural phenomena can be described mathematically, and made a vital contribution to further progress in mechanics. Some other treatises on statical problems were accessible to thirteenth-century students of mechanics; of them, the most important was Jordanus Nemorarius who was perhaps a Dominican; his life-story is still unclear, and several writings have been wrongly ascribed to him.

A book written in fact by Nemorarius and entitled *Liber Jordani de Ponderibus*, underlined the character of statics as a *scientia media*, although this expression is not used explicitly (see p. 199). The preface states that

> Since statics is subordinated to geometry as well as to physics (*philosophia naturalis*) a number of things pertaining to this science must be proved in a philosophical (that is physical) way, but other things by means of geometry.

Here, then is an investigation of physical problems using mathematical methods. Another treatise from about the same period entitled *Liber Archimedis de Insidentibus in humidum* but whose author is unknown, pretends to be a translation of *On floating bodies* by Archimedes. The book is particularly interesting when viewed against the background of the strictly axiomatic structure of the Archimedean prototype. While Archimedes' work begins abruptly with a number of axioms, apparently taken out of the blue (p. 104), the Mediæval author proceeds in a different way. His preface thus states that

> Since the size of certain bodies cannot be found geometrically because of their irregular shape, and since the price of certain goods is proportional to their sizes, it was necessary to find the ratio of the volumes of bodies by means of their weights in order to fix their definite prices, knowing the volume ratios from the weight ratios.

The author thus clearly regards statics as a science with a practical purpose, and makes such a statement in the treatise, illustrating clearly the different presentations of the same science in Antiquity and the Middle Ages. The preface is followed by a number of definitions which prove to be an account of the construction and use of the balance. Compared with Archimedes, another new feature is that the experimental foundations of the theory are described; it is also notable that the concept of specific weight is consistently used throughout this section, which is followed by six axioms called *petitiones*, then by the deduction of a number of propositions, so that at this point the treatise becomes more like its Archimedean model.

Jordanus and the New Principles of Statics

Jordanus Nemorarius's works *Elementa super demonstrationem ponderum* and *Liber de ratione ponderis* contained new lines of argument which enabled him to solve a problem which the statics of Antiquity had left unanswered. A statement, later called 'the axiom of Jordanus', on the relation between a given weight and the height to which this is lifted by the fall of a counterweight of a given size through a given distance is of particular importance. Jordanus maintained that if the weight P is lifted through a vertical distance h, another weight kP

can be lifted through the distance h/k. The truth of this is a direct result of the conservation of energy found in classical mechanics, since the work done by the counterweight is the same in the two cases, but the notion of energy was unknown to Jordanus, as was that of work in the sense in which it used in classical physics. Nevertheless, he suspected the significance of the inverse proportionality of weight and height, and made consistent use of it in his proofs of theorems in statics, which, in this way, became related to dynamics.

A new concept introduced by Jordanus was that of *gravitas secundum situm* (the gravity of a body relative to its position), which implies that if a body moves in a vertical line, in free fall for example, it will possess its maximum gravity. But if it is forced to move on an inclined plane its gravity *secundum situm* will become smaller as the inclination of the plane to the vertical increases. On two inclined planes of the same height but of different inclinations, the body will have smaller gravity *secundum situm* when placed on the less inclined plane.

Armed with these two principles Jordanus arrived at a condition for the equilibrium of a body on an inclined plane, a problem which Heron, Pappos, and Arab scientists had tackled without success (see p. 117). He considers two inclined planes with the same vertical height $BD = a$, and the lengths $DC = l_1$ and $DA = l_2$ (Fig. 15.1). On DC is a mass of

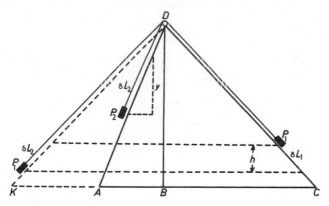

Fig. 15.1. *Figure illustrating Jordanus's proof of the law of the inclined plane.*

weight P_1, and on DA another of weight P_2, connected by the string passing over the pulley at D. Jordanus maintains that the equilibrium condition is $P_1/P_2 = l_1/l_2$.

Jordanus proves this statement indirectly, but without essentially changing the proof it can be expressed as follows. Imagine a third inclined plane DK with the same height and length as DC. On DK is placed a mass of the same weight P_1 as that on DC. If these two masses are connected by a string over the pulley they will be in equilibrium, in modern terms for reasons of symmetry.

The mass on DC is now given a small displacement δl_1 down the inclined plane DC. Because of the equilibrium this implies that the mass on DK is displaced the distance δl_1 upwards along DK. If h is the vertical component of δl_1 one mass is lifted the distance h at the same time as the other is lowered the same vertical distance.

The next step of the proof uses Jordanus's axiom: since the mass on DC is able to lift the mass on DK through the vertical distance h, it will also be able to lift the mass on DA through a vertical distance y determined by the relation

$$\frac{y}{h} = \frac{P_1}{P_2} \quad \text{or} \quad y = \frac{P_1}{P_2} \cdot h \;.$$

The vertical displacement y corresponds to the displacement δl_2 on DA. From similar triangles we then have

$$\frac{\delta l_2}{y} = \frac{l_1}{a} \quad \text{and} \quad \frac{h}{a} = \frac{\delta l_1}{l_1} \;.$$

These equations lead to the relation

$$\delta l_2 = \frac{P_1}{P_2} \cdot \frac{l_2}{l_1} \cdot \delta l_1$$

which, connected with the equilibrium condition $\delta l_1 = \delta l_2$, gives

$$\frac{P_1}{P_2} = \frac{l_1}{l_2} \;,$$

which was the statement to be proved.

This elegant argument is interesting in more ways than one. First, it agrees with the solution found by means of the principle of the parallelogram of forces developed by seventeenth and eighteenth century physicists. For if the angles between the inclined planes and the horizontal are α_1 and α_2 we have

$$\frac{l_1}{l_2} = \frac{\sin \alpha_2}{\sin \alpha_1} \;,$$

which means that the condition of equilibrium found by Jordanus can be written in the form $P_1 \sin \alpha_1 = P_2 \sin \alpha_2$, that is, for both particles the component of gravity parallel to their respective inclined planes is the same.

Next, the proof shows how the concept of *gravitas secundum situm* should be interpreted. For if DA is vertical the condition of equilibrium is reduced to $P_1 \sin \alpha_1 = P_2$, that is, the gravity P_1 on the inclined plane is equal to the full weight of P_2.

Lastly, Jordanus's proof is interesting because he determines the equilibrium by means of imagined displacements δl_1 and δl_2 subject to the condition $\delta l_1 = \delta l_2$, thus expressing the physical condition that the string connecting the masses must remain taut. In the seventeenth and eighteenth centuries similar conditions led to the general mechanical principle of virtual displacements, and served as a very general basis for very general formulations of the laws of mechanics.

Jordanus went on to prove the law of the lever from the principle of *gravitas secundum situm*. Consider a lever AB with the fulcrum C.

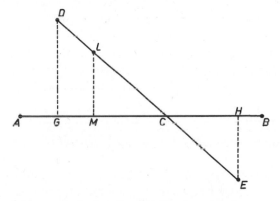

Fig. 15.2. *Jordanus's proof of the law of the lever.*

$CA = a$ and $CB = c$, and the weights suspended at A and B are P_a and P_b respectively (Fig. 15.2). Supposing

$$\frac{P_a}{P_b} = \frac{b}{a} \quad \text{or} \quad P_b = P_a \cdot \frac{a}{b},$$

Jordanus could then prove that the lever will be in equilibrium, so that P_a and P_b have the same *gravitas secundum situm*. Assume that there is no equilibrium, but that P_b is able to lift P_a from A to D, itself descending from B to E. The vertical components of these displacements are HE and GD, for which (from the geometry of the figure)

$$\frac{HE}{GD} = \frac{b}{a} \quad \text{or} \quad GD = \frac{b}{a} \cdot HE .$$

But if P_b descending is able to lift P_a the vertical distance GD, it will—according to the axiom of Jordanus—be able to lift the weight P_b a distance given by

$$GD \div \frac{P_b}{P_a} = GD \cdot \frac{P_a}{P_b} = GD \cdot \frac{b}{a} = ML ,$$

if L is a point on CD so that $CL = CB = b$. But a weight P_b placed at L would therefore not be lifted at all, since equal weights P_b at L and P_b at E are in equilibrium. The original assumption of P_a being lifted to D is thus wrong, and the lever must be in equilibrium.

Another formulation of the condition of equilibrium made by Jordanus was that there is a reciprocal proportion between the weights and their distances from the axis, which shows that Jordanus was very close to the concept of static momentum, although he never introduces the product of weight and distance as such, presumably because, in the traditional way, he usually employed only proportions and ratios.

The Importance of Mediaeval Statics

The *De ratione ponderis* is interesting because it contains a great many propositions on mechanical problems solved by means of the law of the lever, or by even more advanced methods. The determination of the weight with which a beam is pressing on a man lifting it from the ground to an arbitrary position, belongs to the first category while several theorems on the motion of heavy bodies through resisting media belong to the second. Jordanus tried to investigate the way in which velocity depends on various properties of a body such as resistance, weight, and shape. The fact that a freely falling jet of water converges downwards is explained on the explicit assumption that the motion is accelerated, and on the tacit assumption that water is an incompressible fluid. The problem of how a body rebounds from an intercepting surface is also tackled although less satisfactorily. Mediæval mechanics was as yet unable to cope with problems involving the general laws of motion, but this does not diminish the value of the results achieved and handed on to the physics of the Renaissance when statics took its next leap forwards. And the development of Mediæval statics as a particular branch of mathematical physics was of immense importance to physics in general, for it enabled Mediæval scientists to liberate themselves from too close a connection with prevailing philosophical doctrines. Thus statics became, to an even greater extent than optics, a paradigm for the mathematical description of nature.

Chapter Sixteen

The Establishment of Kinematics

THE science of motion entered Mediæval philosophy of nature as a distinct physical discipline in its own right through several channels. Most important were the translations of Aristotle's *Physics* and various commentaries on it by Simplikios, Johannes Philoponos, and Arab scientists. As a result, from the mid-thirteenth century, Mediæval mechanics was dominated by Aristotelian features (see Chapter 9), among them the preference for studying uniform motions, the distinction between natural and forced motions, and the proportionality of force and velocity as the fundamental axiom of dynamics.

The forceful position of Aristotelian philosophy in Mediæval universities meant that doctrines such as these were basic elements in the teaching of physics well into the seventeenth century, although the inherent weakness of the Aristotelian system was made clear much earlier than this. In fact, no sooner had the philosophy of Aristotle been introduced into university teaching than a critical reaction set in, and as early as the thirteenth century philosophers and scientists tried to dispel some major absurdities. The history of scholastic mechanics is thus not only an account of how Aristotelian theory was repeated again and again, it is also the history of a critical movement gaining more and more strength until Galileo and the physics of the Renaissance administered its deathblow. Here, and in the following chapter, only a few typical aspects of this development will be considered. A detailed analysis is made difficult by the close and numerous connections between logical and philosophical problems on the one hand, and physical considerations on the other.

The Concept of Velocity

One of the most peculiar features of Aristotelian scholastic science was the conception of the instantaneous velocity of a moving body. In general, velocity was conceived of as the 'intensive quality' of the motion on a par with its 'extensive quality', that is, the distance traversed. It was natural, therefore, to describe the motion of a body as 'swift' or 'slow' in precisely the same way as, without further elaboration, one talked about its 'whiteness' or 'heat'. It was generally assumed that intensive qualities such as these could somehow be reduced to quantities and expressed in numbers, although no one in fact tried to identify them with measurable quantities. There seemed no need for definitions, as intensities were always dealt with in a purely speculative way. Intensities were considered to be dependent on both space and time. If a body is rotating about an axis, it was argued, the intensity of the motion (that is the velocity of the body) will be different at different points. During free fall, velocity increases with time and the distance traversed, so that the intensity of the motion can be considered relative to both space and time.

Originally, Mediæval kinematics was solely concerned with uniform motions, although it was known that in nature many important movements take place with a non-uniform velocity. This is the case with the free fall, for which Jordanus Nemorarius had already proved an increasing velocity, or acceleration, from the convergence of a falling jet of water. The most important problem in kinematics was thus to find a method of describing the varying velocity of such motions.

As early as the first half of the thirteenth century the problem of varying velocity was tackled by Gerald of Brussels, whose *Liber de motu* was the first Mediæval treatise entirely devoted to kinematic questions. In the theory of motion it holds a similar position to Jordanus's writings in statics, and in the manuscripts both writings are often connected. It is significant that Gerard's treatise quotes Archimedes, whose influence on Mediæval mechanics seems to be greater than historians of science have previously acknowledged.

In the first book of *De motu* Gerard adopted a method that was later to become common in Mediæval kinematics. He tried to find ways to replace a given, non-uniform motion with a uniform motion in some way equivalent to it. The meaning of this can be seen from proposition I, 10, which deals with a line rotating around one end-point, the other end-point describing a circle. The proposition states that the motion of this line is equivalent to that of the middle point, so that if the line performed a purely translatory movement with the same velocity as that of the middle point during rotation, it would describe the same area in the same time. In this way, the rotatory velocity of the end-point was connected with the translatory velocity of the mid-point.

Proposition II, 1 deals with a more complicated problem of a similar nature. Here Gerard considers a circle rotating about a diameter and thus producing the surface of a sphere. The problem then is to find a translatory motion of a circle producing a body with the same volume as a sphere. In the proofs the surface is thought to consist of line elements, very much as in Archimedes' treatise *On method* (see p. 110), which was unknown to Gerard. His solution is incorrect because of an error in computation.

The Merton School

The work of Gerard of Brussels was only a small fore-taste of the great development in kinematical studies which was one of the most distinguishing features of fourteenth century science. In the period from about 1320 to 1350 Merton College, Oxford, was the home of a true mechanical school, with Thomas Bradwardine (*c.* 1290–1349), William Heytesbury (*c.* 1350), Richard Swineshead (*c.* 1340–1350), and John Dumbleton (*c.* 1330–1350) among its most prominent members. There is no doubt that their efforts sprang from the general philosophical problem of the *intensio et remissio formarum*, in other words how the variation of qualitative forms can be described in quantitative terms. The fact that the velocity of a non-uniform motion was a most obvious example of a varying intensity explains why the preoccupations of the Merton school became particularly fruitful in the realm of mechanics.

One preliminary result of the work of the Merton School was the emergence of a clear distinction between kinematics, the description of how a motion proceeds in space and time, and dynamics, the investigation of moving forces. Before this philosophers discussed whether motion in itself demanded criteria other than space and time; for example, was a retarding medium necessary around the moving body?

In the thirteenth century, the kinematic viewpoint had been stressed by Thomas Aquinas in disagreement with another Dominican, Aegidius Romanus, and at the beginning of the fourteenth century the emerging nominalist movement in philosophy had attempted to give the problem of motion a new basis. The nominalists believed only in the existence of individual substances, and refused to consider general or abstract concepts as anything more than mere names. As a result they were, in the field of physics, sceptical about the concept of force, and their leader William of Ockham (c. 1300–c. 1350) stressed that to enquire after the causes of motion (the acting forces) is quite different from describing motion as something perceptible to the senses.

The nominalists had more followers in the field of logic than in natural philosophy; yet they do seem to have stimulated the Merton scholars to treat kinematics and dynamics separately, and to make them fully conscious of the fundamental difference between the two points of view. Thus, in Thomas Bradwardine's *Tractatus de proportionibus* (1328), Chapter 3 deals with the proportions of velocities of motion as related to the forces of the moving and the moved body (dynamics), while Chapter 4 considers the same proportions related to the size of the moved body and the traversed distance (kinematics). Another Merton treatise, *Tractatus de motu*, distinguishes between the determination of the velocity of a body *quo ad causam* (dynamically) and *quo ad effectum* (kinematically).

The fundamental problem on the determination of the velocity of a body at a given instant is explicitly attacked in William Heytesbury's *Regule solvendi sophismata* (1335), in which part 6 treats of *motus localis*, or motion in space. Here he first explains that

> a motion is called uniform when equal distances are traversed in equal times with the same velocity,

which may be taken as a definition of a constant velocity. As for the instantaneous velocity, Heytesbury proposed the following definition:

> In a non-uniform (*difformis*) motion the velocity at a given instant of time may be conceived as the distance which the body would traverse

[in relation to the time spent] if, in a certain interval of time, it was moved uniformly with the velocity it had at the given instant.

Finally, Heytesbury defines uniformly accelerated motion:

Any motion is uniformly accelerated (*uniformiter intenditur*) when the velocity is increased with equal amounts in arbitrary, but equal, intervals of time.

The definition of uniform velocity is correct, it can be noticed, only if the distances are equal when traversed in arbitrarily chosen, equal, intervals of time; this necessary condition is explicitly stated in Richard Swinehead's brief treatise *De motu*. The definition of constant acceleration is totally satisfactory but the attempt to define instantaneous velocity is more difficult. Of course it is obvious that the instantaneous velocity v_0 of a body at a point P_0 and at the time t_0 is equal to the uniform velocity the body would have if, starting from P_0, it performed a uniform motion through a finite interval of time with the same velocity v_0. But this leads to no conclusions, and it is hardly conceivable that a logician of Heytesbury's status would be satisfied with a definition of this kind. His proposal is a testimony to the insurmountable difficulties of giving a logical definition of instantaneous velocity before the introduction of the differential calculus and the concept of limits.

Here one cannot help wondering why the Merton scholars ignored astronomy as a possible source of inspiration for the new kinematics. Mathematicians of the fourteenth century usually knew the Ptolemaic theories of planetary motion which displayed a surprisingly high degree of kinematical ingenuity. For example, Ptolemy was able to calculate the instantaneous, apparent velocity of a planet to any degree of accuracy by numerical methods based on tabulated functions of time (page 97). Many Merton mathematicians and other Mediæval scientists must have been familiar with these methods, but nothing indicates that they ever tried to apply them to mechanical problems, possibly because astronomical kinematics was neither exact nor developed into a systematic theory.

Although the Merton scholars could not give a logical definition of instantaneous velocity this did not prevent them from achieving several results of far-reaching importance to the development of mechanics. First of these was a proposition on uniformly accelerated motion, usually called the Merton relation, and stating that

> If the velocity of a body is increased uniformly from v_0 to v_1 during the time interval t the distance traversed in that time will be

$$s = \left[v_0 + \frac{v_1 - v_0}{2} \right] \cdot t$$

There are many different proofs of this theorem in the various works of the Merton scholars, who must have been aware of its fundamental character. If the velocity v_0 is assumed as zero, the above expression is reduced to

$$s = \tfrac{1}{2} v_1 \cdot t .$$

As the velocity increases uniformly with time we have $v_1 = g \cdot t$ which leads to

$$s = \tfrac{1}{2} g t^2 .$$

Formally, this is the law of free fall, with g interpreted as the acceleration due to gravity, so that it becomes clear that in the first half of the fourteenth century scientists already possessed the tools necessary for the description of fall in a homogeneous field of gravitation in exactly the form applied later by Galileo.

Oresme's Graphical Method

In about 1350, Nicole Oresme (c. 1320–1382) took a new step forward in Mediæval kinematics (Fig. 16.1). At that time he was teaching in the University of Paris, and was strongly influenced by the Merton School, as well as by his own teacher Johannes Buridanus. Oresme's main achievement was the development of a graphical method for representing the variation of qualities. Suppose that the velocity of a body during a given interval of time t_0 is to be visualized, an arbitrary unit is chosen by means of which a line of length t_0 called the *longitudo* is drawn. A point of time t inside the interval t_0 is represented by a point P on the *longitudo*. At P a line is drawn perpendicular to the *longitudo* with a length equal to the velocity of the body at the time t

Fig. 16.1. Nicole Oresme at work. From a manuscript of his French Livre du ciel et
du monde from A.D. 1377, now in the Bibliothèque Nationale, Paris.

(measured in another arbitrary unit). This line is called the *latitudo*. The set of all *latitudo* lines covers a definite area of the plane and defines a figure which Oresme calls the 'configuration' of the motion.

A uniform motion is seen to correspond to a rectangular configuration. If the motion has a constant acceleration the configuration is a triangle (if the initial velocity is zero) or a trapezium. In this way any velocity distribution has its characteristic configuration which immediately reveals a number of features of the motion in question. From a mathematical point of view Oresme's method was extremely important as it was, in the modern sense, one of the first examples of graphical representations. It rapidly became known in the universities, and integrated with the common mathematical pensum. All the indications are that in the 17th century the method contributed both to the development of analytical geometry and to the evolution of the concept of a function.

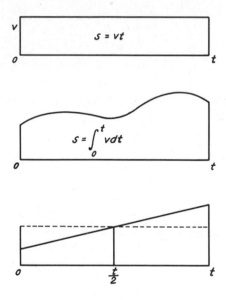

Fig. 16.2. *Configurations of motion according to Nicole Oresme. The configuration at the top corresponds to a motion with* ation *constant velocity* (motus uniformis). *Below is a motion with irregular velocity* (difformis), *and at the bottom with constant acceleration* (uniformiter difformis).

The importance of Oresme's method in kinematics arose mainly because it now became possible to arrive at a geometric determination of the distance traversed by the moving body (Fig. 16.2). If the motion is uniform, the velocity v through the time t, the distance is $s = vt$. But this equals the area of the rectangular configuration of the motion. If the motion is non-uniform, the configuration is not a rectangle, but Oresme assumed that in this case its area will also represent the distance. There is no real proof of this assumption, which Oresme seems to have taken for granted in an intuitive way, but although his method is based on a deficient mathematical foundation it is obvious that in kinematics it was able to serve the same purpose as integral calculus did later. Thus the arbitrary function $v = v(t)$ will determine the upper boundary of a configuration the area of which is the definite integral

$$s = \int_0^t v(t)\, dt$$

which is the distance traversed by the body.

Oresme used his method in many different ways. In his proof of the Merton relation for a motion with constant acceleration (p. 222), for example, the configuration is a trapezium with an area equal to that of a rectangle with the same base and a height equal to the velocity of the middle point. The distance,

$$s = \frac{v_0 + v_1}{2} \cdot t,$$

is thus equivalent to the formula above.

Oresme's Criticism of Aristotle

Oresme also makes use of his method to point out errors and mistakes in Aristotle's many semi-mathematical arguments. Aristotle often used the assertion that if a body moves during an infinite time it will also traverse an infinite distance. Oresme proved that this is not necessarily so by considering the special case of a body which has the velocity v during the first day, $v/2$ during the second day, $v/4$ during the third day, and so on. In this case the configuration becomes a succession of

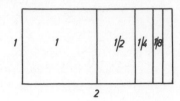

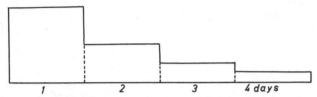

Fig. 16.3. Oresme's proof that a motion through an infinite time may traverse a finite distance.

rectangles as shown in Fig. 16.3, and the total distance is expressed as the sum of the infinite series

$$s = v + \frac{v}{2} + \frac{v}{4} + \cdots$$

$$= v(1 + \tfrac{1}{2} + \tfrac{1}{4} + \cdots)$$

That this sum is finite is proved geometrically by means of a rectangle with sides of 1 and 2 units. From this rectangle a square is cut with side 1 unit, representing the first term of the series. Another square remains, half of which represents the second term. Half of the remaining rectangle corresponds to the third term, and so on. The entire, infinite configuration can thus be contained inside a rectangle of two square units and Oresme concludes that the distance traversed by the body in an infinite time will be finite and equal to $2v$.

Oresme did not restrict himself to applying his method to kinematical problems. In fact he seemed to regard it as universally valid and useful for the description of non-mechanical aspects of nature, including psychology and even theology.

Kinematics and Experimental Science

Oresme's graphical method is a good example of the general attempt in the Middle Ages to give the description of natural phenomena a mathematical basis. This attempt seems to have been one of the main preoccupations of fourteenth-century philosophers who are sometimes acclaimed—with less than justice to Archimedes—as the founders of mathematical physics. But characteristically, quantitative formulations of the laws of nature were never followed up by measurements, despite the fact that all Mediæval philosophers shared the general belief in an underlying mathematical structure of the universe—to the extent that even concepts like goodness, grace, or piety were supposed, in principle, to be expressible in numbers. But exact definitions and concise descriptions from measurements were usually absent, even from mechanical concepts like velocity, force and resistance.

The reasons for the difficulties experienced by physics in the Middle Ages are practical ones. Adequate mathematical tools had not yet been developed, and the techniques of experimental physics were still at an early stage. Precision instruments were unknown except in astronomy and geodesy, and the idea of a scientific laboratory had not yet occurred to university authorities, living as they did in complete isolation from the world of artisans and engineers who might have provided some of the apparatus needed by experimental philosophers.

While acknowledging the empirical basis of natural science, Mediæval scholars therefore had to rely almost exclusively on everyday experience as had scholars in the days of Aristotle, with exception of a few optical and magnetic instruments. Despite this, they were still able to give valuable, logical analyses of the scanty material at their disposal, and to frame hypotheses preparing the way for later physicists. To quote a single example: even in the middle of the fourteenth century the laws of the free fall of bodies were discussed in a hypothetical way. One school maintained that the velocity increases proportionally with time on the basis of the formula $v = gt$, while another supposed it to follow the relation $v = gs$, or to increase proportionally with the distance. Oresme assures his readers that the former hypothesis was generally believed to be true. No proof of this happy intuition was possible, but it is worth noticing that Galileo began his fruitful investigation of free fall with a discussion of precisely the two hypotheses first stated by his fourteenth century predecessors.

227

The Method of the 'Calculators'

The history of science has often, too hastily, discarded the hypothetical mathematical speculations of late Mediæval scientists on laws of nature as mere fantasies. Even without experimental support, a precise hypothesis stated in mathematical terms can prove useful to physics. Take the enormous *Liber calculationum* of the 'Calculator' Richard Swineshead (or Suiseth). His *calculationes* were not wild numerical computations, but a real method of investigating the validity of hypothetical laws. His general procedure was to state the hypothesis as a mathematical relation between a number of variables. Each variable is then given a numerical value taken not from experience, but from physical commonsense. When these particular values are substituted in the equation the result may prove absurd, that is, contrary to commonsense, in which case the hypothesis must be discarded. While the 'method of calculations' thus cannot verify a true hypothesis it may well be able to prove an erroneous one to be false. This explains why Galileo occasionally used *calculationes* of a Mediæval type with great success, and why Leibniz recommended a new edition of Suiseth's book.

Although the method of calculation may have had a real value, its abstract nature was undesirable. In the long term it became unsatisfactory to test a hypothetical law of nature with numerical values taken not from nature itself but from the imagination of the scientist, however clever he might be. But by applying this method, Mediæval scientists were forced to look for better means of providing numerical data. In this light the outburst of experimental activity in the physics of the Renaissance is perhaps more closely connected with late Mediæval science than one might suppose.

Chapter Seventeen

Force and Motion in the Middle Ages

MEDIÆVAL scientists, although they arrived at lasting results in the field of kinematics, achieved less success in the realm of dynamics; yet the literature which survives shows that much more thought was expended on dynamics than on kinematics. One of the reasons for this was undoubtedly the unequal status of the two branches of mechanics in the university curriculum. Kinematics had been liberated from its disadvantageous connection with the general course in philosophy, and established itself as an independent *scientia media*, decidedly mathematical in character (p. 191). The centre of this development, Merton College, had something of the character of a specialized research institute. Dynamics, on the other hand, was inextricably bound to Aristotle's physics, which was taught and commented on by almost every professor of philosophy. It never gained independence and the dynamical questions discussed from the thirteenth century onwards were largely by-products of general, philosophical enquiries into the relation of cause and effect in the realm of natural science. Yet even the supporters of this philosophical mechanics were aware that many of Aristotle's dynamical opinions seemed incomplete, inconsistent, or even absurd; thus Aristotle's presumed authority in philosophy did not prevent his Mediæval followers from sharp criticism in matters of science.

Force Is Necessary

One fundamental Aristotelian principle, the axiom that any motion necessarily presupposes the existence of a moving force, was left undisputed by almost all Mediæval natural philosophers. A famous dictum in the work of Thomas Aquinas stated it: '*omne quod movetur ab alio movetur*' (whatever moves is moved by something else). In the following century even Nicole Oresme, a fierce critic of Aristotle admitted that '*tout mouvement a cause efficiente et final autre que ce meïsme mouvement*' (any motion has an efficient and final cause different from the motion itself). Thus motion was conceived of not as a natural state of bodies, but as an effect which is the result of a cause, and it is easy to understand why Aristotelian philosophers were unable to arrive at the concept of inertial motion as a state in which the body moves along a straight line with constant velocity independent of any external force. In other words, the physical notion of force was connected too closely with the metaphysical notion of cause in the minds of scholars to make a fruitful development of dynamics possible.

The one person to whom force was not an obvious essential to motion was William of Ockham. He and some of his Nominalist followers considered the concept of motion to presuppose nothing but a material body which, at different times, is found at different places. To them, particular forces are unnecessary as causes of motion; a body moves, they argued, simply because it is moving, and there is nothing more to be explained. This extreme opinion has been interpreted as the first intuition of the law of inertia, but the text does not support such a conclusion. Ockham was not prepared to accept an infinite motion, just as force was to him unnecessary to all kinds of motion, whether uniform or non-uniform. This so-called Nominalist doctrine of motion found no great support among natural philosophers of the fourteenth century, nearly all of whom endorsed the traditional view that force is a necessary condition of motion. Their main preoccupation was not to prove the necessity of force, but to find out what kind of motion follows from a given force, a many-sided problem of which we shall first consider the mathematical aspects.

The Fundamental Relation of Dynamics

In Mediæval mechanics, therefore, motion was invariably analyzed in terms of cause and effect, the cause being some physical force in nature, the effect the motion itself, described in kinematical terms. As a result the fundamental question to mechanics was the establishment of a relationship between the dynamic and the kinematic aspects of motion. In classical mechanics this relation is Newton's second law, which implies also the principle of inertia (Newton's first law): if the velocity remains the same no force is needed, a statement that would obviously have been unacceptable to Mediæval scientists. For them, the necessity of force to motion could only be expressed through a relationship in which force is some function of velocity itself.

Initially, thirteenth-century philosophers took over the Aristotelian law of force without any modification. Aristotle postulated (p. 122) that in order to produce a motion the moving force F must be greater than the resistance R, and that the resulting velocity will be proportional to F and inversely proportional to R. Consequently, the fundamental relation was

$$v = \text{const} \times \frac{F}{R} \quad \text{for} \quad F > R$$

$$v = 0 \qquad\qquad \text{for} \quad F \leqslant R$$

As a fundamental presupposition of mechanics this law was used continuously until finally replaced by Newton's laws at the beginning of the eighteenth century.

The Aristotelian solution did not go entirely unchallenged, however. Mediæval philosophers could not close their eyes to the fact that the resulting relationship gave rise to certain difficulties, mainly because of the particular way in which the resistance is introduced in the formula above. For if a body is moving under the action of a constant force through a medium which becomes more and more rarified, the resistance will tend towards zero, and the velocity must increase without limits. In agreement with the general philosophical reluctance to admit the existence of infinite quantities this was considered an absurdity, and as in Antiquity, the problem was solved by denying the possibility of an extended vacuum, that is, a medium without resistance. The philosophers of the Middle Ages thus upheld the ancient mechanical argument for the *horror vacui* of Nature.

By implying discontinuities in the physical world, the relationship between force and motion offended another philosophical principle. If a motion is considered with constant force and gradually increasing resistance, the velocity will drop suddenly from a finite value to zero at the moment when the resistance becomes greater than the force; this was against the dictum *natura non saltat*.

Difficulties such as these led the philosophers to consider alternative forms for the fundamental relation of dynamics. In the thirteenth century Thomas Aquinas quoted a law which presumably dates back to Johannes Philoponos (p. 124), and can be stated:

$$v = \text{const} \times (F - R) \quad \text{for} \quad F \geqslant R$$
$$v = 0 \qquad\qquad\qquad \text{for} \quad F < R$$

The law implies no discontinuity, since the velocity will tend towards zero when R tends towards F. On the other hand, it does not exclude motion in a vacuum, since the velocity will approach a finite limit if the resistance disappears. This is perhaps the reason why this law was not generally accepted, since the impossibility of a vacuum had acquired an almost dogmatic character. Yet Aquinas maintained, with Philoponos, that if a vacuum could be produced, the velocity of a freely falling body would remain finite within it.

Bradwardine's Law of Motion

In the fourteenth century, a third relationship between force and motion was proposed by Thomas Bradwardine (p. 219), who in his *Tractatus de proportionibus* (1328) not only dealt with problems in kinematics but also tried to remedy the defects of the Aristotelian law. At the beginning of Chapter 3 Bradwardine says that 'the velocities of motion follow the proportion of the force of the moving and the force of the moved body'. Here the 'force of the moving body' is the moving force previously denoted by F, while the 'force of the moved body' is the resistance R. Therefore the relation seems to contain nothing more than the Aristotelian assertion that the velocity is proportional to the ratio F/R.

This simple interpretation is, however, incorrect for the term 'follows the proportion' (*sequitur proportionem*) had a particular significance in Mediæval mathematics. When two given ratios a/b and c/d were combined in the new ratio ac/bd they were said to be 'added' together, the term 'add' being used in the special sense of 'combine by multiplication'. As a result, the 'double' of a given ratio a/b is not $2a/b$, but the ratio $(a/b)^2$, produced by 'adding' a/b to itself. In the same way the triple ratio is $(a/b)^3$, etc. This implies that the 'ratio of the two ratios' $(a/b)^p$ and $(a/b)^q$ can be expressed by the ratio p/q. Bearing this terminology in mind, Bradwardine's law acquires its proper meaning, and can be clearly illustrated by a numerical example found in Oresme: Suppose that the values $F = 3$ and $R = 1$ give the body the velocity v. Then the velocity $2v$ will not be the result of $F = 6$ and $R = 1$, but of $F = 9$ and $R = 1$, the ratio $9/1$ being the double of the ratio $3/1$ on the above terminology.

In general terms, Bradwardine's law postulates that the velocity v is a function φ of the ratio F/R satisfying the equation

$$n \times v = \varphi\left(\left[\frac{F}{R}\right]^n\right)$$

This is, in modern parlance, a logarithmic function, so that the new law of force can be expressed in present-day terminology as

$$v = \text{const} \times \log\left[\frac{F}{R}\right]$$

By using this relation Bradwardine managed to avoid the discontinuity of the velocity function which was one of the stumbling blocks of Aristotelian dynamics, for if F converges towards R the velocity v will approach zero in a continuous way. The law does submit, however, to the dogma of the impossibility of a vacuum, since v increases *ad infinitum* when R tends to zero with a constant value of F. Together, these properties made the law acceptable to numerous natural philosophers of the fourteenth century, and must be regarded as the late Mediæval analogy to Newton's second law. Of course it is obvious that compared with later developments in mechanics Bradwardine's law is useless as a basis for describing mechanical experiments. This was not felt in the fourteenth century, where such experiments were unknown, and the law appeared to be in harmony with the small range of everyday experience on which dynamics was founded.

The discussions of the law of force reveal the critical approach to Aristotelian physics maintained by natural philosophers of the fourteenth century. The debate shows, too, the dauntless development of mathematical physics regardless of either precise definitions or of controlled experiments. The dynamic notion of force is just as intuitive and vague as the concept of velocity and the notion of work is often mentioned without being defined.

The Origin of the Moving Force

Apart from the problem of the equation of motion, Mediæval philosophers were much concerned with the origin of the moving forces.
Aristotle had distinguished between 'natural' and 'forced' motions.
Natural motions are caused by the proper 'nature' of the moving body,
that is, the moving force is a tendency to motion residing in the body
itself as part of its substantial 'form'. Aristotle considered these natural
tendencies to be gravity and levity, to which philosophers of the Middle
Ages sometimes added magnetism. Although this doctrine of natural
motion was maintained by most Mediæval scientists, it is possible to
discern increasingly intense efforts to liberate physics from the foundations laid by Aristotle, made particularly by a group of natural
philosophers working in Paris at about the middle of the fourteenth
century and including Jean Buridan and Nicole Oresme.

The Theory of Gravitation and Natural Place

Aristotle had defined gravity as a natural tendency inherent in heavy
bodies to move towards the centre of the world. Oresme maintained,
however, that the centre of the world is a purely mathematical point
incapable of exerting any physical influence. He explained gravity by
assuming that heavy bodies have a natural tendency or inclination to
unite with each other, taking it to be a particular instance of the general
scholastic principle of *inclinatio ad similes*. The tendency to motion
towards the centre of the world (gravity) could then be explained
only by means of a cosmogonical theory: when heavy bodies unite, the
heavier parts will first come together and the less heavy will assemble
around them. On this basis, the universe must be structured in spherical
shells, so that the geometrical centre of the world will be at the centre
of gravity of the Earth.

It is even more interesting that Oresme discards the notion that
gravity and levity are absolute and opposite qualities. Aristotle had
taught that the 'media' water and air have, respectively, no absolute
gravity and levity. He assumed, for instance, that a particle of air
placed in the sphere of fire would move downwards towards its natural
place (the sphere of air), and as a result seem to be heavy. On the other
hand, a bubble of air would appear as light when placed in the

sphere of water. By means of a thought experiment involving a canal stretching from the centre of the Earth through the four elementary spheres to the inner surface of the sphere of the Moon, Oresme takes Aristotle's argument one step further. First the canal is supposed to be filled with water, and a small particle of air placed at the centre. This air will, he said, ascend through the water. According to Aristotle it would stop somewhere inside the sphere of air, but Oresme maintained that it would continue to the upper end of the canal. Similarly, if the canal is filled with fire, a particle of air placed at the top will fall all the way down to the centre of the world. In both cases, Oresme asserted, the motion is 'natural'.

Oresme's thought experiment obviously had very wide implications, as it could undermine the whole concept of natural places, but Oresme did not take it that far. Here, as elsewhere, he ventures on a brilliant, promising idea without following its consequences to the limit (see his consideration of the motion of the Earth, p. 269).

When the traditional concepts of gravity and levity are questioned, it follows that the theory of free fall must also be revised. It was known in Antiquity that fall is an accelerated motion, and Aristotle had explained it by assuming that the gravity of bodies increases as the distance from their natural place diminishes. Oresme assumed (p. 227) that the velocity of the fall is proportional to the time (which Galileo later proved to be correct) but he did not accept the Aristotelian notion of an increasing gravity. For him the natural gravity of a body was independent of the position in which the body is placed, and he was therefore forced to find a new explanation for the increase in velocity. Oresme thought he had achieved his aim by using the concept of *impetus*, which is best considered as part of the problem of violent motion.

The Motion of Projectiles

The motion of projectiles moving along curved paths (as different from the straight lines towards the centre of the Earth which they would follow according to their natural inclination) was, in the Middle Ages, the standard problem in violent motion. All philosophers agreed with Aristotle that when motion is violent the force is not caused by the nature or 'substantial form' of the body. Determination of the origin of this force became particularly important in the 14th century, with the

235

increasing use of artillery in battle. Jean Buridan (*c.* 1290–*c.* 1360) discussed violent motion by referring, in particular, to 'engines throwing large stones', without specifying whether he meant cannons or more traditional catapults.

The problem of the moving force of a missile could be tackled in several ways. First, the traditional Aristotelian theory assumed that the force acting on the projectile comes directly from the surrounding air which, in turn, has received it from the catapult (Fig. 17.1). At a certain point the force is exhausted and the projectile falls to Earth. One result of this theory is that the path of the projectile (the ballistic curve) is composed of two straight lines: the projectile moves upwards in a direction determined by the angle of the catapult or cannon,

ctum rei demoliendæ inferat, aduertimus. Cum itaque globum igneum quis eijcere conatur, & per perpendicularem contendit tangere certum fcopum, oportet illum in primis fcire uim illius iaculatoriæ machinæ, atque deinde diftantiam loci quem cupit igne uexare.

Cognitis his duabus diftantijs, bafi fcilicet & hypotenufa, inclinanda eft machina iuxta affixi quadrantis normulam, ut globus piceus hypotenufæ tramitem recte incedat, atque deficiens in curfu fuo, defcendendo cadat in deftinatu

Fig. 17.1.   *The motion of a projectile on the Aristotelian theory. From Sebastian Münster,* Rudimenta mathematica, *Basileae, 1551.*

and then suddenly vertically downwards. The occasional use of this aypothesis in gunnery can be seen in Fig. 17.1.

Secondly, in the fourteenth century the Aristotelian theory was widely criticised. William of Ockham argued his point from a case in which two arrows are flying close to each other, but in opposite directions. If their motions were explicable as an effect of the surrounding medium then the air must be considered to have motive force in two opposite directions at once, which is absurd. Ockham therefore toyed, at one stage, with the hypothesis that the force on the arrow comes directly from the bow, without the medium between them playing any role. It seems that he is the only Mediæval philosopher to consider action at a distance as a principle of mechanics, but even he later discarded this assumption and adhered to the general philosophical axiom that no agent can have any effect at places where it is not present.

Ockham satisfied himself, thirdly, by declaring that a body moves simply because it is moving. In other words: in order to explain motion it is unnecessary to assume any moving force. At this particular point, Ockham had no followers among Mediæval philosophers. His contemporaries adhered unswervingly to the dogma that force and motion are inseparable.

The Impetus Theory

Philosophers of the fourteenth century largely discarded these three doctrines and took refuge in a fourth hypothesis, the so-called 'impetus' theory which has its origin in the work of Johannes Philoponos (p.125). In the twelfth and thirteenth centuries the impetus theory is occasionally found in the writings of Latin scholars, but it had few thirteenth century supporters. Thomas Aquinas refuted the theory several times, yet made use of it in his explanation of how a ball rebounds when it impinges upon a solid wall. The reason given by Aquinas is that the hand throwing a ball has imparted to it a certain *impetus ad motum*. This impetus is conserved during the impact, and is therefore able to carry the ball away from the wall.

In the fourteenth century the impetus theory became generally accepted, and from about 1320 was taught in the University of Paris. Soon after, Jean Buridan developed it so widely that he must be rated

237

as its main proponent. Buridan's formulation of the theory can best be illustrated by a quotation from his *Quaestiones* on Aristotle's *Physics*:

> When a mover sets a body in motion he implants into it a certain impetus, that is, a certain force enabling the body to move in the direction in which the mover starts it, be it upwards, downwards, sidewards or in a circle. The implanted impetus increases in the same ratio as the velocity. It is because of this impetus that a stone moves on after the thrower has ceased moving it. But because of the resistance of the air (and also because of the gravity of the stone) which strives to move it in the opposite direction to the motion caused by the impetus, the latter will weaken all the time. Therefore the motion of the stone will be gradually slower, and finally the impetus is so diminished or destroyed that the gravity of the stone prevails and moves the stone towards its natural place. In my opinion one can accept this explanation because the other explanations prove to be false whereas all phenomena agree with this one.

The implanted impetus, it will be noted, is caused by the velocity and supposed to be proportional to it. Elsewhere, Buridan considered it as proportional to the weight of the body. In properly chosen units, therefore,

$$\text{impetus} = \text{weight} \times \text{velocity},$$

by which Buridan has given a precise meaning to impetus, a concept which was previously rather vague. From the formal point of view this new concept in dynamics equals the quantity of motion of classical physics, but in reality the two are very different because, as will be shown they play different parts in their respective dynamic theories. The important point is that in the Mediæval sense of the word impetus is a *force* with the same physical status as gravity, levity, magnetism, and so on. Nevertheless, the theory may well have prepared the way for the concept of inertia. The quotation above implies that in the absence of all resistance the impetus would carry the projectile with constant velocity along a straight line to infinity, which is the kind of motion described by the law of inertia. Thus even if the impetus is a force in the Aristotelian sense (a cause of motion) it may well have led philosophers to consider the type of motion described by Newton's first law.

Buridan's application of the impetus theory to the motion of projectiles obviously led to a ballistic curve different from that given by the Aristotelian theory. This problem was investigated more closely by another Parisian scholar, Albertus of Saxonia (c. 1316–1390) who distinguished three different stages in the motion of projectiles. First was an initial stage in which the impetus is dominant, and gravity regarded as insignificant, the result being motion along a straight line. Secondly, Albertus defined an intermediate stage in which gravity recovers, and the path begins to deviate downwards from the straight line; this part of the path was often conceived of as part of a circle. Thirdly, he postulated a final stage where the impetus is completely spent, and gravity alone draws the projectile downwards along a vertical line. Figure 17.2 shows that the impetus theory did result in an

Fig. 17.2. Ballistic curve based on the impetus theory. From Walther Ryff *Der Furnemb-sten . . . Mathematischen und Mechanischen Kunst,* Nuremberg, 1547.

improved form of the ballistic curve, although in a purely qualitative sense, and from which it would have been impossible to deduce range tables of any practical value.

Buridan used the impetus theory in rather an odd way to explain the accelerated motion of the free fall. Initially, he argued, gravity alone acts on the body, which acquires a small velocity. But this velocity entails a small *impetus* which, together with gravity, augments the velocity, which in turn augments the gravity, and so on. This strange theory was later adopted by Oresme. Buridan also dealt with

239

phenomena such as the rebounding of a ball from a wall, a vibrating string, and the motion of pendulums.

That the impetus theory cannot be considered as a direct forerunner of the concept of inertia is abundantly clear from the fact that Buridan applied it also to circular motion. The impetus of a body was thought to be a force tending to uphold the initial motion, whether straight or circular, an idea which survived in physics for at least three centuries, and was even used by Galileo.

The circular impetus certainly delayed the clear understanding of inertial motion as rectilinear, but it did play a very important part in helping Mediæval philosophers to look at celestial motions in a new way. In the thirteenth century the revolutions of the celestial spheres were usually explained on the assumption that each sphere is moved and governed by a separate 'intelligence', a theory which had abandoned the Aristotelian concept of animated spheres. Buridan remarked that such 'intelligences' were never mentioned in Holy Scripture, and that they could be completely discarded if one assumed that at the creation of the World God imparted a certain amount of impetus to each sphere. Since the spheres move without resistance, he went on, their impetus will never diminish, but will keep them in a state of uniform rotation throughout all time. This idea was also taken up by Oresme, who made use of the analogy between the universe and a clock which is wound up once and for all.

This theory of the heavenly motions marks a radical break with the traditional view, in which the belief in a fundamental difference between terrestrial and celestial phenomena was firmly rooted. That fourteenth century philosophers were able to include celestial mechanics in a system developed to describe terrestrial motions is a testimony to their independence and their intellectual ability, although the implications of their theorizing were not realized until Newton's time.

In the centuries that followed, the impetus theory became extremely popular and widely accepted. Oresme tried to give it another interpretation by connecting the impetus with the increase in velocity, but not with velocity itself. This could have resulted in a notion of force similar in some respects to that of Newton, but Oresme's form of the impetus theory was soon forgotten, and it was passed on to succeeding generations in the shape given it by Buridan.

The Resistance to Motion

In many ways, the mechanics of the late Middle Ages succeeded in breaking away from tradition. Numerous new ideas in several different fields of physics and astronomy were proposed, examined, adopted or discarded. But despite their efforts, Mediæval natural philosophers proved unable to create any lasting foundation on which a new natural science could be constructed. One of the reasons for this failure becomes obvious when fourteenth century dynamics is compared with that of Galileo. The most obvious difference is in scientific method. Galileo's understanding of how to support hypothetico-deductive conclusions with experimental proof gave physical theory a far more solid basis than did the simple, qualitative, everyday experiences from which Mediæval science started.

Equally important was the fact that Galileo introduced the principle of the mutual independence of the motive forces, which enabled him to break up a complicated problem of motion into a number of simpler components. One particularly important consequence of this principle was Galileo's method of finding approximate solutions by disregarding for example, the resistance to motion offered by the air when dealing with free fall. Such approximations were foreign to Mediæval physics, which always took the resistance into account, and distinguished between various resistant forces acting on the moving body. First, there was the obvious resistance, or friction, from the surrounding medium. Next, a force like gravity could be conceived of as a kind of resistance, namely to a violent motion as long as the impetus was still active. Finally, some Mediæval physicists assumed a kind of resistance unknown to Aristotle. It was called *resistentia intrinseca*, and thought of as an inherent tendency seeking to bring the moving body back to its natural state of rest. But regardless of the particular type of resistance operating on the body, the combined effect of the moving force and the resistance were always considered. In fact, these two forces were inseparable not only in Aristotle's law of motion, but also in Bradwardine's, and in both cases the velocity was a function of the ratio of force to resistance. The Galilean approximation of putting the resistance equal to zero was therefore mathematically inapplicable to these theories and as a result it was impossible to discuss motion as ideally separated from any resistance, and to take the resistance into account as only a separate small correction to the ideal solution. This seems to be the most characteristic difference between Mediæval mechanics and the fundamental assumption of the independence of forces from which classical mechanics started in the sixteenth and seventeenth centuries. Despite their many brilliant ideas and daring innovations, a wide gulf separated the natural philosophers of the fourteenth century from their successors.

Chapter Eighteen

Mediæval Astronomy

A T the beginning of the Middle Ages astronomy was the only science to emerge from the cultural decline without being completely deprived of its former glory. Throughout the Dark Ages, astronomy was regarded as the most precious discipline of the quadrivium, and for centuries afterwards retained this lofty position, although the reasons for this esteem varied.

Astronomy was the most developed branch of science, and as such served as a model for a mathematical description of nature in other fields. This became particularly important during the thirteenth and early fourteenth centuries, when theoretical astronomy had again reached a maturity which the pioneers in mathematical statics and kinematics strived to attain. Astronomy was also interesting to philosophers, partly as material for epistemological discussions on the nature of scientific hypotheses and theories, partly as a basis for metaphysical disquisitions on the ultimate structure of being, dependent, in turn, on a supreme cause, just as the motion of the planetary spheres was caused by the *primum mobile* outside the fixed stars. This viewpoint was often reflected by poets like Geoffrey Chaucer who, in *The knight's tale*, stressed the contrast between the tumultuous and unstable world below and the eternal order and harmony above, to which men should look if they are to learn the true principles of human existence. Similar reflections are to be found in Dante's *Divina commedia*.

Astronomy and Astrology

During the second half of the Middle Ages astronomy gained a universal significance as the technical basis for astrology. Before the middle of the thirteenth century, at least outside Italy, the Church attacked astrology in an attempt to quell astrological 'prognostications' and more occult forms of divination. After the anxious years of the Black Death around A.D. 1350 the struggle ended in a complete victory for astrology. Kings and princes kept court astrologers in honoured positions, and commoners asked for the astrologers' advice

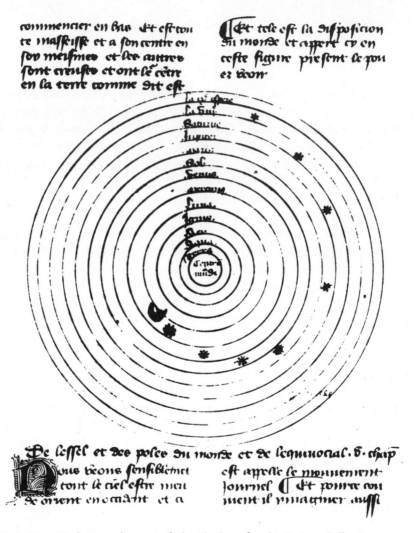

Fig. 18.1. Nicole Oresme's system of celestial spheres from his Traictie de l'espère, c 1375. *From a manuscript in Bibliothèque Nationale, Paris.*

on even the most intimate details of life, such as the most auspicious date and hour for the conception of a child.

Many sober-minded philosophers and scientists tried to dam this stream. A particularly ingenious argument against astrology was put forward by Nicole Oresme, who wrote several tracts against divination. He realized that astrological predictions could be possible only if the planets returned to the same relative positions within regular intervals of time. If not, it would have been impossible to deduce the laws by which they are supposed to influence the lower spheres of the universe (Fig. 18.1). Now the planets do not return to the same positions relative to each other, unless their periods of revolution are commensurable and can be expressed in what we would call rational numbers. In a treatise *On the incommensurability of celestial motions* Oresme shows that there are very few rational, compared with irrational, numbers. It is therefore very improbable that planetary motions should be commensurable, and unlikely that the universe will ever return to exactly the same order.

This combination of mathematical insight and an argument of probability undermined one of the fundamental assumptions of astrology, but had no practical consequences. The old superstition was irresistible and the Renaissance period greatly enforced it, even reducing the *Almagest* of Ptolemy to a mere introduction to Albumasar and other astrological authors. Nevertheless, the increasing interest in the positions of the planets was a stimulus to astronomy. Evidence of this is the many attempts at improving astronomical tables made in the fifteenth century, which again meant a renewed interest in both observation and planetary theory.

Latin Astronomy of the Early Middle Ages

From the works of Isidore of Seville (p. 188) and the Venerable Bede (p. 188) it is possible to gain an impression of the work of rescue performed in the early Middle Ages to save the astronomical and cosmological knowledge of Antiquity. Secondary Latin authors provided the more elementary astronomical concepts: the spherical shape of the Earth; Eratosthenes' measure of its circumference; the diurnal rotation of the heavenly sphere; the axis of the world and its

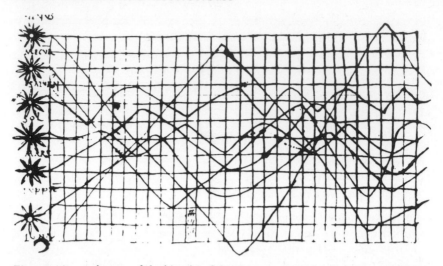

Fig. 18.2A. *A diagram of the latitudes of the planets as functions of time, using a kind of co-ordinate system. It is found in an eleventh-century manuscript now in the Bayerische Staatsbibliothek, Munich.*

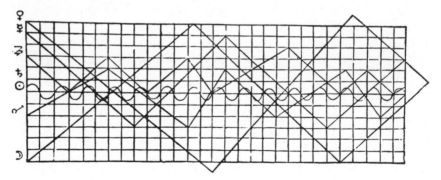

Fig. 18.2B. *The same diagram found in Johannes Hervagius's edition of Bede's works, Basle 1563. The diagram proves that graphical representation of functions was not completely unknown in the Middle Ages.*

poles; the principal circles of spherical astronomy; the direct motion of the planets eastwards among the fixed stars; their approximate periods; the annual movement of the Sun, and the cause of eclipses and the phases of the Moon. Last, but not least, the elements of the theory of the calendar were retained from which a particular Mediæval science developed, called the *Compotus*, on which Bede wrote an extremely competent manual *De ratione temporum*. All this was genuine astronomy, marked by an unhesitating rejection of judicious astrology branded by St. Augustine as impious superstition.

The meagre knowledge of the heavens formed a slender, but solid basis for spherical astronomy. Planetary theory, on the other hand, had almost no links with the past. Ptolemy's *Almagest*, in Greek, could no longer be read, and an early Latin translation by Boëthius (p. 188) was soon lost. As a result astronomers of the early Middle Ages were ignorant both of Hipparchos's solar theory and of the epicycle theories of the planets. For a general conception of the universe they had to rely on such ancient and primitive sources as Chalcidius's translation of the main part of Plato's *Timaios*.

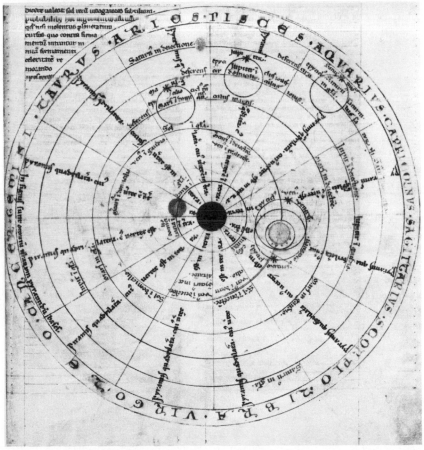

Fig. 18.3. The geo–heliocentric system. From a thirteenth–century astronomical manuscript now in the Royal Library, Copenhagen.

The Carolingian Renaissance took a decisive step forwards. A number of manuscripts dating from the eighth to the twelfth century contain a figure showing the universe with the Earth at its centre, and the Moon, Mars, Jupiter and Saturn revolving around it, but with Mercury and Venus moving in circles around the Sun (Fig. 18.3). This was the geo-heliocentric system, dating from Heracleides of Pontos (p. 62), which, in the early Middle Ages enjoyed a widespread popularity propagated by late classical commentators, such as Martianus Capella. The same author had also assumed a diurnal rotation of the Earth about its axis, but there is no trace of this in the astronomy of this period. Instead, astronomers accounted for the daily revolution of the heavens by supposing that the sphere of the fixed stars outside the planets actually performed a rotation with a period of one stellar day.

The Astrolabe and the Awakening of Astronomy

As far as is known, the astronomers of the early Middle Ages made very few observations and knew of no astronomical instruments in the true sense. An illustration in a tenth century manuscript from St Gallen shows a monk gazing at a star through a long tube, which served only as a means of keeping his sight fixed on a particular star. And as far as instruments in general were concerned, only some simple types of sundials seem to have been in use.

At the turn of the millenium A.D. 1000 a new era of astronomy began, connected at first with Gerbert of Aurillac (c. 930–1003, alias Pope Sylvester II, 999–1003) who had frequented Spanish monasteries and was acquainted with at least some of the astronomical works of the Muslims. It seems that in one sense the new development in Europe began with the introduction of a new and ingenious instrument, the plane astrolabe described by Islamic astronomers in the ninth century and now made known in Latin by a short treatise written by the Benedictine Hermann of Reichenau (1013–1054).

The astrolabe was a small brass instrument shaped as a circular disc (*mater*), which could be held in a vertical position on the thumb by means of a ring. On one side of the disc (the *dorsum*) the limb was divided into 360 degrees and provided with a diopter or *alhidade* which, when adjusted to the rays from the Sun or a star, showed the altitude of the object on the limb (Fig. 18.4). The other side of the instrument,

To knowe the senyth of the altitude of the sonne et

This is no more to don but any tyme of the day tak the altitude of
the sonne ã by the azymut in which he shall stonde... ...son in
which pns of the firmament he is ã the same will minyton
sen by the myle of any stone which: the stere stite est or west or
north or any pns ... ... the name of the azymut in which is
the stere ... for the more declaratoñ is her the figure

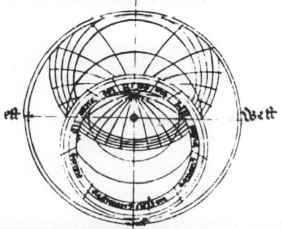

To knowe sothly the degree of the longitude of the mone or of
any planete y hath no latitude. and for the tyme fro the eclip
til first ...

Tak the altitude of the mone ã phis this altitude up among thine
almikindes on which side that the ynis stonde ã see ther a prikke
ã tak thanne anon jibe up on the mones side the altitude of any
stere fix which y thou knowest ã set his certis up on his altitude
among thin almykindes ther the stere his fotindo ã Venus than
no which degree of the zodiac to which the prikke of the altitude of
the mone ã tak ther the degree in which the mone standith ã this
conclusion is veray soth yif the sterres in this astrolobie stonden aft

Fig. 18.4

249

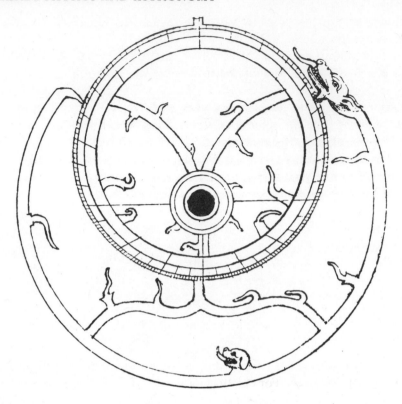

þy zodiak of þin aftrelabye ys ſchapen as a com
pas blyndly þat conteneþ a large brede as after the
fmtite of þin aftrelabye. in enſample þat þe zodiak in
hcuene ys ymagyned to ben a ſurface conteinynge a

Fig. 18.4-18.5. Two drawings from Chaucer's Treatise of the astrolabe *showing the
tympanon and the rete. From manuscripts in the University Library, Cambridge and the British
Museum, reproduced from R. T. Gunther,* Early Science in Cambridge, *Oxford, 1937.*

the *facies,* had the graduated limb raised a little so that the *mater* could
include another circular disc (the *tympanon*) on which a stereographic
projection of the various circles of the heavenly sphere was engraved
according to the geographical latitude at which the instrument was to
be used (Fig. 18.5). Above the *tympanon* was another thin plate or disc
(the *rete*) on which the ecliptic and the positions of a number of bright
stars were engraved in the same projection as before. From this disc

as much material as possible was cut away, so that the circles on the *tympanon* were visible below it.

A typical way of using the astrolabe was to determine the altitude of a star by means of the *alhidade* at the *dorsum*, and then to turn the *rete* at the *facies* around its centre until the point marking the star in question fell upon the altitude circle of the *tympanon* corresponding to the measured value. The hour could then be read off on the limb when the date was known, or vice versa.

In the course of time, the astrolabe was put to many other uses. Its dual character of measuring instrument and computer made it useful for solving many problems of spherical astronomy or geodesy. The enormous popularity of the astrolabe is obvious from the 1,500 or so specimens of the period A.D. 1000–1700 still preserved in museums, and from the numerous surviving treatises which describe its use, of which Chaucer's Middle English *Tretis of the astrolabie* written in A.D. 1391 is among the most detailed and lucid.

The Revival of Theoretical Astronomy

The introduction of the astrolabe sparked off a minor revolution in Mediæval astronomy. For the first time since the days of Ptolemy, it was now possible to make astronomical observations resulting in numerical values for the positions of the Sun and the planets. Consequently, a better knowledge of planetary motions became imperative, and again scientists had to turn their attention to the Arabs. Adelard of Bath (p. 189) translated the astronomical tables of al-Khwārizmī (p. 176) into Latin, together with their *Canones*, or directions for use. At about the same time, in the middle of the twelfth century, the famous *Toledo tables* were translated and adapted to the longitudes of various localities in Western Europe, such as Marseilles, London and Toulouse. These tables gave Mediæval astronomy a secure foundation since it now became possible to compute planetary positions for a given point of time, and to compare the derived values with observations. There was still an insufficient knowledge, however, of the theoretical models on which the tables were founded, and the next step was logically an effort to reconquer the theoretical astronomy of the ancients.

In the field of theoretical astronomy the *Almagest* of Ptolemy was by far the best manual. It had been translated into Latin from Arabic or Greek in the twelfth century, but the rare surviving manuscripts of the *Almagest* show that it was never extensively used as a textbook in Mediæval universities. No doubt it was too long, and initially too difficult from a mathematical point of view. More elementary manuals for students were necessary, and from about the middle of the thirteenth century several such introductory texts began to appear. A beginning was made by Johannes de Sacrobosco (John of Holywood), an English-man who taught astronomy at Paris about 1240 to 1254 and who wrote a *Tractatus de sphaera*, a *Compotus*, and an *Algorismus* which together formed a set of short, clear texts adapted to the needs of the liberal arts student. The first gave the elements of spherical astronomy, the second was an introduction to time reckoning, the third an exposition of the arithmetic used in astronomical computations. The three texts still exist in hundreds of manuscripts, and were immensely popular during the remainder of the Middle Ages and long after the invention of printing.

The *Tractatus de sphaera* or the *Sphaera* (Fig. 18.6) was divided into four 'books'. The first treated the heavenly sphere, its revolution, the spherical shape of the Earth and its immovable position in the middle of a spherical universe. The second book dealt with the celestial equator; the ecliptic; the zodiac and its constellations; the meridian; the altitude of the pole, and the division of the Earth into geographical zones by means of the tropics and the polar circles. Book three described the rising and setting of the heavenly bodies; the length of day and night corresponding to different geographical latitudes and 'climates', that is zones bounded by circles parallel to the equator and defined by the length of the longest day of the year. All this was excellent material, and could be used today as a first introduction to astronomical geo-graphy. The fourth book, however, was less satisfying. It gave a very brief description of the motion of the Sun and the planets, and an outline of the theory of eclipses; but the treatment was obviously much too thin to be of any use at a time when the precise Ptolemaic models had become known to professional astronomers. Thus the *Sphaera* had to be supplemented by a more detailed manual of theoretical astronomy. Several texts of this sort came to light during the second half of the thirteenth century, the most widely used being the anonymous *Theorica planetarum;* this work is conserved in almost as many manu-scripts as Sacrobosco's three textbooks, which it usually follows in the astronomical codices of the later Middle Ages. It has been ascribed both

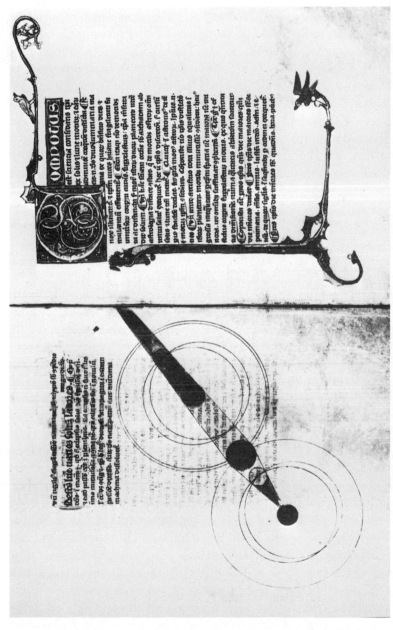

Fig. 18.6. Two pages from a Sacrobosco manuscript showing (left) the end of the Tractatus de sphaera with a figure illustrating the theory of eclipses, and (right) the beginning of the Compotus, that is, his treatise on time reckoning. From a manuscript in the Royal Library, Copenhagen.

to the translator Gerard of Cremone (p. 189) and to the astrologer Gerard of Sabionetta, but seems to be the work of an unknown teacher, who, at some time between 1260 and 1280, tried to remedy the defects of the last book of the *Sphaera*.

The *Theorica planetarum* (Fig. 18.7) begins with an exposition of the Hipparchian theory of the Sun, followed by the Ptolemaic theories of the Moon and the outer planets. Next come chapters on the motion of Venus and Mercury, and the direct and retrograde movements of the planets. The various geometrical models are described briefly, but carefully, and the important definitions are clearly stated. Less successful are the final sections on the motion in latitude, eclipses, the use of astronomical tables, and planetary aspects. The author seems, at this point, to have come to the end of his resources, and later astronomers supplemented (as he had supplemented Sacrobosco) his work with special treatises on these subjects. Thus a copious collection of astronomical standard works gradually grew up which together (they were often bound together in the same codex) gave as complete a general survey of astronomy as the Middle Ages were able to provide.

The standard collection of astronomical texts reveals, too, how the more technical development of astronomy took place. From the start it usually contained a calendar, which to begin with was computed by Robert Grosseteste (p. 193) for a 76-year period ending in 1296, after which it was replaced by a somewhat better calendar due to Petrus Philomena de Dacia (Peter Nightingale) and computed in Paris in A.D. 1292. At about the same time the old *Toledo tables* were discarded and replaced by the new *Alphonsine tables* which were then used for 300 years. The most remarkable feature, however, is without doubt the great interest in new astronomical instruments, of which many are described in later versions of the collection—a testimony to the rapid spread of the art of instrument-making all over Europe. The 'old quadrant' (Fig. 18.8) described in the voluminous Castilian encyclopedia *el Libros del saber*, compiled at Toledo by the astronomers of King Alfonsus X, was known in France as early as the first half of the 13th century, but towards the turn of the century it was supplemented by the 'new quadrant' (Fig. 18.9) invented by the Jewish astronomer Jacob ben Mahir (Profatius Judaeus) at Montpellier, and made known in Latin by Peter Nightingale. It was an ingenious combination of a quadrant and an astrolabe, the circles of the astrolabe being folded upon the quarter circle of the quadrant.

254

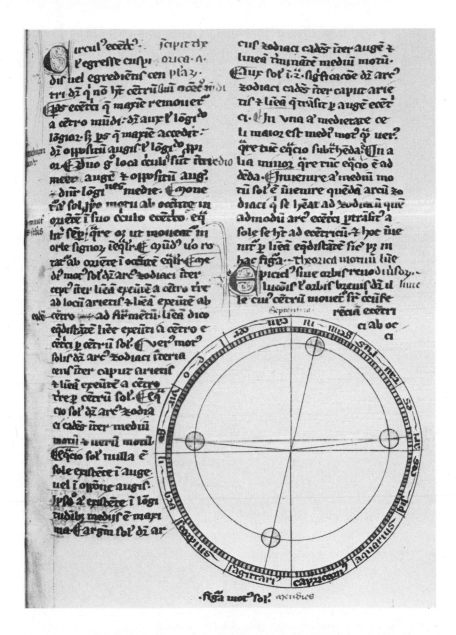

Fig. 18.7. *The first page of the Theorica planetarum with a figure showing the theory of the Sun on Hipparchos's model. From a manuscript in the Royal Library, Copenhagen.*

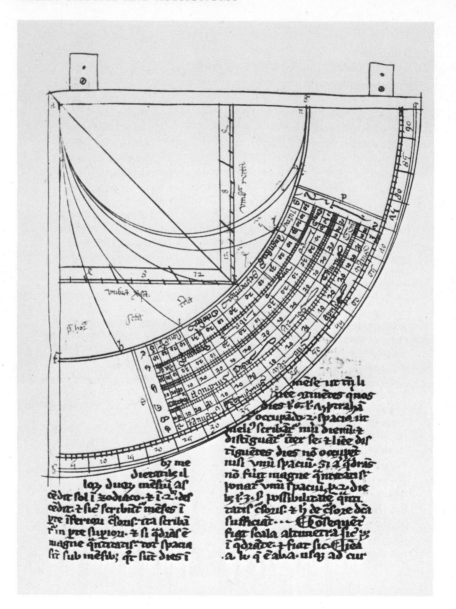

Fig. 18.8.   The 'Old quadrant' which was made known in Europe during the first half of the thirteenth-century. It served primarily for measuring altitudes, but could also be used as a sundial because of the engraved, curved hour lines and the adjustable 'cursor' or scale representing the annual motion of the Sun along the ecliptic. From a manuscript in the Royal Library, Copenhagen.

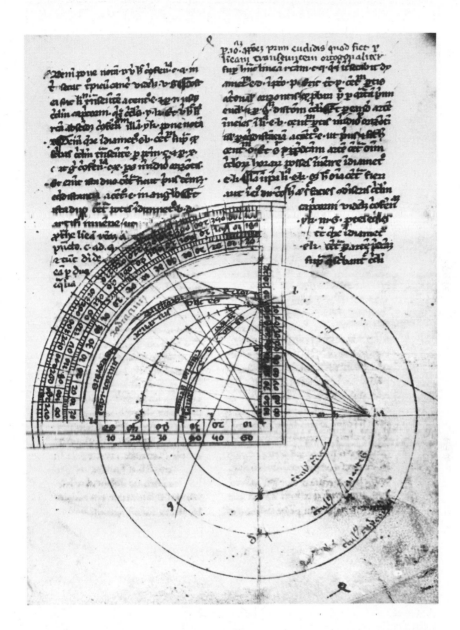

Fig. 18.9. A page from Petrus de Dacia's Tractatus quadrantis *explaining the construction of the 'New quadrant' of Profatius Judaeus. From a manuscript in the Royal Library, Copenhagen.*

257

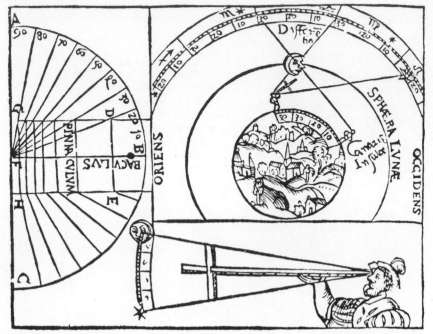

Fig. 18.10. The construction and use of the cross staff. From Petrus Apianus, Cosmographia *Antwerp, 1531.*

During the years of the fourteenth century the Provençal Hebrew astronomer Levi ben Gerson (1288–1344) invented the cross staff, an extremely simple device for measuring angles which became widely used for navigational purposes (Fig. 18.10). A highly sophisticated instrument for measuring and demonstrating angles in all three celestial systems of co-ordinates (the equatorial, ecliptic, and horizontal) was the *torquetum,* or 'Turkish instrument' which together with the armillary sphere became a common aid for teaching spherical astronomy (Fig. 20.1)

Astronomical Computing Instruments

The *equatorium* or *volvella,* of which a great number of forms were developed from about A.D. 1270 onwards, was an analogue computer. With it, the longitude of a planet at a given time could be found mechanically and cumbersome work with astronomical tables and manual computations largely avoided. In this field too, Latin astronomers were preceded by Muslims like al-Zarqālī, whose *azafea* was

258

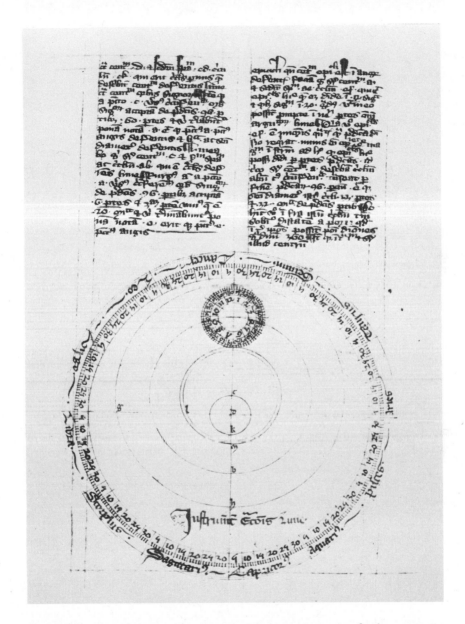

Fig. 18.11.   A page from Campanus de Novara showing an equatorium for the computation of positions of the Moon (see Fig. 7.15.). From a manuscript in the University Library, Cambridge.

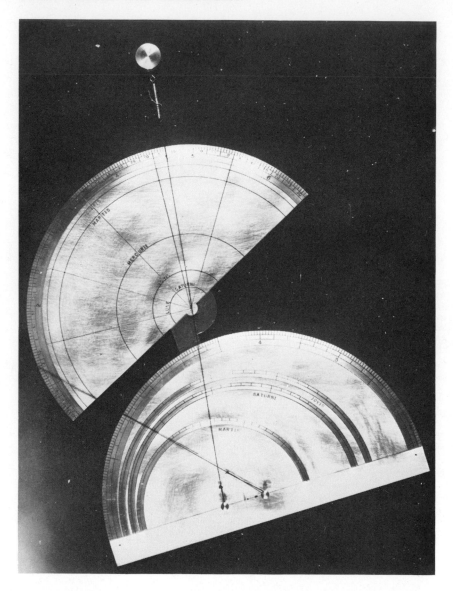

Fig. 18.12. The semissae *of Petrus de Dacia. The lower half circle represents the ecliptic and the equants of the superior planets. The equants of the other planets are engraved on the back. The upper half circle represents the epicycles. The parameters are adjusted so that all the planets have the same deferent represented by a pin on a rotating ruler of which a part is seen beneath the two discs. The photograph shows a modern reconstruction of the instruments. No Mediæval specimen has survived.*

mentioned in Chapter 13 (p. 180). The first instrument of this kind to appear in Latin Europe was the *theorica planetarum* of Campanus of Novara (Fig. 18.11) who used the principle of analogy in a very direct way. The various circles in the geometrical model of the motion of a planet were cut out of parchment or brass and placed correctly relative to each other, a thread from the equant centre to the epicycle centre marking the mean longitude. By means of another thread from the centre of the instrument (representing the Earth) through the planet, the true longitude of the planet could be read off the ecliptic scale. In this way the relation

true longitude = mean motus ± aequatio centri ± aequatio argumenti

could be dealt with mechanically (p. 97) in as much as the difficult corrections *aequatio centri* and *aequatio argumenti* were introduced automatically by the instrument itself, without involving trigonometric computation or tables. All that was needed was a table of the mean longitude of the planet and another of the mean longitude of the Sun.

Campanus had to construct a special instrument for each particular planet, each instrument having three fully-graduated circles. In A.D. 1292 Peter Nightingale invented a modified form (called the *semissae*) in which a single instrument served for all the planets, thus reducing the construction considerably (Fig. 18.12). The same principle was later followed by astronomers such as Johannes de Lineriis in Paris, and by Geoffrey Chaucer. In the fifteenth century, inventors like Johannes Fusoris and others succeeded in developing *equatorea* without threads, and with concentric circles only; a great many types of such *volvellae* are still preserved in museums and libraries. The accuracy of these computing instruments was not great unless they were made on a very big scale. Perhaps they served primarily for teaching purposes, or for rapid computations which could later be carried out more exactly by hand. Their historical importance is that they prepared the way for the idea of mechanical computation and for the invention, in the seventeenth century, of the slide rule.

Development of Observational Astronomy

Observatories in the modern sense were, in the Middle Ages, very rare outside Muslim centres such as Toledo, Cairo, Baghdad, and Samarkand, where large instruments were built and placed in permanent positions. In Latin Europe astronomers made observations from their own houses, usually with much smaller instruments. One of the first records of a systematic series of observations was made by an anonymous astronomer at Roskilde in Denmark; in A.D. 1274 he measured the altitude of the Sun at noon every day in the year by means of an astrolabe. The data were used for a computation (by trigonometrical methods newly taken over from the Arabs) of the length of the day, which was noted daily in a calendar. Most later Mediæval calendars by Peter Nightingale, and William of St Cloud—himself a keen observer—retained this feature. Occasionally, special instruments were constructed for specific purposes, such as the large sector with which, in 1318, Johannes de Muris determined the vernal equinox at Evreux; it had a radius of 15 feet and a graduated arc of 15 degrees.

The Mechanical Clock

The mechanical clock, which appeared in Europe in the last decades of the thirteenth century must be counted as an astronomical instrument and as one of the most important inventions of the Middle Ages. Its highly developed form tells the tale of a long evolution, yet the prehistory of the clock is not yet known in detail, although water clocks, astrolabes with gear mechanisms and mechanical Chinese timepieces seem to be among its ancestors.

One of the first clockmakers in Europe was the Italian Giovanni de' Dondi (1318–1389), whose astronomical clocks were real masterpieces of mechanical skill and astronomical ingenuity. The central part of the mechanism of any clock is the escapement, and in those of the Middle Ages is always of the verge and foliot type. Via a gear mechanism, a heavy weight turns the vertical crown wheel (Fig. 18.14) which is a cylinder with pointed teeth. In front of this wheel is the verge – a vertical axle with two small pallets which engaged the crown wheel at the top and bottom respectively. The axle bears a horizontal lever, or foliot, with two adjustable weights, an arrangement implying that the crown wheel can move one tooth only when, say, the upper verge turns the foliot a given angle to one side. When the tooth is free of the upper verge, the lower one takes over and stops the wheel, which will then move a further step forwards only after the foliot has swung to the opposite side. Verge and foliot thus oscillate to and fro, and the motion becomes slower if the adjustable weights are removed to a greater

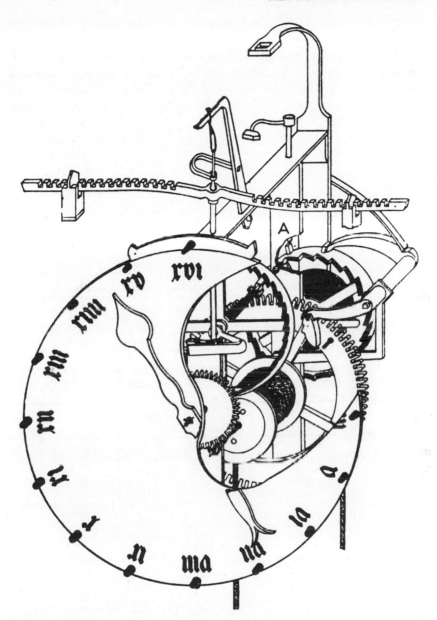

Fig. 18.13. Mediæval clock from c. A.D. *1390 now in the Germanisches Museum, Nuremberg. The verge escapement with the balance rod and crown wheel was used in all Mediæval clocks. This specimen also has an adjustable striking chain which meant it could be used as a monastic alarm clock. From* A History of Technology, *III, p. 652, Oxford, 1954.*

distance from the axle. In this way, the oscillations can be controlled and the pointers turned at the correct speed.

In this kind of astronomical clock the period of oscillation is determined not only by the moment of inertia of the foliot, but by the complete mechanism. The whole clock takes part in time-keeping, and the result is a lack of accuracy and reliability (Fig. 18.14). Mediæval

H. CHRIS NEWTON, A.R.P.S.

THE ANCIENT CLOCK. SALISBURY CATHEDRAL
CIRCA 1386

Fig. 18.14. The ancient clock of Salisbury Cathedral (c 1386) restored to its original condition. The balance rod is seen at the top, the going train to the right, and the striking train to the left.

clocks had to be adjusted at frequent intervals by means of a sundial; this explains why the ancient art of sundial-making now spread rapidly in the wake of the mechanical clock, giving rise to an enormous amount of literature on gnomonics (dialling) between the fourteenth and the eighteenth centuries. The first clocks were used mainly as public or domestic time-keepers, but were of no use for scientific purposes. Reliable astronomical clocks became possible only in the 17th century when Galileo, Huygens, and others replaced the pendulum through which the time-keeping device was separated from the rest of the foliot by a mechanism which served only to provide the pendulum with enough energy to maintain its oscillations, and to couple it with the pointers. This was a great advance in accuracy, and from that point clocks made their way into observatories and laboratories. The verge escapement itself was still used until the eighteenth century when it was replaced by the anchor escapement.

The Attack on Ptolemy

The upsurge of interest in both observational and theoretical astronomy during the second half of the Middle Ages led to more general discussions among scientists, including both physical and philosophical questions of the nature of astronomical theories. Scholars had access once more to the highly developed planetary theory of Ptolemy, which by means of somewhat artificial geometrical models, tried to describe the movement of the planets in a purely mathematical way. Yet the universe as a whole was still described in Aristotelian terms, as a structure of real, 'physical' spheres located in space (p. 100). Even the philosophers of Antiquity had discussed how these two kinds of description were related, and the argument, now resurrected, gained such great force that it divided Mediæval thinkers into two hostile camps, as it had the early Muslim astronomers in the east (p. 182). But it was not until astronomy began to flourish in Moorish Spain that the battle really sprang up in the west.

In his great manual *De astronomia libri IX* (translated by Gerard of Cremona) the Seville astronomer Jābir ibn Afflah (early twelfth century) criticized Ptolemy on many specified points: his trigonometry was too intricate, he introduced the equant point without stating his reasons, many of his numerical parameters were wrong, he placed Venus and

Mercury below the Sun but ought to have placed them above it, and so on. But Jābir's astronomy still followed the main lines of the *Almagest*, which he described as a 'great gift from the Lord'.

A more radical attack on Ptolemy came from the Aristotelian philosophers, headed by Ibn Rushd (1126–1198), also called Averroës and the 'Commentator' by the Latins and father of the very strictly Aristotelian school of Latin 'Averroists'. Ibn Rushd was not diverted by technical details, but tried to prove that in order to be consistent with Aristotelian physics astronomy can allow neither eccentric circles nor epicycles. The basic reason, he said, is that a circular motion cannot be natural, and therefore not perpetual, unless it takes place around a heavy body. The existence of an epicycle or eccentric circle must, therefore, presuppose the existence of heavy bodies outside the centre of the world. But this is impossible, since the latter is the only natural place of heavy bodies.

Concentric Planetary Theories

Averroës seems to have been aware that an attack on Ptolemaic astronomy based on 'physical' premises would make no great impression on astronomers without the proposition of an astronomically acceptable alternative. The only attempt at creating a mathematically well-defined concentric theory was made by the Spanish-Arabic astronomer al-Biṭrūjī of Cordova (*c.* 1100–1185). He was known to the Latins as Alpetragius, from his main work which, in A.D. 1217, was translated by Michael Scot under the title *De motibus coelorum*. The planetary system he proposed has very much in common with Eudoxos's theory (p. 74), the basic principle being (Fig. 18.15) that around the pole N of the celestial equator a point C_1 rotates on a concentric sphere with a constant velocity, completing a revolution in the mean tropical period of the planet; the arc $NC_1 = \varepsilon$ is equal to the obliquity of the ecliptic. Another point C_2 has a distance $C_1 C_2 = \beta_{max}$ equal to the maximum latitude of the planet. It rotates around C_1 with the mean anomalistic velocity of the planet. The latter is at P where the arc $C_2 P = 90°$ rotates with the same velocity in the opposite direction. This model satisfied Aristotelian philosophers in its use of concentric spheres alone; but its mathematical consequences were inferior to

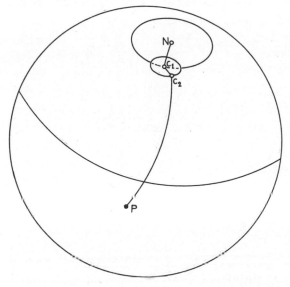

*Fig. 18.15. Figure illus-
trating the concentric planetary theory of al-Biṭrūjī.*

those of Ptolemaic astronomy. In particular, the retrograde motions
could not be adjusted to the individual planet in a satisfactory way.

The later Middle Ages led to a considerable number of treatises
directed against Ptolemaic astronomy, most of them stemming from
al-Biṭrūjī, or directly from Aristotle. They were rarely written by
professional astronomers, but by philosophers to whom concentric,
and material, spheres were essential. A notable exception is the *De
reprobatione ecentricorum et epiciclorum* by Henry of Hassia (or Langenstein, 1325–1397) who was a competent scientist and, like Nicole
Oresme, an eager adversary of astrology.

The Status of Scientific Hypotheses

Other thinkers dealt more carefully with matters concerning the planets,
and pointed out that astronomical theories of a mathematical nature
cannot be regarded simply as 'pictures' of the physical universe. These
theories aimed at a mathematical deduction of celestial phenomena.
Their purpose was to 'save' phenomena, and this might be possible in
more than one way. Thomas Aquinas thus underlined the fact that
neither the Aristotelian spheres, nor the Ptolemaic circles are visible:

267

> According to Ptolemy the heavenly bodies are not fastened to the spheres, but possess a special motion different from that of the spheres [. . .]. But in Aristotle's opinion the stars are fixed into the orbits and do not move except together with the motion of the orbits. However, according to what really is the case, our senses do not perceive the motion of the spheres but only that of the heavenly bodies.

Following this, Aquinas mentions two different types of scientific explanations or 'reasons', the first corresponding to 'physical' theories of an Aristotelian nature, the second to abstract mathematical descriptions like that of Ptolemy:

> The reason for a certain matter can be set forth in two ways. In the first way in order to give a sufficient proof for a certain principle—as when one in natural science introduces a reason which is sufficient to prove that the rotation of the heavens takes place with uniform velocity. In the other way the reason is not introduced because it proves the principle in a sufficient manner, but because it shows that some previously accepted principle agrees with the ensuing effects—as when in astronomy the eccentric circles and the epicycles find their reason in the fact that the observable celestial phenomena can be saved by this assumption. Nevertheless, this method is no sufficient proof because they [that is the phenomena] may perhaps be saved in another way.

Thomas Aquinas's statement implies a clear insight into the hypothetical nature of astronomical theories. They are designed to 'save' phenomena, but are insufficient as descriptions of the physical nature of the universe.

No solution to the problem was found in the Middle Ages, but the discussion was a valuable one because it allowed scientists to become familiar with the distinction between physical and phenomenological descriptions, and taught them to proceed by means of hypotheses contrary to traditional views. Even Copernicus was influenced by the phenomenological point of view, although he ultimately aligned himself with the 'physicists' (p. 315). It was Newtonian mechanics which first fulfilled the dream of deducing planetary motions from the laws of physics. Then the Aristotelians' programme was carried through—but only at the cost of Aristotelian physics itself.

The Diurnal Motion of the Earth

Some of the physical problems which concerned astronomers of the later Middle Ages can now be discussed. First of these is the daily motion of the Earth as a possible explanation of the apparent rotation of the celestial sphere, a hypothesis which was known in Antiquity, but usually discarded because Aristotle had proved the Earth to be immovable from its elementary nature. Nevertheless, the critical philosophers of the fourteenth century gave it serious reconsideration and Nicole Oresme examined it in minute and careful detail in his French *Livre du ciel et du monde* of 1377. Like many other commentators on Aristotle (in this case the *De caelo*), Oresme tried to revise some of the outdated notions of the philosopher. He was fully aware that all celestial phenomena will appear in precisely the same way whether the apparent rotation of the heavens is due to a rotation of the sphere of the fixed stars (and thus of the other heavenly spheres), or to a diurnal rotation of the Earth about its own axis. He also pointed out that it is more simple to assume a rotation of the Earth, and was mainly concerned with the many objections which might be hurled at such a view.

The Earth, said the Aristotelians, cannot rotate because if so a strong wind would, contrary to experience, blow constantly from the East. Oresme explained that the assumption needed is that the atmosphere also takes part in the rotation. Another objection based upon the doctrine of elements tried to prove that the Earth is incapable of natural, circular motion: it consists, the argument ran, mainly of a heavy element with a natural motion downwards towards the centre of the universe; since Aristotle had shown that no element has more than one natural motion, its supposed rotation would be a movement contrary to nature, and therefore not eternal. This was one of the strongest arguments of the Aristotelians, which even Galileo was unable to cope with in a completely satisfactory way. Oresme answered the objection with a very long and ingenious argument, and one of great importance because it implies a conscious break with the doctrine of natural place.

Finally, Oresme asked whether the motion of the Earth must be rejected because the Bible speaks of it as immovable and of the Sun as moving. He answered that the language of the Bible is adapted to the usage of ordinary people and everyday modes of expression from which no astronomical consequences can be drawn, an attitude founded upon the authority of St Augustine and which at the time met with no opposition from the Church, in contrast to what befell Galileo when, 250 years later, he put forward the same argument. After this clear investigation of the problem one cannot help wondering why Oresme finally announced that, although the motion of the Earth is possible, and even probable, he himself did not believe in it. Perhaps he found it too difficult to deny the immediate testimony of the senses.

The Size of the Universe

Philosophers of the late Middle Ages were much occupied with the question of the extent of the universe. The traditional opinion inherited from Antiquity (p. 102) was that the world as such is a sphere of finite size beyond which nothing exists—not even empty space. The first to break with this view seems to have been Thomas Bradwardine (d. 1349), whom Oresme followed in assuming that the outermost sphere of the heavens is bounded, not by nothingness, but by an infinite, empty space. This means that the notion of space had changed since Aristotle to whom space was an entity filled with matter.

Oresme argued: the last sphere rotates about its axis with a period of one (stellar) day. It is possible to imagine that this 'sphere' is not spherical in that it could have an excrescence at some point. As this excrescence must take part in the rotation, we must admit the existence of an outer space which sometimes is filled with matter (the excrescence when passing) but will usually be empty. Now there is no evidence of such an irregularity in the highest heaven, but this objection can be met with the answer that God in His omnipotence could have created the world in this way, although in fact He did not. To deny the possibility of an empty space outside the heavens therefore implies the rejection of the omnipotence of God. Furthermore, God is Himself infinite, and He is everywhere present in the universe. A finite universe therefore seems incompatible with the theological conception of God.

Quasi-theological arguments such as these were often used by Mediæval critics of Aristotle who made use of the freedom and omnipotence of God to show that many of Aristotle's statements were not necessary or binding conclusions. Archbishop Etienne Tempier's condemnation, in A.D. 1277, (see p. 192) of several Aristotelian and other doctrines is the point of departure for this train of thought.

The concept of an infinite universe was too radical to win general acceptance during the Middle Ages. It is mentioned by Nicolaus of Cusa (p. 280) and after him by Giordano Bruno (died 1600), but Galileo still regarded the world as finite in size. Newton was the first person who could make the infinite universe acceptable to scientists in the form of his 'absolute space', which was supposed to exist independently of material bodies—a view exactly opposite to Aristotle's.

The Problem of Uniform, Circular Motion

Some interesting developments in planetary theory took place in the later Middle Ages among Muslim astronomers in the eastern part of Islam. Although the general attack on Ptolemy from philosophical quarters had been tended off, the theory of epicycles could still be attacked, and for reasons more liable to impress professional astronomers. Ibn al-Haitham had already accused Ptolemy of betraying his own principles, and making his system inconsistent—a reference to the obvious fact that Ptolemy violated the axiom that any natural, circular motion should be uniform as seen from its own centre. This was clearly seen (Fig. 7.18) in the motion of the epicycle centre C on the deferent circle around the centre D. For this motion was not uniform seen from D, but only from another point E (the equant centre). A number of astronomers therefore felt the importance of 'saving Ptolemy from himself' by introducing changes to his models such that all circular components of motion became uniform as seen from their own centres.

In this context, it must be remembered that Ptolemy first tried to work out a model in which C moved on an 'equant' circle with uniform velocity around the centre E of the same circle. But, presumably in order to account for the observed variations in the velocity, or the size, of the epicycle, he had to diminish its distance TC from the Earth at the apogee, and augment it at the perigee. This was achieved by choosing the deferent circle as the path of the epicycle centre C. The circularity of the motion of C was thus maintained—but at the cost of its uniformity, since E was retained as the 'centre of uniform motion'.

A way out of this dilemma was proposed by the great Persian scientist Nāṣir al-Dīn al-Tūsī (1201–1274), who made the important discovery that a rectilinear movement can be produced by a combin-

271

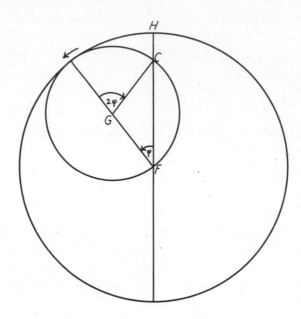

Fig. 18.16. Nāsir al-Dīn's theorem.

ation of two circular motions. He considered (Fig. 18.16) a circle with centre G rolling inside the periphery of another circle with centre F and a radius twice as long as that of the rolling circle. He was then able to prove that any point C on this rolling circle will move upon a straight line FH through F. It follows that the motion of C can be produced by a linkage of two vectors $\overrightarrow{FG}$ and $\overrightarrow{GC}$, of which $\overrightarrow{GC}$ rotates about G with double the angular velocity of $\overrightarrow{FG}$, which rotates about F in the opposite direction.

With a device of this kind it was possible to avoid introducing the deferent. In Fig. 18.17 the circle around D represents the Ptolemaic deferent with radius R and eccentricity $e = TD = DE$, where T and E are the centres of the Earth and the equant respectively. The Ptolemaic equant circle is the locus of the point F which rotates about E so that $EF = R$, and the angle $AEF = c_m$ (the mean centrum) is a linear function of time. The epicycle centre is at C.

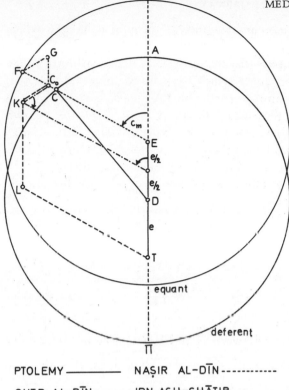

PTOLEMY ——————— NAṢIR AL-DĪN - - - - - - - - - - -

QUṬB AL-DĪN – - – - – - IBN ASH-SHĀṬIR – - – - – -

Fig. 18.17. Determination of the epicycle centre C in Ptolemy, Nāṣir al-Dīn, Quṭb al-Dīn, and Ibn ash-Shāṭir. Re-drawn from E. S. Kennedy, 'Late Mediæval planetary theory', Isis, 57, (1966), 367.

Nāṣir al-Dīn's model can now be characterized by the decomposition of the vector $\overrightarrow{TC}$ from the Earth to the epicycle centre according to the vector relation

$$\overrightarrow{TC} = \overrightarrow{TE} + \overrightarrow{EF} + \overrightarrow{FG} + \overrightarrow{GC},$$

where $\overrightarrow{TE}$ is a constant vector of length $2e$ pointing towards the apogee; $\overrightarrow{EF}$ is a vector of constant length R rotating with uniform angular velocity around E so that the angle AEF (called the mean centrum c_m of the planet) increases as a linear function of time; $\overrightarrow{FG}$ and $\overrightarrow{GC}$ are

273

vectors of equal and constant lengths $e/2$ forming a linkage which makes C oscillate on EF.

This construction implies that the epicycle centre C will move on a closed, but non-circular, curve which has only the points A ($c_m = 0°$) and Π ($c_m = 180°$) in common with the Ptolemaic deferent. Thus it places the epicycle at the correct distances from the Earth at both apogee and perigee. In between it will be displaced only a little relative to the deferent. This has been achieved by a model in which all component circular motions are uniform about their own centres.

A variant of Nāṣir al-Dīn's model was proposed by his disciple Quṭb al-Dīn al-Shīrāzī (1236–1311), who noticed that the vector $\overrightarrow{GC}$ in Fig. 18.17 always remains parallel to TA. This leads immediately to a more economical model (involving one circle less) characterized by the relation

$$\overrightarrow{TC} = \overrightarrow{TH} + \overrightarrow{HK} + \overrightarrow{KC},$$

where K is the mid-point between D and E, $\overrightarrow{HK}$ is equal and parallel to EF, and $\overrightarrow{KC}$ equal and parallel to FG. It follows that the two models are equivalent.

An interesting variation of these models was introduced by the Damascus astronomer Ibn ash-Shāṭir (1304–1376) and is based on the decomposition

$$\overrightarrow{TC} = \overrightarrow{TL} + \overrightarrow{LK} + \overrightarrow{KC}.$$

From the figure it can be seen that this is the exact equivalent of the two former models. But it has the new feature that the vector $\overrightarrow{TL}$ (of length R) is rotating around the Earth. If this model is transformed from the vector representation to the geometrical language of epicycles and deferent (Fig. 18.18) it can be said that Ibn ash-Shāṭir used a concentric deferent of radius R on which the point L performs a direct motion with uniform velocity equal to the mean daily motion. Around L is a first epicycle of radius $3/2\ e$ on which K performs an indirect motion with the same angular velocity, so that the direction LK remains parallel to TH. K is the centre of a second epicycle with a radius of $e/2$. On it is the centre C of the third epicycle carrying the planet.

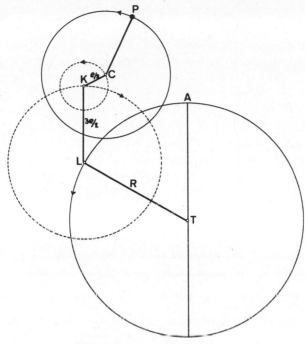

Fig. 18.18. Ibn ash-Shāṭir's theory of the superior planets.

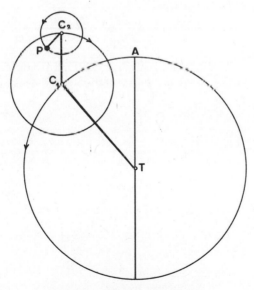

*Fig. 18.19. Ibn ash-Shāṭir's theory
of the motion of the Sun.*

275

These models can be applied to the motions of the three superior planets. It is not surprising that they had to be more complicated in the case of Mercury, or simpler in the the case of the Sun. Ibn ash-Shāṭir's solar theory is shown in Fig. 18.19. It has no less than two epicycles, with centres C_1 and C_2. The vector $\overrightarrow{C_1C_2}$ remains parallel to the apsidal line. The Sun is at P.

Since he himself does not mention it, one can only speculate as to why ash-Shāṭir found it necessary to introduce two epicycles into the theory of the Sun when Apollonius had already shown that one was enough to describe the only anomaly of the Sun's motion. But perhaps he wanted to make a theory as similar as possible to his theory of the Moon.

Ibn ash-Shāṭir's lunar theory is particularly interesting to the subsequent history of astronomy, and is illustrated in Fig. 18.20. Here $\overrightarrow{TS}$ is the vector from the Earth to the mean Sun. The centre of the first lunar epicycle is defined by the vector $\overrightarrow{TC_1}$, forming the uniformly increasing angle ε (the mean elongation of the Moon) with $\overrightarrow{TS}$. The centre of the second epicycle C_2 is determined by the vector $\overrightarrow{C_1C_2}$

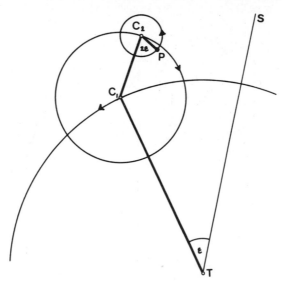

Fig. 18.20. Ibn ash-Shāṭir's theory of the Moon.

rotating about C_1 in the retrograde direction. The Moon is at P, where $\overrightarrow{C_2P}$ forms the angle 2ε with C_1C_2. The parameters are

$$R = TC_1 \ = 60; 0$$
$$r_1 = C_1C_2 = \ 6; 35$$
$$r_2 = C_1C_2 = \ 1; 25$$

One significant fact is revealed by these models, namely that theoretical astronomers in late Mediæval Islam had acquired a very high degree of sophistication and ability to handle the traditional tools (epicycles and deferents) of planetary theory. The freedom to choose between equivalent models must have become very obvious to them and apart from the desire to use only uniform circular motion, the driving force behind this fascinating model-making must have been the longing for mathematical simplicity. But sooner or later the question of whether the models were possible from a physical view-point invariably arose. This was the case with Copernicus who was one of the last of the Mediæval model-makers.

Chapter Nineteen

The Change Called Renaissance

The End of the Middle Ages

IT is a matter of semantics when the Middle Ages ended and the new science began, but fifteenth century science obviously represents a period of transition which is marked in other fields by the Renaissance in art and literature. In the history of science the use of the word Renaissance gives rise to several questions, since it is impossible to make the development of science conform to the traditional scheme of Antiquity as the supremely creative period, the Middle Ages as a dark interlude when nothing happened, and the Renaissance as a re-birth of the outlook of Antiquity. In fact, it is dubious whether such a division is valid in any field. The historian will wish to explain any given period by what happened immediately before it, not by receding ten or twelve centuries, and the study of the history of science as such has proved how difficult it is to apply the traditional term 'Renaissance' to the transition from Mediæval to 'classical' science. This is best understood if we consider—as a simplification—the science of Antiquity to be dominated by three currents connected with the names of Plato, Aristotle, and Archimedes, in order to ask which of these was resuscitated in the fifteenth and sixteenth centuries.

Unfortunately, fifteenth century science has been investigated only minimally, and much less is known of it than of the science of the preceding centuries, although a complete, reliable survey of the physics or astronomy of this period is a most urgent historical task. Until the period has been studied in detail the only acceptable way of dealing with it is to take a few examples from the names which have survived as outstanding representatives of their time.

Nicolas Cusanus

One very conspicuous figure in fifteenth century philosophy was Nicolas Cardinal Cusanus (1401–1464), whose many writings touch upon almost every general idea animating the first half of the century. Cusanus was an eclectic, convinced of a fundamental harmony between Plato and Aristotle, but was also inspired by such different philosophers as the Pre-Socratics, the Pythagoreans, the Greek fathers, the German mystic Eckhart (c. 1260–1329), and the late Mediæval critics of Aristotle. Many of Cusanus's opinions are important to natural philosophy, or to the history of mathematics. In his book *De docta ignorantia* he stresses the necessity of mathematics for all kinds of knowledge, even in the sphere of theology, and several minor works reveal his efforts to come to terms with a number of specific mathematical problems. Yet, Cusanus was convinced that natural science must be founded not only on experience, but on quantitative experiments, which alone can provide numerical data for the mathematical description of nature. Thus his work *Idiota* included a dialogue on statical experiments proposing concrete methods of finding quantitative results by means of simple apparatus like balances and water clocks.

Cusanus knew that any substance has a specific weight which can be determined by one general method. He recommended systematic measurements of the specific weights of metals and alloys in order to determine their composition. What he has in mind is a kind of analytical chemistry similar to that already in use among miners and metallurgists, the central purpose of which would be the unmasking of fraudulent alchemists 'This would serve to expose the sophistic nature of alchemy and show how far the alchemists are removed from truth'. Cusanus was also aware that the specific weight of water is not always the same, and that in blood and urine it is influenced by age or illness. Accordingly, he argued, such measurements would be useful to medicine. It seems that Cusanus was the first person to demonstrate practical methods for finding the strength of a magnet, the humidity of the air, and several other quantities by means of a balance.

It is unlikely that Cusanus found time to conduct more than a few experiments himself, but his methods were occasionally used by scientists of the following centuries. Thus Galileo's way of determining small intervals of time by weighing the water flowing from the water clock was originally put forward by Cusanus, who thought of measuring the time of fall of a body in order to find the resistance of the air. Another of his ideas was to weigh a given amount of earth and let plants grow in it without adding anything but water. The plants were then burned and the ashes weighed together with the earth. This would show how much of the element water had been transformed into the element earth, which was the main component of the solid plant substance. The same experiment was later performed with great accuracy by Van Helmont, and played an important role in the anti-alchemistic argumentation in Robert Boyle's *The sceptical chymist.*

In the field of theoretical mechanics Cusanus shared the ideas of fourteenth century scientists, but combined them with intuitions of his own. He was a supporter of the impetus theory, and applied it to the motion of projectiles in the same way as Albert of Saxony (p. 239). In his book *De ludo globi* (1463), he dealt with the motion of a spherical ball, maintaining the idea of an impetus which gradually wears out; yet he said that a perfectly round sphere would never of itself stop its motion, being unable to influence its own behaviour. This would be the case of a sphere rolling upon a perfectly smooth horizontal plane, or of the heavenly spheres, rotating about an axis through the centre. Cusanus obviously still believed in the impetus of a circular motion, although he was prepared to accept a kind of inertial motion of a sphere upon a plane. But when Cusanus then identified the impetus of the heavenly spheres with the Platonic soul of the universe he was dangerously near to reintroducing the 'intelligences' as causes of motion—precisely the pitfall Buridan and Oresme had tried to avoid.

Cusanus's ideas on the system of the world were most remarkable. He adopted the hypothesis of an infinite universe developed by Bradwardine and Oresme, although he indicated Eckhart as its first exponent. In the *De docta ignorantia* he emphasized that the world is an all-embracing unity in which the all necessarily is in everything. It follows, he said, that the universe has no fixed, immovable centre. The world cannot be said to be infinite, yet it is not finite because it has no boundaries. Thus the Earth is neither the centre of the universe, nor immovable. Cusanus regarded the Earth merely as 'a noble star', its own specific light, heat, and influence making it different from other stars. He also toyed with the idea that observers could be placed upon the Sun, Moon, Mars or other stars, maintaining that any observer would see the universe turning around himself. Only the poles would be different from those seen from the Earth. It was Cusanus who coined the much-repeated phrase 'the whole machine of the universe has, as it were, its centre everywhere, but its circumference nowhere'. Such audacious opinions make Cusanus the first supporter of the cosmological principle that everywhere, roughly speaking, the universe must have very much the same structure.

Leonardo da Vinci

Another prominent figure in Renaissance science was Leonardo da Vinci (1452–1519) who is divorced from the scholastic tradition, but in a different way from Cusanus. Leonardo had no regular education and was a very disorderly and unsystematic thinker. He must have acquired some knowledge of Aristotle, Archimedes, Jordanus Nemorarius, and 14th-century Paris scientists over a long period, although he often misinterpreted their ideas. Leonardo's Latin was bad, and he was unable to make a serious study of the great works of Mediæval or Ancient mathematical or scientific authors. He planned several works, but finished none, and his very copious notebooks are haphazard collections of a wealth of ideas expressed in words or drawings but devoid of any system. As a painter he achieved fame in his life time, although he finished only a dozen pictures, and both his paintings and his manuscripts reveal his extraordinary powers of observation. He rarely approached problems through books and study, but through his own keen perception.

Leonardo's artistic work often led him to scientific studies. The representation of the human body was followed up by anatomical observations of a quality never known before. The problem of perspective was new and essential to Renaissance painting, so he threw himself into the study of both geometrical and physical optics. He made himself acquainted with the structure of the eye, the function of the iris and lens, just as he knew the optical axis and conceived the eye as a kind of *camera obscura*. Stereoscopic vision gave rise to a number of his drawings, as well as the formation of images in mirrors and lenses. He was the first person to draw the caustic curve produced by the reflection of light in a spherical mirror (Fig. 19.1). The mathematically difficult 'problem of Alhazen' (by means of which rays do you see a candle in a spherical mirror?) is attacked and solved mechanically by a linkage.

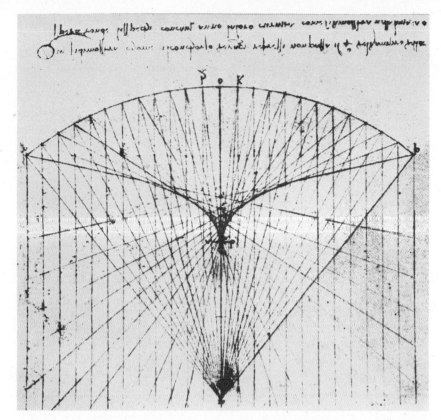

Fig. 19.1. Leonardo da Vinci's construction of the caustic curve produced by reflection of parallel rays of light in a cylindrical mirror. From the Arundel Codex of Leonardo's drawings.

In physics Leonardo's main interest was in mechanics, a subject on which he has bequeathed an amazing collection of notes and figures touching upon almost every mechanical problem. His most original work is embodied in his many drawings of flowing water, based on extremely careful observations of streams, rivers, waterfalls, vortices, and so on. He seems to have realized the principle of continuity of water flowing in a channel of variable cross-section, and the principle of equal circulation in vortex motion (Fig. 19.2). Leonardo also studied waves and understood that a wave transports motion without transporting matter, just as he tried to account for the interference of two waves by the principle of superposition. Ideas such as these were new to physics so here Leonardo opened them into a new field of study. In the

283

Fig. 19.2.   Leonardo da Vinci's drawings of the flight of birds.

more familiar realm of hydrostatics Leonardo stated the law of communicating vessels, and has been credited with the insight that the pressure of water on a surface element is independent of its orientation. He also made propositions for tackling aerodynamic problems by considering the motion of air as analogous to the motion of water.

It is much more difficult to evaluate Leonardo's achievements in other branches of mechanics. The many hundreds of drawings of gear wheels, pulleys and other machine parts plus those of cranes, cannons, and vehicles of various sorts, show his interest in technological problems. Similarly, his sketches of boats with paddle wheels and double boards, of air screws, flying machines (driven by man power), and many other contrivances, have established his reputation as a great inventor, forecaster of the aeroplane, the submarine, and many other technical achievements realized in later eras.

Such absolute praise may, however, be unjustified. What little is known of late Mediæval technology shows that many of Leonardo's mechanical contrivances were nothing more than very careful representations of machinery well known to contemporary engineers sometimes with additions of his own. Nothing indicates that he ever put such imaginary inventions into practice, and in the history of technology it is better to forget Leonardo the Inventor. He is far more important as a faithful and gifted reporter of the achievements of engineers in his own and the preceding centuries.

In the field of theoretical mechanics, Leonardo was interested in the problem of the inclined plane which he tried to solve in a different way from Jordanus (p. 211), by replacing the force and the body by one force perpendicular to the inclined plane, and another parallel to it. However, he did not discover the general law of the parallelogram of forces. Many drawings reveal his interest in ballistics, and show his intuitive grasp of the form of the ballistic curve (Fig. 19.3). Particularly interesting are his speculations on friction, in which he states a law that the friction between a body and a plane, horizontal surface is proportional to the weight of the body, with our modern coefficient of friction equal to $\frac{1}{4}$. He even proposed an experiment to prove it, but never carried it out. Similarly, his many reflections on the strength of materials and the properties of beams and pillars are mere speculations; their value cannot be judged until much more is known about the rules used by Mediæval architects.

Leonardo does not merit the important place in the history of science

Fig. 19.3. Bombardment of a fort. Drawing showing Leonardo's conception of the ballistic curve. From the Windsor Collection of Manuscripts.

awarded him by his most enthusiastic admirers. Apart from the fact that most of his work was hidden and neglected until the end of the nineteenth century, his lack of scientific method and orderly thought robbed his ideas of any true fertility. While he proposed experiments in great numbers, Leonardo left them to be carried out by others who did not understand them. Scientifically, he was simply a man of many interests who spent his life making drawings, revealing himself as one of the keenest observers of nature of all times, and giving testimony to the great variety of problems under consideration towards the end of the fifteenth century.

The Eclectic Attitude

The learned scholar Cusanus and the unsystematic autodidact Leonardo had one feature in common, namely their eclectic attitude. They did not belong to any traditional school, but accumulated ideas at random, subjecting them freely to their own interpretations. This attitude became more and more prevalent towards the end of the fifteenth century. The great philosophical systems were weakened by the penetrating criticism begun in the fourteenth century, and their various elements split up, to reappear in new combinations in the works of individual thinkers. Most of the universities continued teaching along traditional lines, and the established schools of thought lingered on, still expending much of their energy in maintaining their particular philosophical tenets against each other, and against over-radical innovators. Nevertheless, an increasing number of philosophers were imbued with a spirit of independence, a spirit which enabled them to combine new ideas with methods or concepts from the past.

This new and freer attitude towards traditional disciplines does not imply that the Mediæval heritage was discarded, for several of the most fruitful notions of Renaissance physics had their origins in Mediæval science. In his work on statics Cusanus depended on 13th-century results achieved by Jordanus, and their later development. His conception of gravitation was a continuation of ideas first discussed in the fourteenth century, as was his conception of the universe as a whole. Leonardo was greatly influenced by late Mediæval technology and, although in a much more indirect way, by the Parisian critics of Aristotle. Both Cusanus and Leonardo reached a clear understanding that not only is science based on experience, but that regular experiments are necessary. The principle of experience had been stressed by the Aristotelian philosophers of the 13th century, at the time when the experimental method was first practised by students of optics and magnetism. From this aspect, Renaissance science is not so much a fundamental break with the past as a development of its fertile ideas, stimulated by new philosophical insight and a changed social environment.

The Greek Influence

It is not at all easy to evaluate the impact of Renaissance philosophy on science. To many Mediæval scholars Aristotle was *the* philosopher whose ideas formed the general framework of most intellectual discussion, although his absolute authority was not universally upheld. Even to his critics Aristotle was the father of philosophy, and held in an esteem similar to that which a modern physicist accords to Newton. But in the 15th century many philosophers liked to regard themselves as the followers of Plato, or even of Pythagoras, a movement closely connected with the steady influx of Greek scholars from threatened Constantinople, starting in about A.D. 1400 and reaching a climax after the fall of the city in 1453. Among them must be mentioned both the Aristotelian, George of Tredizont (1396–*c*. 1486), and the Platonist Bessarion (*c*. 1395–1472), who settled in Italy, bringing with them numerous Greek manuscripts and reviving interest in the Greek language. The result was a lively study of Greek authors in general, and of Plato in particular. New translations were made, and the very expression 'Middle Ages' was now coined by scholars who earnestly believed themselves to be the direct perpetuators of the philosophy of Antiquity.

This early humanism can be thought of as a new ideal of education opposed to that of the scholastics, stressing the importance of the *trivium* (the 'humanities') and furthering it through the study of Greek while, at the same time, relegating the traditional disciplines of the *quadrivium* to a less prominent position. Scientific problems were discussed with more emphasis on their philosophical application than on a detailed analysis of their mathematical aspects. For students of Plato it was easier to cultivate general ideas than hard, factual details. Sometimes even the wilder speculations of ancient Platonists and neo-Pythagoreans were revived, a trend hinted at in Nicolas Cusanus's adherence to the old doctrine of a world soul. Leonardo had a completely animistic notion of the Earth. He thought it was possessed of a vegetative soul, and therefore as much a living creature as a plant. The skeleton of the Earth was to him the mountains, the soil is its flesh, rivers and fountains its arteries, and the heat of the Earth its soul, views more primitive even than those of the *Timaios*. Such notions may possibly have played a role in dynamical geology, since a living being is subject to change, and thus geological processes could be construed as visible signs of the life of the Earth.

Although, traditionally, the Aristotelians were supporters of the principle of experience the possibility cannot be excluded that the revival of interest in Platonism may have influenced the development of a more experimental philosophy. This apparent paradox was a logical outcome of the Platonic theory that the objects of science are unchangeable, immaterial forms, which are not objects of the senses. Consequently, sensible experience must necessarily be approximate. Aristotle had taught that a measure must be exact, so the inevitable uncertainty connected with all measurements could prejudice good Aristotelians against quantitative experiments. But even in the thirteenth century a reader of William of Moerbeke's translation of Proclos's *Commentary* on the *Timaios* could learn that exact knowledge is a privilege of the Gods, while mere mortals must satisfy themselves with approximations. This, too, Aristotle would have admitted. But as Platonism progressed, philosophers realized more and more that although approximations were unavoidable they were no scandal to science. Cusanus strongly underlined the fact that even if the entire truth cannot be grasped it can be approached more and more closely, just as figures can be drawn representing the ideal circle more and more skilfully, without perfection ever being reached. The uncertainty of measurements is thus, he argued, a natural condition of any kind of experience. The task of the scientist is not to despair of the imperfection of knowledge, but to find better and better methods of approximation.

The Separation of Research and Teaching

Another feature common to Cusanus and Leonardo is that their efforts in science were made outside the universities. It is true that Cusanus had a regular university education, whereas the autodidact Leonardo was a self-taught, had only a slight and superficial knowledge of scientific literature and found his inspiration in technology and art. But neither they nor other scientific pioneers held teaching posts. An historical gulf now began to open between the innovators on the one hand and the old centres of learning on the other. In the thirteenth century, the universities had been able to attract all the best scholars of the time. In the fourteenth century they were the obvious centres for critical discussions and alternative solutions to traditional problems. But in the fifteenth century the intellectual standard of the universities began to decline. Their growing numbers thinned the ranks of scholars inside the individual institutions, national and religious divisions became more and more apparent, and new subject matters were ignored, mainly because the curriculum was usually defined in the university statutes and therefore not easily changed.

The disastrous consequences of this gradual decay appeared during the sixteenth and seventeenth centuries when many of the protagonists of science simply left the universities to seek support from kings and princes, either on a private basis or as members of scientific academies. Characteristically, neither Copernicus nor Tycho Brahe occupied university chairs. The situation reached a climax when, in 1610, Galileo left his chair at Padua to become a court mathematician of the Grand Duke of Tuscany. From the point of view of the universities the trend had catastrophic results. Scientific research was separated from the teaching of science, and the ordinary student no longer became acquainted with the most recent ideas. Yet science itself certainly profited from a new contact with technology which was able to pose new problems and to teach scientists new methods. If the Mediæval universities had founded polytechnic faculties, however, this development might well have been avoided, and the rift between scientific research and humanistic scholarship been less deep and less lasting.

The Invention of Printing

The separation of university teaching and progressive science did not result in dismal consequences for the development of science, which not only got new social connections, but also profited from the invention of printing in the middle of the fifteenth century. Printing made it possible to distribute books much more widely, and also to propose new opinions for discussion independent of established curricula. Problems could be raised and examined by an international forum irrespective of lectures and disputations. It thus became possible for research to be carried out on a private basis, and each scholar had the opportunity of providing himself with a library of much greater variety and scientific value than the minute collection of books possessed by the Mediæval professor. The extent to which the new art of printing served the interests of science is easily seen from the distribution of the first printed books on various subjects. Among the incunabula (books printed before 1500) there were more than 3,000 editions of about 1,000 scientific and technological works by about 650 different authors, distributed as follows (edition numbers are approximate).

| Medicine | 850 editions |
| Technology, alchemy | 300 editions |

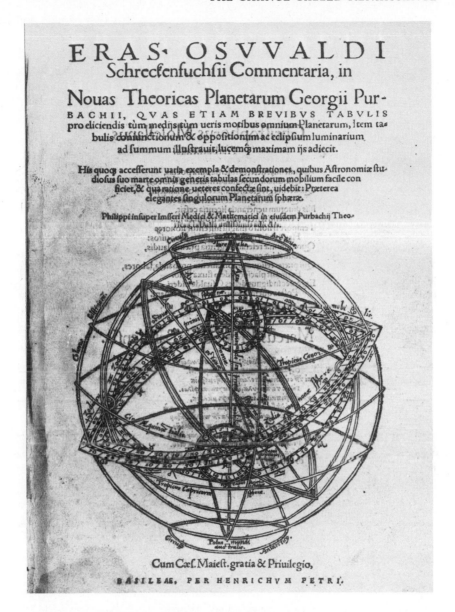

Fig. 19.4. Regiomontanus's misgivings about the art of printing as a means of perpetuating old errors is illustrated by this title page to a commentary on his own Theoricae novae planetarum *showing a complicated armillary sphere destined to reproduce mechanically the trepidation of the equinoxes. From Erasmus Schreckenfuchs,* Commentaria in Nouas Theoricas Planetarum Georgii Purbachii, *Basel, 1556.*

291

Natural philosophy, natural history	400 editions
Astronomy, astrology	200 editions
Mathematics, geometry, physics	100 editions
Miscellaneous	800 editions

Among the first books to be printed were Pliny's *Historia naturalis*, published in Venice in 1469, Strabo's *Geographia*, published in Rome in 1469, and the alchemical (Hermetic) writing *Pimander*, which appeared at Treviso in 1471. Several works appeared in numerous editions: thus the *Sphaera* of Sacrobosco was printed 30 times in Latin, as well as in a Portuguese translation. Aristotle's *Problemata* appeared 28 times, and his collected works 8 times in Latin and once in Greek, compared with the collected works of Plato, which appeared only twice before 1500; so despite the growing interest in Plato among the humanists, Aristotle was still much more in demand. A medical treatise on gynaecology called *Secreta mulierum* by Albertus Magnus could be obtained in no less than 56 printings, before 1500, and there were many later ones. A number of late Mediæval astrological treatises and prognostications enjoyed an immense popularity.

Despite its obvious advantages, printing was not always regarded as an unqualified blessing. The books were usually hastily edited by the printer himself with little critical competence, and as well as propagating new ideas, printing also served, to some extent, to perpetuate old errors. This was the reason why Regiomontanus questioned the value of the invention in a preface to a book in which the old *Theorica planetarum* (Fig. 19.4) was subjected to a devastating criticism, the new printed versions of it being regarded as the most serious hindrance to the reform of astronomy.

Renaissance Science

The preceding sketch of the efforts of Cusanus and Leonardo, and the milieu in which they worked, is by no means exhaustive as an attempt to visualize the scientific change that marked this period. But it is sufficient to show that the expression 'Scientific Renaissance' covers a very complex phenomenon. Many of its important aspects had been evolving for centuries while others were merely features of the fifteenth century. In the 15th century, therefore, the criticism of Aristotle's physical views was continued, but his principle of the primacy of sense experience was retained. Platonism was discussed as a possible alternative to Aristotelian philosophy, but its denigration of concrete phenomena was not accepted. Science broke free from the old university tradition and established fertile contacts with artists, engineers, and architects, conserving the scholastic ethos of science as an international activity. The pioneers worked more as individuals, yet printed literature established a free communication between them. It is impossible to speak of the Renaissance of science as a rebirth of the science of Antiquity. On the one hand, too much had happened to natural philosophy during the Middle Ages; on the other, the new movement still lacked the most decisive influence which ancient science was able to provide, through the works of Archimedes, whose force was not really felt until the following century.

For these reasons, the expression 'Renaissance of Science', is best avoided and the dialectic situation outlined above described simply as Renaissance Science. Even the period which followed it was, in common with this Renaissance, a scientific movement with deep roots in the past, but was increasingly imbued with a wealth of new ideas. This feature unites the semi-Mediæval work of Copernicus, and the typically Renaissance achievements of Tycho Brahe, with those of Galileo, the most influential of the founders of modern physics.

Chapter Twenty

The Reform of Astronomy

Viennese Astronomy

THE new trends in late Mediæval science are nowhere more apparent than in the field of astronomy. Despite the appearance of revolutionary figures like Cusanus, who conceived of an infinite universe and of the Earth as a star among other stars, a long time was still to pass before such radical opinions found general support. In the realm of practical astronomy, however, problems such as the growing need for a reform of the Julian Calendar gave astronomy a new social relevance. Even astrology, with its continuously spreading sphere of influence, contributed to an increased interest in planetary theory, the immediate result being the construction of numerous new collections of tables. Finally, the art of making astronomical instruments was carried to a degree of perfection previously unknown.

From about the middle of the 15th century more conscious efforts at reshaping the traditional astronomy over a wide front can be discerned. The pioneer figure in the field was the Austrian astronomer George Peurbach (1423–1462), who in many ways is typical of this period of transition. He was a protégé of the two cardinals Cusanus and Bessarion, and influenced by their respective Latin and Greek backgrounds, so that he had one foot in the camp of traditional astronomy, the other in that of the humanists whose ideal was the rediscovery of the Greek sources of science. At Bessarion's request, Peurbach undertook a new translation of the *Almagest*, and finished the first six books before his untimely death. In 1460, he published his Vienna lectures under the title *Theoricae novae planetarum*, and in them gave a very clear and reliable exposition of planetary theory with the aim of making the old *Theorica planetarum* of the thirteenth century unacceptable as a university textbook, and of cleaning from astronomy the many aberrations and errors which, since Ptolemy's time, had been perpetrated in the works of both Arabic and Latin astronomers.

Peurbach did not have complete success, for he believed in the peculiar concept of the precession which went back to Thābit Ibn Qurra (p. 183), but which had been given a distinct form by the Alphonsine astronomers of the thirteenth century. According to them, the sphere of the fixed stars performs a uniform rotation in the course of 49,000 years, while at the same time the equinoctial points, in the course of 7,000 years, perform an oscillatory motion or *trepidatio*, to and fro around a mean position, with an amplitude of 9°.

An interesting new detail introduced by Peurbach was his analysis of the motion of Mercury. He demonstrated that as a result of the Ptolemaeic theory, the epicycle centre of Mercury must move on an oval curve which very much resembles an ellipse. Al-Zarqālī (p. 176) knew this, but there is no proof that Peurbach relied on his treatise. Peurbach's main achievement was his computation of many new astronomical tables and ephemerides which were meant to replace the *Alphonsine tables*, whose inexactness had now become obvious. He also performed significant work as an instrument-maker; he constructed an apparatus for mechanical determination of the planetary motions, called the *speculum planetarum*, and a new geodetic measuring instrument, the *quadratum geometricum*.

Regiomontanus and the Nuremberg Institution

Peurbach's efforts, interrupted by his early death, were continued by his pupil Johannes Müller (1436–1476), also known as Regiomontanus, who in 1461 took over Peurbach's Chair in Vienna. The following year he went with Bessarion to Italy where he stayed for a number of years, collecting manuscripts. He continued the translation of the *Almagest* which was not printed until 1496, under the title *Epitome in Almagestum*, indicating that it was not an exact translation, but rather an incomplete paraphrase. Regiomontanus, too, stood by the old astronomy. His importance to science is not in his work on planetary theory or his astronomical tables but in the fact that he settled in Nuremburg in 1471 and established there a complete astronomical

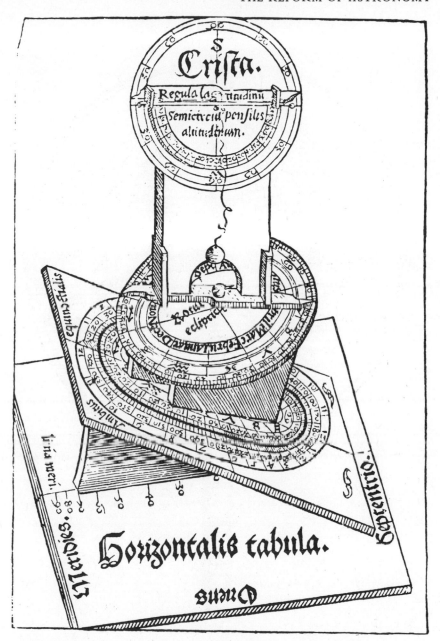

Fig. 20.1A (opposite). Among the astronomical instruments of the later Middle Ages was the torquetum by which horizontal, equatoreal, and ecliptic co-ordinates could be measured. The wood-cut is reproduced from Regiomontanus, Scripta de torqueto, Nuremberg 1544.

297

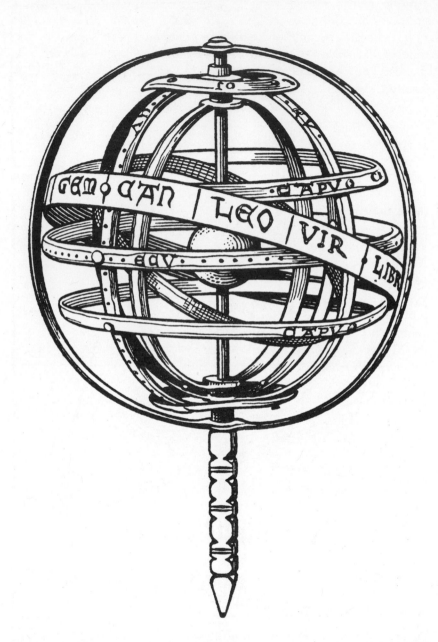

Fig. 20.1B shows a small armillary sphere for teaching purposes. Now in the Whipple Museum, Cambridge. From A History of Technology, *III, 613, Oxford, 1954.*

institute with an observatory, a workshop for instrument making, and a printing office, all paid for by a rich citizen, Bernhard Walter (1436–1504) who also became one of Regiomontanus's pupils. For the first time in Europe a scientific institute planned on generous lines was founded outside a university, an illustration of the first parting between scientific research and old seats of learning.

Nicolaus Copernicus

'The reform of astronomy' is a phrase often found in the sixteenth and seventeenth-century works to express justifiable esteem for the achievements of Peurbach and Regiomontanus. In historical perspective, however, it would be far more correct to use the word 'reform' to describe the work of Nicolaus Copernicus (1473–1543). Although Copernicus had many of his roots in the Middle Ages and Antiquity, it was he who started the development which in the centuries after his death finally changed the world picture. Yet his conception of astronomy was so closely tied to that of Ptolemy that it is appropriate to finish an account of the exact sciences in Antiquity and the Middle Ages with an examination of his achievements.

Copernicus was born in Torun and educated at the University in Cracow, from whence, in 1496, he went to Italy. In 1503 Copernicus finished his Italian visit by taking his doctor's degree in canon law at the University of Ferrara. The rest of his life was chiefly spent in Frauenberg, where he held a canonry at the cathedral. Contrary to popular opinion, he never became a priest, but was a successful jurist, doctor, administrator, and private astronomer, free from any university association. It is not easy to follow the development of his astronomical ideas in detail, but at the very beginning of the sixteenth century Copernicus must have begun to concentrate on his new theories, which spread but slowly through the learned world.

Copernicus's first communication appeared in a minor treatise *Commentariolus* originating before 1530 and circulating only in manuscript copies. In 1536, Cardinal Nicolaus von Schönberg asked Copernicus to publish his ideas in a larger work, but although such a manuscript was ready, Copernicus hesitated. In 1540, however, a *Narratio prima de libris revolutionibus* came out, written by the Lutheran mathematician George Joachim Rheticus (1514–1576) who had visited

299

Copernicus the year before. Now that the ground was prepared, Copernicus finally decided to publish his work. It was printed at Nuremberg in 1543, under the title *De revolutionibus orbium coelestium libri sex.*

In *De revolutionibus*, Copernicus described very briefly the arguments which led him to the new astronomical theory. The most important passage is his dedication to Pope Paul III:

> When for a long time I speculated about the uncertainty of the mathematical traditions it began to prey on my mind that the philosophers—who sometimes examine the smallest details of the universe so closely—had been unable to give a reliable explanation of the movements of the world machine which had been created for our sake by the most perfect and regular of all master builders. Therefore I set to read all the philosophers whose books I could procure, in order to see whether any of them postulated a concept different from that taught at the schools by the professors of mathematics. First, I found in Cicero, then, that Nicetas [that is Hicetas of Syracuse, see p. 62] had believed in the motion of the Earth. Thereupon I found in Plutarch that others had been of the same opinion [. . . whereupon he mentions the Pythagoreans]. This was the occasion which made me also think over the motion of the Earth. As I knew that others before me had felt free to assume arbitrary circular motions for the deduction of coelestial phenomena I thought that I, too, was permitted to search after a deduction of the coelestial phenomena more trustworthy than the one used up till now, by assuming a motion of the Earth, even if this view appeared absurd, too.

Copernicus's dedication has several important features, the first of which is his spontaneous impulse to go back to the Greek sources to find clues for his astronomical ideas. In this, he proves himself a child of Renaissance and humanism: astronomy must be reformed by a return to the original thoughts on the subject.

In the rest of the work, therefore, Copernicus quotes only classical and Arabian authors, besides Jacob ben Mahir, but no Latin astronomers from the period between Martianus Capella and Peurbach, although he was familiar with the main works of Mediæval astronomers. It is not certain, however, whether he knew of Oresme, whose discussion of the diurnal rotation of the Earth might well have formed a basis for Copernicus's arguments in its favour; but as Oresme wrote on the rotation of the Earth in French, there is little probability that Copernicus was directly influenced by him. Despite this, some of Copernicus's ideas are not original in the fine sense. Copernicus was primarily, as the preface to *De revolutionibus* shows, a theoretical astronomer. He was ultimately searching for a model to explain or 'save' observable phenomena in a better way than the traditional one. As far as the phenomena proper were concerned, he usually relied on the ancient astronomers. Copernicus generally accepted Ptolemy's values for the different astronomical constants and parameters, although, for instance, he preferred the sidereal to the tropical year; he was not, unlike

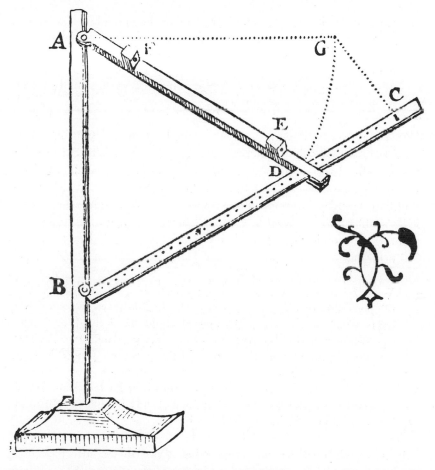

Fig. 20.2. Copernicus's parallactic ruler (see Fig. 7.16.). The instrument later came into the possession of Tycho Brahe, but has since been lost. From the third edition of Copernicus, De revolutionibus orbium coelestium, Amsterdam, 1617.

Tycho Brahe, imbued with a zeal for exact and copious measurings observations but contented himself with adding some of his own to the Ptolemaic list (Fig. 20.2).

In the introduction to the *Commentariolus*, Copernicus's theoretical motives stand out more clearly. He points out that in Antiquity the planetary theories of Eudoxos and Calippos were rejected in favour of Ptolemy's theory which corresponded better with the observed phenomenon. Yet, although Ptolemy's system fits the numerical data, Copernicus maintains that it still presents several difficulties. For

> these theories, that is Ptolemy's theories of the individual planets, were inapplicable unless the existence of certain equant circles, too, was assumed. It then became apparent that a planet did not move uniformly, neither on its deferent nor around its epicycle centre. A system of this kind, consequently, does not appear sufficiently perfect or sufficiently in agreement with common sense.

Copernicus evidently did not object to the Ptolemaic system because of its disagreement with experience. Like his predecessors in Antiquity he ignored this disagreement (see p. 101), and instead raised the objection that in planetary theory Ptolemy did not adhere to the old principle of uniform circular motions alone—the very objection from which Nāṣir al-Dīn and his successors started their investigation of alternative theoretical models. It is clear that Copernicus intended to follow the old principle:

> When I had become aware of these disadvantages I often reflected on the possibility of working out a more reasonable arrangement of circles from which every seeming anomaly could be derived, and where everything would move with uniform velocity around its own centre, thus corresponding with the idea of perfect ['absolute'] motion. After I had approached this very difficult and almost insoluble problem I found out, at last, how it could be solved by means of far fewer and simpler constructions than those hitherto applied, if only I could start from certain conditions which are called axioms.

It is worth noticing, however, that Copernicus also had more rational motives for adhering to uniform circular motions. In referring to the various apparent irregularities of the planetary motions, he states in the *De revolutionibus*:

Nevertheless, these motions must be presumed to be circular or composed of circular motions because it is seen that these irregularities are subjected to a definite law of unvarying recurrence, which would be impossible without the assumption of circular motions. For only the circle can bring back the bygone.

This implies the condition that every arbitrary, periodical, and plane movement can be decomposed into a series of uniform, circular motions. It is mathematically correct, although neither Ptolemy nor Copernicus knew of a formal proof.

The Fundamental Principles

The Copernican system was not the result of any new experience. It was in every respect a theoretical construction based upon three fundamental principles: first, a metaphysical principle of the perfection of circular motion; secondly, a mathematical principle of the possibility of resolving every periodic motion into a series of uniform, circular motions; thirdly an epistemological principle of the greatest possible simplicity of mathematical theories of the description of nature.

Copernicus was in a sense a link in the long chain of theoretical astronomers and constructors of the universe which started with the Pythagoreans. His originality emerges as his three axioms are considered more closely. Here they are reproduced from the *Commentariolus*, where they are found as seven individual statements.

1. There is no common centre of all the heavenly circles or spheres.

This conflicts with the Aristotelian view, according to which all planetary spheres were centred in the Earth centre, and is explained in more detail in the following two axioms:

2. The centre of the Earth is not the centre of the Universe, but only a centre of gravity and the lunar sphere.

3. All spheres rotate about the Sun as their central point, and therefore the Sun is the centre of the Universe.

NICOLAI COPERNICI

net, in quo terram cum orbe lunari tanquam epicyclo contineri diximus. Quinto loco Venus nono menſe reducitur., Sextum deniᶜⱳ locum Mercurius tenet, octuaginta dierum ſpacio circū currens, In medio uero omnium reſidet Sol. Quis enim in hoc

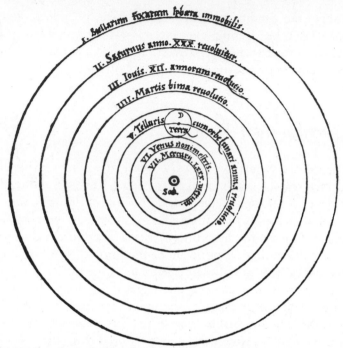

pulcherimo templo lampadem hanc in alio uel meliori loco po neret, quàm unde totum ſimul poſsit illuminare? Siquidem non inepte quidam lucernam mundi, alij mentem, alij rectorem uo‑ cant. Trimegiſtus uiſibilem Deum, Sophoclis Electra intuentē omnia, ita profecto tanquam in ſolio re gali Sol reſidens circum agentem gubernat Aſtrorum familiam. Tellus quoᶜⱳ minime fraudatur lunari miniſterio, ſed ut Ariſtoteles de animalibus ait, maximā Luna cū terra cognationē habet. Concipit interea à Sole terra, & impregnatur annuo partu. Inuenimus igitur ſub hac

Fig. 20.3   *The first printed representation of the Copernican system. From the first edition of* Copernicus, *De revolutionibus orbium coelestium,* Nuremberg, *1543.*

304

Axiom 3 implies particularly that the Earth also moves around the Sun. This implies, in turn, that the fixed stars must necessarily have a yearly parallax, as the line of sight to a star must change as a result of the movement of the Earth. The objection could then be raised that such a parallax had never been observed in the case of the fixed stars, an objection answered in the fourth axiom:

4. The ratio between the distance of the Earth from the Sun, and the height of the firmament [that is the distance of the fixed stars from the Sun] is so much smaller than the ratio between the radius of the Earth and its distance from the Sun, that the distance of the Earth to the Sun is as nothing to the height of the firmament.

This fourth axiom implies that the motion of the Earth will evoke no appreciable change in the line of sight to the fixed stars.

Axioms 2, 3 and 4 lay the groundwork for an explanation of the most fundamental astronomical phenomena. First to be considered is the diurnal revolution of the celestial globe:

5. Every movement observed on the firmament is caused by the motion of the Earth and not by that of the firmament. Together with its surrounding elements, the Earth during a diurnal motion performs a complete revolution round its fixed poles, while the firmament and the highest heavens remain motionless.

The apparent yearly motion of the Sun is thereafter explained away:

6. That which to us appears as movements of the Sun is not due to any motion of the latter, but to a motion of the Earth and our own sphere, during which we rotate about the Sun like any other planet. The Earth has consequently more than one motion.

7. The apparent retrograde and direct motion of the planets is due not to their [proper] motion but to that of the Earth. The motion of the Earth only is therefore sufficient to explain so many diversities in the heavenly phenomena.

The last part of this seventh axiom points out that the new theory is simpler than the old one, which had to introduce a special device for each planet to explain the second anomaly (see page 99).

Finally, concerning the order of the planets or 'spheres', the *Commentariolus* states that 'Highest of all is the immovable sphere of the fixed stars which contains all things and gives them their right place'. After this follow Saturn, Jupiter, Mars, the Earth with the lunar sphere, the latter 'revolving round the Earth and keeping up with it as an epicycle', Venus, and lastly Mercury next to the Sun. The fundamental features of the system had now been laid out, and Copernicus's essential task was then to elaborate the details and to choose numerical values for the parameters in such a way that the observed phenomena could be deduced from the model. In the *Commentariolus* this section of the theory is mentioned only briefly, but is worked out with great care and still greater simplicity in *De revolutionibus*.

The clear exposition of the foundations on which the Copernican system were laid, states precisely where the break with the traditional conception occurs, and which of the old elements were to be retained. The most important of the new features was the heliocentric character of the system: the centre of the universe is not the Earth, but the Sun. The Earth is one of several planets with motions of their own. The break with traditional and with popular views was founded on a purely astronomical basis: the planetary theories became more simple. Later on it became clear that Copernicus had other, more philosophical, or mystical reasons for his theory. For instance:

> But in the centre of the Universe is the seat of the Sun. For could one give this light another and better place in this magnificent temple [that is the Universe] than that from whence he at the same time can illuminate everything equally? Not unreasonably is he by some people called the light of the world, by others its soul or master. Trismegistos calls him the visible God, and Sophocles's Electra calls him the all-seeing. Thus the Sun rules, as if he were sitting on a royal throne, the whole family of stars revolving about him.

In this poetic reflection on nature Copernicus followed an exceedingly old tradition which still flourished, for example, in Keplers time.

In his treatment of the actual structure of the universe Copernicus was obviously affected by earlier conceptions of it. He adopts the heavenly spheres of the old astronomy which according to Aristotle, carry the planets with them during their movements (Fig. 20.4). It is impossible to decide, however, whether Copernicus conceived of the

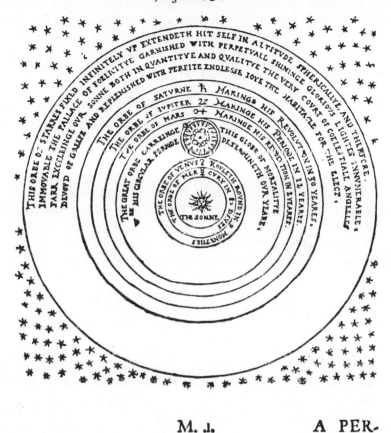

Fig. 20.4.   The Copernican system embedded in an infinite universe of fixed stars. From Thomas Digges, A Perfit Description of the Caelestiall Orbes, *London, 1576.*

spheres as material bodies, but obviously the notion of planets moving in purely mathematical orbits in an otherwise empty space was foreign to him. Only Kepler took this decisive step on the path towards the modern world view. Next, it seems from his axioms that in keeping with tradition Copernicus regarded the universe as both bounded and spherical. The fixed stars are, for him, still embedded in a sphere, and axiom 5 hints at a still higher sphere or *primum mobile.* The finite aspect

307

is also implied in Copernicus's description of the Sun as placed at the centre of the universe, for an infinite universe would have no centre. In this respect Nicolaus Cusanus, to whom the infinitely extended universe had its 'centre everywhere and its surface nowhere' was far more radical than Copernicus.

The Annual Motion of the Earth

De revolutionibus consists of six books, of which the first deals with the spherical shape of the Earth and the universe, the motion of the Earth, and a number of lemmas from plane and spherical trigonometry. *Book II* is on spherical astronomy. It is obvious that when writing his work Copernicus had the *Almagest* very much in mind. In *Book III* of the *Almagest*, Ptolemy had given an account of Hipparchos's theory of the Sun's motion. In a similar way Copernicus treated the motions of the Earth in his third Book. Figure 20.5 shows the mechanism in schematic form. The position of the Earth T relative to the immovable Sun S can be broken down according to the relation.

$$\overrightarrow{ST} = \overrightarrow{SC_1} + \overrightarrow{C_1C_2} + \overrightarrow{C_2T}.$$

Here $\overrightarrow{C_2T}$ is a vector with the arbitrary length $R = 60$ units, revolving

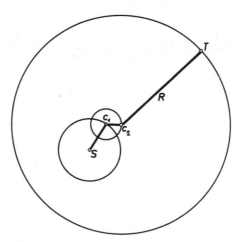

Fig. 20.5.   *The motion of the Earth* T *around the Sun* S *on Copernicus's model.*

directly with uniform angular velocity around C_2 and completing a revolution in one year. It explains the annual revolution.

$\overrightarrow{C_1 C_2}$ is a small vector with the length $r_2 = 0;16, 55, 12$ units, performing a uniform, retrograde revolution in 3,434 years.

$\overrightarrow{SC_1}$ is a vector with the length $r_1 = 2;12, 28, 48$ units. It performs an extremely slow revolution with a period of 53,000 years.

If a brief span of years is considered the centres C_1 and C_2 can be regarded as fixed. The Earth will then move on a circle around C_2, with the eccentricity $e = SC_2$. Thus the model is geometrically equivalent to Hipparchos's theory. But because of the slow motions of C_1 and C_2 this eccentricity will have a secular change between the values $(r_1 + r_2)$ and $(r_1 - r_2)$; the same motions are responsible for the precession of the equinoxes and the movement of the aphelium of the Earth's orbit.

The Theory of the Moon

Copernicus's lunar theory, put forward in *Book IV* of *De revolutionibus* has, of course, to be geocentric. It is illustrated in Fig. 20.6 which shows a deferent of radius R, and a primary and a secondary epicycle of radii r_1 and r_2 respectively, of which r_2 is carrying the Moon P. The vector $\overrightarrow{TC_1}$ rotates towards the east with a uniform angular velocity ω equal to the mean sidereal motion. $\overrightarrow{C_1 C_2}$ has a uniform, retrograde motion with the angular velocity $\omega = 13°;3,53,56$ per day equal to the Ptolemaic mean anomalistic motion. Finally $\overrightarrow{C_1 C_2}$ rotates twice around C_1 during each lunation.

The three radii of the model have the numerical values

$$R = TC_1 = 60;0$$
$$r_1 = C_1 C_2 = 6;35$$
$$r_2 = C_2 P = 1;25$$

The values are extremely interesting. A comparison with the lunar theory of Ibn ash-Shāṭir (p. 277) shows that not only is the general structure of the two models the same, but the geometrical parameters are also identical. It would be very tempting to explain this agreement

309

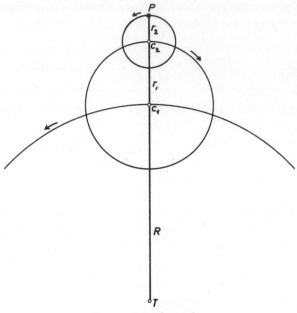

*Fig. 20.6. The circles
of the Copernican lunar
theory.*

on the assumption of direct influence. If this is true, the line of transmission is unknown, but it is not impossible that an account of ash-
Shātir's work might have reached the west, though there is no other
evidence to support such an assumption. The hypothesis is not even
necessary, for Copernicus—like ash-Shātir,—could well have calculated his parameters in such a way that the model gave the maximum
and minimum equations known from the *Almagest*, which in fact it
does. A final solution to the problem is a matter for further research
into the history of astronomy in the 15th century.

   While retaining the Ptolemaic equations, Copernicus improved the
lunar theory on one particular point. The parameters quoted above
imply a maximum Earth-Moon distance of

$$R + (r_1 + r_2) = 68;0,$$

and a minimum distance of

$$R - (r_1 + r_2) = 52;0,$$

which means that the apparent size of the Moon varies during a revolution as 3:4. This is still too large, but a great improvement on the ratio of 1:2 implied in the *Almagest*.

The Planetary Theories

Finally, the theories of the planets can be considered. In Fig. 20.7.

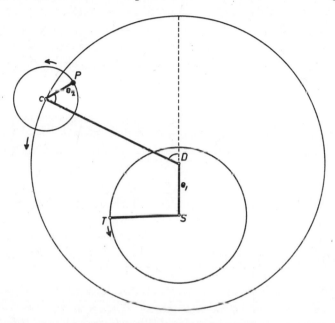

Fig. 20.7. *The Copernican theory of the superior planets.*

S is the Sun, around which the Earth T is moving along a path which, for the sake of simplicity, is taken as circular. A fixed point D on the apsidal line, distant e_1 from S, is the centre of an eccentric circle with radius $DS = R$. A small epicycle with radius $CP = e_2$, carrying the planet P rides on this circle. The vector $\overrightarrow{DC}$ rotates around D with a uniform direct motion and the sidereal period of the planet. CP has a direct motion on the epicycle with the same period. When C is on the apsidal line the vectors $\overrightarrow{SD}$, $\overrightarrow{DC}$ and $\overrightarrow{CP}$ are on a straight line, with $\overrightarrow{CP}$ pointing towards S. In general, we have the relation

$$\overrightarrow{TP} = \overrightarrow{TS} + \overrightarrow{SD} + \overrightarrow{DC} + \overrightarrow{CP}.$$

It is obvious that the common link between the various planetary models is that the orbit of the Earth is the same in all of them. This means that the annual component of the motion of the planet is explained in a more rational and economic way by Copernicus than by Ptolemy who, quite arbitrarily, had to postulate this component in each one of his models (in the superior planets as the motion on the epicycle, in the inferior planets as the motion on the deferent); this is one of the real advantages of the Copernican theory.

Another advantage of the theory is that Copernicus settles the old question of the order of the planets in an objective way. Ptolemy had determined the ratio $q = r/R$ of the radius of the epicycle to the radius of the deferent; being unable to find r and R separately he had chosen $R = 60$ units in all models and determined the radii of the epicycles from the ratios $q_1 = r_1/R$, $q_2 = r_2/R$, ... etc. Since Copernicus puts $r_1 = r_2 = ... r =$ the radius of the terrestrial orbit,

$$q_1 = \frac{r}{R_1} , q_2 = \frac{r}{R_2} \ldots \text{etc.}$$

so that the order of R_1, R_2 ... etc. is determined.

Copernicus and Ptolemy

The Copernican models are sometimes said to be simpler than Ptolemy's because they contain no equant circles. This is true, but the small epicycles appear instead, whose significance is not obvious. To investigate their implication, the Copernican model in Fig. 20.7 can be transformed into an equivalent, geocentric model with the Earth T at rest. This leads to the arrangement in Fig. 20.8 where the position vector

$$\overrightarrow{TP} = \overrightarrow{TD'} + \overrightarrow{D'C'} + \overrightarrow{C'C''} + \overrightarrow{C''P}.$$

If
$$\overrightarrow{TD'} = \overrightarrow{TD}$$

$$\overrightarrow{D'C'} = \overrightarrow{DC}$$

$$\overrightarrow{C'C''} = \overrightarrow{CP}$$

$$\overrightarrow{C''P} = \overrightarrow{TS},$$

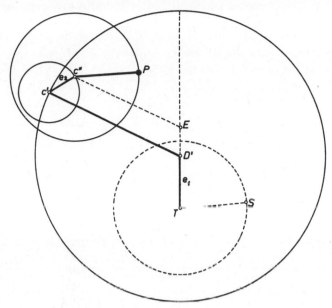

Fig. 20.8. The Copernican theory of the superior planets in a geocentric arrangement.

the two models are seen to be equivalent. The path of the Earth, it will be noticed, now reappears as a secondary epicycle with the radius *r* and a centre O'' revolving on the primary epicycle.

This transformation suggests that the Copernican models were not quite equivalent to the models of the *Almagest*. The reason is that the path of the centre C'' of the planetary epicycle is no circle, because C'' has a direct motion upon the small primary epicycle; had the motion been indirect, the path would have been circular, as known from Apollonios. The reason for the introduction of the small epicycle is not quite clear. It can be better understood if we ask for the conditions that C'' moves on a closed curve which approximates to the Ptolemaic deferent as nearly as possible.

A first condition of this is that C'' revolves with the same constant angular velocity as $\overrightarrow{D'C}$ around an equant point *E* on the apsidal line of the transformed model. This is the case if $\overrightarrow{C''E}$ is parallel to $\overrightarrow{D'C'}$, that is, if $D'EC''C'$ is a trapezium which—because the angle $ED'C'$ equals the

313

angle $D'C'C''$—is equilateral with $D'E = e_2$. This gives the equation

$$TE = 2e = e_1 + e_2,$$

where e is the eccentricity of the Ptolemaic deferent.

A second condition is that the apogee of the deferent of the transformed model coincides with that of the Ptolemaic. This gives the equation

$$e_1 + R - e_2 = R - e,$$

or

$$e_1 = e + e_2.$$

The two equations are solved by

$$e_1 = \tfrac{3}{2}e \quad \text{and} \quad e_2 = \tfrac{1}{2}e.$$

But these are precisely the values given by Copernicus to the eccentricity $SD = TD' = e_1$ of his model, and to the radius $CP = C'C'' = e_2$ of the small epicycle. It is worth noticing that here too Copernicus is in complete agreement with Ibn ash-Shāṭir (see p. 274), apart from the fact that Ibn ash-Shāṭir keeps to the geocentric model. This means that although Copernicus had no equant circles or equant points in his models, one must conclude that he did, to some extent, maintain the equant mechanism. The only difference is that while Ptolemy halved the eccentricity, that is, placed the deferent centre D in the middle of the line TE, Copernicus distributed the total eccentricity of the equant $2e$ differently, giving $\tfrac{3}{2}e$ to his deferent and concealing the rest of it in the small epicycle. Seen from this viewpoint the equivalence between the Copernican and the Ptolemaic models becomes patently clear.

Copernicus must therefore be considered as the last of the 'old' astronomers in that he simply took over all the mathematical apparatus of Ptolemaic astronomy and merely rearranged its elements in another way. His work resembles that of the late Arabic astronomers Nāṣir al-Dīn, Quṭb al-Dīn, and Ibn ash-Shāṭir, who tried to stress the principle of uniform, circular motion more consistently than Ptolemy had cared to do.

Physical Implications of the Theory

This view does not overlook the fact that Copernicus produced something new and original. The idea of transferring the centre of the universe from the Earth to the Sun had been put forward many times, but since Aristotle's time, had usually been refuted. Even a scientist like Oresme could not convince himself of the motion of the Earth, although he was unable to disprove it. Thus it was an intellectual *tour de force* when Copernicus not only proposed and defended a heliocentric theory but also proved that it was mathematically consistent and possible, and did so in a great work whose scientific standard every serious astronomer and mathematician was forced to respect. Copernicus is therefore the first of the 'new' astronomers. He did what neither his Arabic or Latin predecessors had done, and it was now impossible to discard the heliocentric universe and the motion of the Earth as a mere fantasy.

Copernicus had proved that his system was a possible hypothesis with no implied contradictions, and that it would describe observed facts with the same precision as the Ptolemaic theories, and with some real simplifications. The author of the anonymous preface to the first edition of *De revolutionibus* had stressed the purely hypothetical character of its content, adding that it was unnecessary to regard such hypotheses as true; it was sufficient that they agreed with observations. Did Copernicus share this view, or did he regard his system as more than a mere hypothesis? Was it just a play with eccentrics and epicycles, without any relation to reality? Or was it a description of the actual structure of the physical universe?

It was fairly obvious that the unsigned preface was not written by Copernicus himself and later Kepler was able to show that it came from the pen of Osiander, who had seen the work through the press at Nuremberg. Copernicus's own thoughts are not explicitly stated in *De revolutionibus*, but there is some circumstantial evidence that he did regard his astronomical models as a true picture of the solar system. If that were not the case it is difficult to understand why, in his own dedication to Pope Paul III, he should use the phrase 'even if this opinion (of the motion of the Earth) seemed absurd (*et quamvis absurda opinio videbatur*)'. A mere hypothesis would not become absurd if it offended against an invisible cosmological structure, but only if it proved unable to account for observable fact.

Copernicus's own support for the 'physical' interpretation of his system was obvious to the readers of his *Commentariolus*, where he said:

> Let nobody believe that I have maintained the motion of the Earth from arbitrary reasons like the Pythagoreans: a strong proof is found in my account of the circles.

An account of the 'circles' as such can establish no more than the mathematical possibility of the theory. What Copernicus presumably means is that in his account he had also given physical arguments in support of his general idea. This is true and the fact that he entered on a discussion of the physical implications of his theory shows that he was unable to satisfy himself with a mere mathematical hypothesis.

Aristotle and Ptolemy had proposed a number of 'physical' reasons for the siting of the Earth at the centre of the universe in an absolutely immovable position. Some of these reasons are dealt with in Book I of *De revolutionibus*, but in a curiously condensed and unsystematic way. One suspects that Copernicus had no great admiration for traditional physics, a suspicion upheld by his remark that he leaves the problem of the infinity or finiteness of the universe to the physicists.

A single example is enough to show how careless Copernicus was in trying to prevent objections to his theory. A classic argument against the daily rotation of the Earth is found in Book I of the *Almagest*, where Ptolemy maintains that if the Earth were rotating everything on its surface would be flung away in the opposite direction to the motion. Copernicus goes even further, saying that the terrestrial globe itself would explode, and wondering why Ptolemy does not fear that the heavens would do the same, being so much larger and subject to much stronger forces during the rotation allotted to them by Ptolemy himself.

On the basis of the traditional conceptions of physics and cosmology, Copernicus's argument is not very convincing. According to both Aristotle and Ptolemy the rotation of the heavens is a 'natural' form of motion, characteristic of the fifth element, or ether, of which the heavenly bodies are formed. The rotation, as 'natural' to heavenly bodies, would not make them disintegrate. Copernicus could even be faced with the counter-argument that the only 'natural motion' of the Earth is towards the centre of the universe, and that as a result a daily rotation would be 'unnatural' and 'violent' and therefore impossible.

Either Copernicus was ignorant of Aristotelian physics, or cared for it so little that he did not trouble himself with putting forward his reasons in such a way that obvious objections were anticipated. It is very unlikely that Copernicus did not know of Aristotle's physics, so what were his own physical opinions?

In *De revolutionibus* Copernicus states unambiguously that if the Earth rotates around its own axis this must be a natural motion. In other words, he acknowledges the Aristotelian distinction between 'natural' and 'violent' motion. But he does this in a way which no Aristotelian philosopher would consider legitimate, and in fact gives some of the main concepts of Aristotelian physics a new interpretation. To Copernicus the natural motion of heavy or light bodies is downwards, or upwards, only if such bodies are outside their natural place. At this place itself, their natural motion is circular—a postulate quite unacceptable to traditional physics and cosmology. Copernicus's adoption of it is a clear break with established principles, and excludes any real dialogue with the Aristotelian opponents of the heliocentric astronomy. In return it sets Copernicus free to venture upon a new conception of the relation between physics and astronomy, of which the crucial point is obviously the problem of gravitation.

Copernicus on Gravitation

Once the Earth is conceived of as a planet moving in an orbit far away from the centre of the universe, the Aristotelian doctrine of gravity becomes impossible. But heavy bodies still tend to move towards the centre of the Earth. This is an observable fact. The corollary is that they do not tend to move towards a centre of the universe which is far away and in an ever-changing direction.

Copernicus was bound, therefore, to give another explanation of the phenomena on which any theory of gravitation is founded. He found his explanation in the late Mediaeval idea of gravitation as an *inclinatio ad similes* inherent in all bodies. A necessary result of this is that not only the Earth but all the other planets act as centres of gravitation. Copernicus states very explicitly that

> gravity is nothing but a natural urge which the Creator of all things in His Divine Providence has implanted in the individual parts in order that they may unite to attain unity and wholeness by adopting a spherical form. It must be assumed that this property is found even in the Sun, the Moon, and the other planets in such a way that their observed, unchangeable, spherical form is assured, at the same time as they, nevertheless, perform their several revolutions.

317

The idea of any celestial body being an individual centre of gravity was put forward by Nicolas Cusanus (p. 287). It is not known whether Copernicus was directly influenced by him, or if he drew up his own conclusions independently from premises well known to many natural philosophers in the late Middle Ages, although adopted, in their full implications, by only a few before, well into the eighteenth century, Newton's general theory of attraction was universally accepted.

Copernicus took great pains in developing the implications of his system for the theory of gravitation, which shows, that he not only realized the necessity of a break with traditional views, but also that he regarded his astronomical theory as important to the views of the physical structure of the universe.

After Copernicus

The publication of *De revolutionibus* in 1543 was a landmark in the history of science, and in the two centuries that followed much of the debate in astronomy and physics was centred on it. To understand the fate of the Copernican system two important facts must be remembered. The first is that it was possible to ignore Copernicus's more personal opinions on the physical status of his theory. It could be conceived as a purely mathematical structure from which planetary positions at a given point of time could be calculated completely regardless of what is at the centre of the world; this holds true because all calculated co-ordinates must, ultimately, be transformed to a geocentric system of reference in order to be comparable with observations. This was possible in the Copernican system, too. Professional astronomers were satisfied with the mathematical structure of the Copernican theories. This is seen from the new *Prussian tables* calculated by Erasmus Reinhold (1511–1553) and published in 1551. They were based on Copernican methods and parameters and destined to replace the now obsolete *Alfonsine tables*. The fact that Copernicus used the same mathematical technique as Ptolemy made such applications of the new system easy to competent calculators.

Confronted with astronomers of such a mathematical persuasion the Copernican system had to stand or fall on its practical merits, in particular its degree of simplicity. It was soon discovered that in most cases the heliocentric theory led to the same numerical results as the Ptolemaic models, and that the new system was simpler, at least in some respects. Thus the total number of uniform, circular motions was 55 in the *Almagest*, but only 34 in the *De revolutionibus*. Very probably, a book containing nothing but mathematical astronomy would have made all practising astronomers familiar with heliocentric views without provoking any antagonism.

Secondly, Copernicus deliberately chose to present his system as a physical theory. He realized the impossibility of concealing the fact that it was something more than a mere kinematical description. The motion of the Earth was 'unnatural' in the Aristotelian sense and could not be maintained without a serious break with prevailing physical opinions. This break was started with the Copernican theory of gravitation, but in a sketchy way which left the main structure of Aristotelian physics unaffected. His postulate that all planets are their own centres of gravitation was unable to convince those who searched in vain for physical proofs of the theory. Galileo's lifelong efforts were mainly an attempt to provide such proofs but he did not succeed, and his personal fate reveals how serious the situation had become less than a hundred years after Copernicus's death.

Copernicus's decision to stress the physical implications of his theory immensely increased the number of his potential adversaries. Not only could mathematically competent astronomers take part in the discussion, but the armies of professors of philosophy in all the universities of Europe could rally under the colours. Unable to grasp the technical side of the Copernican or Ptolemaic systems, they rushed to the defence of Aristotle, prepared to fight the battle in the physical field alone. Copernican astronomy stands therefore not only as one of the starting points for the reform of astronomy, despite its old-fashioned apparel of eccentrics and epicycles but also as one of the most powerful incitements to the introduction of a new mechanics.

Bibliography

I General Works

1. THE STUDY OF THE HISTORY OF SCIENCE

Enriques, F. *Signification de l'histoire de la pensée scientifique*, Paris, 1934
Kuhn, T. S. *The structure of scientific revolutions*, Chicago, 1962.
Sarton, G. *The study of the history of mathematics and the study of the history of science*, New York, 1957.
Stuewer, R. H. (ed.) *Historical and philosophical perspectives of science*, Minneapolis, 1970. (Minnesota Studies in the Philosophy of Science, Vol. 5.)

2. BIBLIOGRAPHICAL WORKS

Annual critical bibliography of the history of science, published in each volume of *Isis*.
A biographical dictionary of scientists, ed. T. I. Williams, London, 1969.
Chambers's dictionary of scientists, ed. A. V. Howard, London, 1961.
Dictionary of scientific biography, ed. C. C. Gillispie, vol. 1 ff, New York, 1970 ff.
The early history of science. A short handlist, London 1950.
Ferguson, E. S., *Bibliography of the history of technology*, Cambridge, Mass., 1968.

Histoire des sciences et des techniques (Section 522 of the Bulletin Signa-
letique) Paris, 1946 *ff.*

Houzeau, J. C. et Lancaster, A. *Bibliographie générale de l'astronomie
jusqu'en 1880.* New edition by D. W. Dewhirst, 1–2 London, 1964.

Mayerhöfer, J. *Lexicon der Geschichte der Naturwissenschaften*, 1 *ff*, Wien
1959 *ff.*

Olschki, L. *Die Literatur der Technik und der angewandten Wissenschaften
vom Mittelalter bis zur Renaissance*, Leipzig, 1919.

Poggendorff, J. *Biographisch-literarisches Handwörterbuch zur Geschichte
der exacten Wissenschaften*, 1–2, Leipzig 1863, reprint Amsterdam,
1965.

Russo, F. *Histoire des sciences et des techniques*—Bibliographie, Paris,
1954, new ed. Paris, 1969 (best short bibliography).

Sarton, G. *Introduction to the history of science* vols. 1–3, Baltimore—
Washington 1927–47 (a standard work for the period until A.D.
1400).

Thorndike, L. and Kibre, P. *A catalogue of incipits of mediæval scientific
writings in latin*, revised ed., London 1963 (Mediæval Academy
of America Publ. No 29).

3. PERIODICALS

Abhandlungen zur Geschichte der Mathematischen Wissenschaften, Leipzig
1877–1913.

Ambix. Journal of the society for the study of alchemy and early chemistry,
London, 1937, *ff.*

Archeion. Archivio di storia della scienza, Roma, 1919–1943.

Archive for history of exact sciences, Berlin, 1960, *ff.*

Archiv für die Geschichte der Naturwissenschaften und der Technik, Leipzig
1908–31.

Archives internationales d'histoire des sciences, Paris, 1947, *ff.*

Bullettino di Bibliografia e di Storia delle Scienze Matematiche e Fisiche,
Roma 1868–87, reprint, New York, 1964.

*Centaurus. International magazine of the history of mathematics, science, and
technology*, Copenhagen, 1950, *ff.*

Chymia. Annual studies in the history of chemistry. Philadelphia, 1948, *ff.*

Isis. Revue internationale d'histoire des sciences, Bruxelles, later Cambridge,
Mass., 1913, *ff.*

Janus. Revue internationale d'histoire des sciences, Amsterdam, later Leiden, 1896, *ff.*

Lychnos (annual), Uppsala, later Bruges, 1936, *ff.*

Osiris (annual) Cambridge, Mass., 1936, *ff.*

Quellen und Studien z. Gesch. d. Mathematik, Astronomie und Physik, Berlin, 1931–42.

Revue d'histoire des sciences, Paris, 1947, *ff.*

Zeitschrift für Mathematik und Physik, historisch-literarische Abteilung. Leipzig 1856–1900.

II   *Works on Various Disciplines and Periods*

1.   MISCELLANEOUS WORKS

Clagett, Marshall (ed.) *Critical problems in the history of science*, Madison, Wisconsin, 1959.

Dampier, W. C. *A history of science and its relations with philosophy and religion*, Cambridge, 1929.

Dannemann, F. *Die Naturwissenschaften in ihrer Entwicklung und in ihrem Zusammenhange dargestellt*, vols. 1–2, Leipzig, 1920–21.

Darmstaedter, L. *Handbuch zur Geschichte der Naturwisschenschaften und der Technik*, Berlin, 1908 (chronological tables).

Daumas, M. (ed.) *Histoire de la science* (Encyclopédie de la Pleïade). Paris 1957.

Dijksterhuis, E. J. *Die Mechanisierung des Weltbildes*. Berlin, 1956. (English transl., *The Mechanisation of the World Picture*, Oxford 1961).

Duhem, P. *Essai sur la notion de théorie physique de Platon à Galilée*, Paris, 1908.

Forbes, R. J. and Dijksterhuis, E. J. *A history of science and technology*, vol. 1, Ancient times to the seventeenth century. vol. 2, The eighteenth and nineteenth centuries, London, 1963.

Hall, A. Rupert and Hall, Marie Boas, *A brief history of science*, New York, 1964.

Hull, L. W. H. *History and philosophy of science*, London, 1959.

Mieli, A. *Panorama general de historia de la ciencia*, vols. 1–12, Buenos Aires, 1945–61.

Russo, F. *Histoire de la pensée scientifique*, Paris, 1951.

Singer, Ch. *A short history of scientific ideas to 1900*, Oxford, 1959.

Taton, R. (ed.) *Histoire générale des sciences*, tôme 1, *La science antique et médiévale*. Paris, 1957, 2 ed. 1966.

Taylor, F. Sherwood, *An illustrated history of science*. London, 1955.

Taylor, F. Sherwood, *A short history of science*, London, 1939.

Taylor, F. Sherwood, *Science past and present*, London, 1949.

Thorndike, L. A. *History of magic and experimental science*, vols. 1–8, New York, 1923–58.

Whewell, William, *History of the inductive sciences*, 3 vols. London 1837–47, reprint London, 1967.

2. PRE-GREEK SCIENCE AND ANTIQUITY IN GENERAL

Frankfort, H. A., Wilson, J. A. and Jacobsen, T. *Before philosophy*. London 1954, (Egyptian and Babylonian mythology; picture of the world).

Garland, H. and Bannister, C. O. *Ancient Egyptian metallurgy*, London, 1927.

Kugler, F. X. *Die babylonische Mondrechnung*, Freiburg, 1900.

Kugler, F. X. *Sternkunde und Sterndienst in Babel*, vols. 1–2, Münster 1907–1924.

Levey, M. *Chemistry and chemical technology in ancient Mesopotamia*, Amsterdam, 1959.

Neugebauer, O. *Astronomical cuneiform texts*, Newhaven, 1945.

Neugebauer, O. *The exact sciences in antiquity*, Copenhagen, 1951 (mostly on astronomy and mathematics), Rev. ed. New York, 1962.

Neugebauer, O. and Parker, Richard A. *Egyptian astronomical texts*, vols. 1–3, London, 1960–69.

Parker, R. A. *The calendars of ancient Egypt*, Chicago, 1950.

Rey, A. *La science dans l'antiquité*, vols. 1–5, Paris, 1930–48.

Rutten, M. *La science des chaldéens*, Paris, 1960 (a brief introduction).

Schiaparelli, G. *Astronomy in the Old Testament*, Oxford, 1905.

Thompson, R. C. *A dictionary of Assyrian chemistry and geology*, Oxford, 1936.

Waerden, B. L. van der, *Science awakening*, Groningen, 1954, New York, 1963 (mostly on mathematics).

Waerden, B. L. van der, *Anfänge der Astronomie*, Groningen, 1965.

3. THE GREEKS

Baccou, R. *Histoire de la science grecque de Thalés à Socrate*, Paris, 1951.

Brunet, P. et Mieli, A. *Histoire des sciences, Antiquité*, Paris, 1935.

Burnet, J. *Early Greek philosophy*. London, 1892, 3rd ed. 1920 (on the pre-Socratics with texts in English translation).

Cohen, M. R. and Drabkin, I. E. *A source book in Greek science*, New York, 1948.

Diels, H. *Fragmente der Vorsokratiker*, vols. 1–3. Berlin, 1934–37, (a collection of Greek texts).

Farrington, B. *Greek Science*, vols. 1–2, London, 1944–49.

Farrington, B. *Science in antiquity*, London, 1936, 1950.

Freeman, Kathleen, *The Pre-Socratic Philosophers*, Oxford, 1946, 1966.

Guthrie, W. K. C. *A history of Greek philosophy*, vol. 1, *The earlier presocratics and the Pythagoreans*, New York/Cambridge, 1962.

Guthrie, W. K. C. *A history of Greek philosophy*, vol. 2, *The presocratic tradition from Parmenides to Democritus*, Cambridge, 1965.

Heiberg, J. L. *Geschichte der Mathematik und Naturwissenschaften im Altertum*, München, 1925, reprint 1960.

Hoppe, Edmund, *Mathematik und Astronomie im klassischen Altertum* (Neudruck der Ausgabe von 1911). Wiesbaden, 1966.

Lippman, E. A. *Musical thought in ancient Greece*, New York/London, 1964.

Loria, G. *Histoire des sciences mathématiques dans l'antiquité hellénique*, Paris, 1929.

Mittelstrass, J. *Die Rettung der Phänomene*, Berlin, 1962.

Mugler, Charles, *Les origines de la science grecque chez Homère. L'homme et l'univers physique*, Paris, 1963.

Pauly, August *Real-Encyklopädie der klassischen Altertumswissenschaft*, ed. G. Wissowa, Stuttgart, 1893, *ff.*

Reymond, A. *Histoire des sciences exactes et naturelles dans l'antiquité gréco-romaine*, Paris, 1924, English translation, New York, 1963.

Robin, L. *La pensée grecque et les origines de l'esprit scientifique*, Paris, 1923, reprint 1963 (with bibliography).

Santillana, G. de, *The origins of scientific thought from Anaximander to Proclus*, 600 B.C. to A.D. 300, New York/Chicago, 1961.

Sarton, G. *A history of science*, vol. 1, *Ancient science through the golden age of Greece*, Cambridge, Mass., 1952, vol. 2, *Hellenistic science and culture in the last three centuries B.C.*, Cambridge/Mass., 1959.

Stahl, W. H. *Roman science, origins, development, and influence to the later middle ages*, Madison, 1962.

Tannery, P. *Pour l'histoire de la science hellène*, Paris, 1887.

Theiler, W. *Zur Geschichte der teleologischen Naturbetrachung bis auf Aristoteles*, 2. Aufl., Zürich u. Lpz. 1925, new ed. Berlin, 1965.

West, M. L. *Early Greek philosophy and the orient*, Oxford, 1971.

West, M. L. *Three presocratic cosmologies*, Class. Quart., 57 (1963), 154–76.

4. THE ARABS AND THE HEBREWS

Adnan, A. *La science chez les turcs ottomans*, Paris, 1939.

Arnold, T. Guillaume, A. *The legacy of Islam*, Oxford, 1931.

Carmody, F. J. *Arabic astronomical and astrological sciences in Latin translation*, Berkeley, 1956.

Carra de Vaux, Baron Bernard, *Les penseurs de l'Islam*, vol. 2, *Les géographes, les sciences mathématiques et naturelles*, Paris, 1921, *ff.*

Encyclopedia of Islam, Leiden 1908–38.

Ferrand, G. *Introduction à l'astronomie nautique arabe*, Paris, 1928.

Gandz, S. *Studies in Hebrew astronomy and mathematics*, New York, 1970.

Kennedy, E. S. 'Late mediæval planetary theory', *Isis*, 57 (1966) 365–78.

Mieli, A. *La science arabe et son rôle dans l'évolution scientifique mondiale*, Leiden, 1939, reprint 1966.

Nasr, S. H. *An introduction to Islamic cosmological doctrines. Conceptions of nature and methods used for its study by the Ikhwān al-Ṣafā, al-Bīrūnī, and ibn Sīnā*, Cambridge, Mass., 1964.

O'Leary, D. L. *How Greek science passed to the Arabs*, London, 1949.

O'Leary, D. L. *Arabic thought and its place in history*, London, 1958.

Pines, S. *Beiträge zur islamischen Atomenlehre*, Berlin, 1936.

Rosenthal, F. *Das Fortleben der Antike im mittelalterlichen Islam*, Zürich/Stuttgart, 1965.

Ruska, J. *Arabische Alchemisten*, vols. 1–2. Heidelberg, 1924.

Schmalzi, P. *Zur Geschichte des Quadranten bei den Arabern*, München, 1929.

Schoy, K 'Die Gnomonik der Araber', Berlin-Leipzig 1923 (in Bassermann-Jordan, *Gesch. d. Zeitmessung u.d. Uhren*, Band 1, Heft F).

Sayili, A. *The observatory in Islam and its place in the general history of the observatory*, Ankara, 1960.

Sedillot, L. *Traité des instruments astronomiques des arabes*, Paris, 1841.

Steinschneider, M. *Mathematik bei den Juden*, Berlin-Leipzig 1893–99; reprinted Hildesheim, 1964.

Suter, H. *Die Mathematiker und Astronomen der Araber und ihre Werke*, Leipzig, 1900; Suppl. 1902.

Wiedemann, E. *Aufsätze zur arabischen Wissenschaftsgeschichte*, vols. 1–2, Hildesheim, 1970.

Winter, H. J. J. *Eastern science*, London 1952 (a brief introduction, including Chinese and Indian science).

5. THE MIDDLE AGES

Birkenmajer, A. *Études d'histoire des sciences et de la philosophie du moyen âge*, Warszawa 1970 (Studia Copernicana vol. 1).

Butterfield, H. *The origins of modern science*, 1300–1800, London, 1949.

Clagett, M. 'Some novel trends in the science of the fourteenth century'. p. 275–303, in C. S. Singleton (ed.), *Art, science, and history in the renaissance*, Baltimore, 1967.

Crombie, A. C. *Augustine to Galileo. The history of science*, A.D. 400–1650, London, 1952.

Crombie, A. C. *Mediæval and early modern science*, vols. 1–2, New York, 1959 (new edition of the work above).

Crombie, A. C. *Robert Grosseteste and the origins of experimental science 1100–1700*, Oxford, 1953 (a very comprehensive bibliography).

Crump, C. C. and Jacob, E. F. *The legacy of the middle ages*, Oxford, 1951.

Duhem, P. *Études sur Léonard de Vinci*, vols. 1–3, Paris, 1906–13.

Gunther, R. T. *Early science in Oxford*, vols. 1–14, Oxford, 1921–45.

Gunther, R. T. *Early science in Cambridge*, Oxford, 1937.

Haskins, C. H. *Studies in the history of mediæval science*, Cambridge, Mass., 1924 (especially good on the 11th and 12th centuries).

Klebs, A. C. 'Incunabula scientifica et medica'. Short title list. *Osiris* 4 (1938), 1–359 (printed scientific literature from the 15th century), separate reprint Hildesheim, 1963.

Koch, J. (ed.) *Artes Liberales von der antiken Bildung zur Wissenschaft des Mittelalters*, Leiden, 1959.

Maier, A. *Die Vorläufer Galileis im XIV. Jahrhundert*, Rome, 1949.

Maier, A. *Zwei Grundprobleme der scholastischen Naturphilosophie*, Rome, 1951.

327

Maier, A. *An der Grenze von Scholastik und Naturwissenschaft*, Rome, 1952.

Maier, A. *Ausgehendes Mittelalter. Gesammelte Aufsätze zur Geistesgeschichte des 14. Jahrhunderts*, 2 vols., Rome, 1964–67.

Randall, J. H. *The school of Padua and the emergence of modern science*, Padova, 1931.

Rashdall, H. *The universities of Europe in the middle ages*, vols. 1–3, New ed., Oxford, 1936.

Thorndike, L. *Science and thought in the fifteenth century*, New York, 1929.

Weisheipl, J. A. 'Classification of the sciences in mediæval thought', *Mediaeval studies*, 27 (1965), 54–90.

Wise, J. E. *The nature of the liberal arts*, Milwaukee, 1947.

6. THE RENAISSANCE

Boas, M. *The scientific renaissance, 1450–1630*, London, 1962.

Kristeller, P. O. *Eight philosophers of the Italian renaissance*, Stanford, 1964 (Petrarch, Valla, Ficino, Pico, Pomponazzi, Telesio, Patrizi, and Bruno).

Kristeller, P. O. *Renaissance philosophy and the mediæval tradition* (Wimmer Lecture XV), Latrobe, Penn. 1966.

Lacroix, P. *Science and literature in the middle ages and the renaissance*, New York, 1964.

Sarton, G. *The appreciation of ancient and mediæval science during the Renaissance 1450–1600*, Philadelphia, 1955.

III *Individual Disciplines*

1. MATHEMATICS

Aaboe, A. *Episodes from the early history of mathematics*, New York, 1964

Cantor, M. *Vorlesungen über die Geschichte der Mathematik*, vol. 1, Leipzig 1892.

Hofmann, J. E. *The history of mathematics*. Transl. from the German by Frank Gaynor and Henrietta O. Midonick, New York, 1957.

Neugebauer, O. *Vorgriechische Mathematik*, Berlin, 1934, reprinted 1969.

Schramm, M. 'Steps towards the idea of function: A comparison between eastern and western science of the middle ages'. *Hist. Sci.* 4, (1965), 70–103.

Yuschkewitsch, A. P. *Geschichte der Mathematik im Mittelalter*, Leipzig, 1964.

2. ASTRONOMY

Abetti, G. *The history of astronomy*, London, 1954.

Balss, H. *Antike Astronomie*, München, 1949.

Bauer, G. K. *Sternkunde und Sterndeutung der Deutschen im 9.–14. Jahrhundert* (Germanische Studien 186). Nendeln/Liechtenstein 1967 (reprint of the Berlin, 1937 ed.).

Bigourdan, G. *L'astronomie, évolution des idées et des méthodes*, Paris, 1911.

Boll, F. *Sphaera. Neue griechische Texte und Untersuchungen zur Geschichte der Sternbilder*, Leipzig, 1903, reprint Hildesheim, 1967.

Coleman, J. A. *Early theories of the universe*, New York, 1967.

Couderc, P. *Les étapes de l'astronomie*, Paris, 1945.

Dicks, D. R. *Early Greek astronomy*, London, 1970.

Doig, P. *A concise history of astronomy*, London, 1950.

Dreyer, J. L. E. *A history of astronomy from Thales to Kepler*, New York, 1953.

Duhem, P. *Le système du monde*, vols. 1–10, Paris, 1913–59.

Goedon, P. *L'image du monde dans l'antiquité*, Paris, 1949.

Koyré, A. *From the closed world to the infinite universe*, Baltimore, 1957.

Neugebauer, O. 'The Survival of Babylonian methods in the exact sciences of antiquity and middle ages'. *Proc. Amer. Phil. Soc.* 107 (1963), 528–35.

Orr, M. A. *Dante and the early astronomers*, London, 1913 and 1956.

Pannekoek, A. *A history of astronomy*, London, 1961.

Sticker, B. *Bau und Bildung des Weltalls. Kosmologische Vorstellungen in Dokumenten aus zwei Jahrtausenden*, Freiburg, 1967.

Thiele, G. *Antike Himmelsbilder*, Berlin, 1898.

Toulmin, S. and Goodfield, J. *The fabric of the heavens. The development of astronomy and dynamics*, London, 1961.

Tuckerman, B. *Planetary, Lunar, and Solar Positions, A.D. 2 to A.D. 1649 at Five-day and Ten-day Intervals*, Philadelphia (Memoirs, vol. 59) American Philosophical Society, 1964.

Waerden, B. L. van der, 'Die Astronomie der Pythagoräer. *Verh. nederl. Akad. d. Wetensch.*, 20, 1951.

Zinner, E. *Geschichte der Sternkunde*, Berlin, 1931.

Zinner, E. *Astronomie. Geschichte ihrer Probleme*, München, 1951.

Zinner, E. *Entstehung und Ausbreitung der coppernicanischen Lehre*, Erlangen, 1943.

Zinner, E. *Verzeichnis der astronomischen Handschriften des deutschen Kulturgebietes*, München, 1925.

Zinner, E. *Geschichte und Bibliographie der astronomischen Literatur in Deutschland zur Zeit der Renaissance*, 2. ed. Stuttgart, 1964.

3. ASTRONOMICAL INSTRUMENTS

Cittert, P. H. van, *Astrolabes*, Leiden, 1954.

Destombes, M. 'La diffusion des instruments scientifiques du Haut Moyen Âge au XVe siècle'. *Cah. Hist. mond.* 10 (1966), 2–51.

Drecker, J. *Theorie der Sonnenuhren*, Berlin, 1925.

Edwardes, E. L. *Old weight-driven chamber clocks, 1350–1850*, vol. 1: *Weight-driven chamber clocks of the middle ages and renaissance*, Altrincham, 1965.

Gunther, R. T. *The astrolabes of the world*, vols. 1–2, Oxford, 1932.

Hartner, W. *The principle and use of the astrolabe. Oriens-Occidens*, Hildesheim, 1968, 287–318.

Kiely, E. R. *Surveying instruments, their history and classroom use*, New York, 1947.

Lloyd, A. *Some outstanding clocks over seven hundred years 1250–1950*, London, 1958 (particularly on Mediæval clock works).

Maddison, Fr. 'Early astronomical and mathematical instruments. A brief survey of sources and modern studies'. *Hist. Sci.* 2 (1963), 17–50.

Michel, H. *Scientific instruments in art and history*. Transl. R. E. W. Maddison and Francis R. Maddison, New York/London, 1967.

Michel, H. *Traité de l'astrolabe*, Paris, 1947.

Price, D. J. *The equatorie of the planetis*, Cambridge, 1955.

Repsold, J. A. *Zur Geschichte der astronomischen Messwerkzeuge* vols. 1–2, Leipzig, 1908–14.

Schmidt, F. *Geschichte der geodätischen Instrumente und Verfahren im Altertum und Mittelalter*, Neustadt, 1935.

Stevenson, E. L. *Terrestrial and celestial globes*, vols. 1–2, New Haven, 1921.

Zinner, E. *Deutsche und niederländische astronomische Instrumente des 11–18 Jahrhunderts*, München, 1956.

Zinner, E. *Alte Sonnenuhren an europäischen Gebäuden*, Wiesbaden, 1964.

4. TIME RECKONING

Bassermann-Jordan, E. *von, Die Geschichte der Zeitrechnung und der Uhren*, Berlin, 1920–25.

Drecker, J. *Zeitmessung und Sterndeutung in geschichtlicher Darstellung*, Berlin, 1925.

Ginzel, F. K. *Handbuch der mathematischen und technischen Chronologie. Das Zeitrechnungswesen der Völker*, vols. 1–3, Leipzig, 1906–14, reprint 1958.

Krauss, J. *Vom Messen der Zeit im Wandel der Zeiten*, Wolfshagen, 1950.

Kubitschek, W. *Grundriss der antiken Zeitrechnung*, München, 1928.

Wijk, W. E. van, *Origine et développement de la computistique médiévale*, Paris, 1954 (Conférences du Palais de la Découverte No. 29).

5. CARTOGRAPHY

Brown, A. *The story of maps.* New York, 1949 (important bibliography.)

Destombes (ed.) *Mappemondes A.D. 1200–1500.* Catalogue préparé par la Commission des Cartes Anciennes de l'Union Géographique Internationale. Amsterdam, 1964.

Durand, D. B. *The Vienna—Klosterneuburg map corpus of the fifteenth century: A study in the transition from mediæval to modern science*, Leiden, 1952.

Kretschmer, K. *Die italienischen Portolane des Mittelalters*, Berlin, 1909.

6. ASTROLOGY

Bauer, K. *Sternkunde und Sterndeutung der Deutschen im 9–14 Jahrhundert unter Ausschluss der reinen Fachwissenschaft*, Berlin, 1937.

Boll, F. *Sternglaube und Sterndeutung. Die Geschichte und das Wesen der Astrologie*, Berlin, 1918.

Bouche-Leclercq, A. *L'astrologie grecque*, Paris, 1899.

Cumont, F. *Astrology and religion among the Greeks and Romans*, New York, 1912.

Gundel, W. and Gundel, H. G. *Astrologumena. Die astrologische Literatur in der Antike und ihre Geschichte* (Sudhoffs Archiv. Beiheft 6). Wiesbaden, 1966.

Klibansky, R., Panofsky, E. and Saxl, Fr. *Saturn and melancholy. Studies in the history of natural philosophy, religion and art.* London, 1964.

Kroll, J. *Die Lehren des Hermes Trismegistos*, Münster, 1914.

Wedel, T. O. *The mediæval attitude toward astrology particularly in England*, New Haven, 1920.

7. PHYSICS

Boyer, C. B. *The rainbow. From myth to mathematics*, New York, 1959.

Daujat, J. *Origines et formation de la théorie des phénomènes électriques et magnétiques. 1. Antiquité et moyen-âge*, Paris, 1945.

Drachmann, A. G. *Ktesibios, Philon and Heron. A study in ancient pneumatics*, Copenhagen, 1948.

Gilbert, O. *Die meteorologischen Theorien des griechischen Altertums*, Leipzig, 1907.

Heller, Àgost, *Geschichte der Physik von Aristoteles bis auf die Neueste Zeit.* 1, *Von Aristoteles bis Galilei;* 2, *Von Descartes bis Robert Mayer*, Wiesbaden, 1965.

Hoppe, Edm. *Geschichte der Optik*, Leipzig, 1926, reprinted Wiesbaden, 1967.

Jammer, M. *Concepts of mass in classical and modern physics*, Cambridge, Mass. 1961.

Jammer, M. *Concepts of force*, Cambridge, Mass., 1957.

Jammer, M. *Concepts of space*, Cambridge, Mass., 1954.

Kemble, Ed. C. *Physical science, its structure and development.* vol. I, *From geometric astronomy to the mechanical theory of heat*, Cambridge, Mass./London, 1966.

Klemm, Fr. 'Vom Perpetuum Mobile zum Energieprinzip'. *Abh. Ber. deut. Mus.* 1966.

Lejeune, A. *Euclide et Ptolemée, deux stades de l'histoire de l'optique géometrique grecque*, Louvain, 1948.

Lippmann, E. O. von, *Geschichte der Magnetnadel bis zur Erfindung des Kompass*, Berlin, 1932.

Mottelay, P. F. *Bibliographical History of Electricity and Magnetism*, London, 1922.

Ronchi, V. *Histoire de la lumière*, Paris, 1956.

Sambursky, S. *The physical world of the Greeks*, London, 1956, New York, 1962.

Sambursky, S. *Physics of the Stoics*, London, 1959.

Sambursky, S. *The physical world of late Antiquity*, London, 1962.

Schuck, A. *Der Kompass*, vols. 1–3, Hamburg, 1911–18.

Urbanitzky, A. von, *Elektricität und Magnetismus im Alterthume*, Wien, 1887, reprinted Wiesbaden, 1967.

Vollgraff, J. A. *De leer van het licht voor Huygens*. 1, *De optika in de oudheit*, Leiden, 1910.

8. MECHANICS

Borel, E. *L'évolution de la mécanique*, Paris, 1942.

Callahan, J. F. *Four views of time in ancient philosophy*, Oxford, 1948.

Clagett, M. *The science of mechanics in the middle ages*, Madison, Wisc., 1959.

Cornford, F. M. *The laws of motion in ancient thought*, Cambridge, 1931.

Dijksterhuis, E. J. *The mechanization of the world picture*, Oxford, 1961.

Dijksterhuis, E. J. *Val en Worp. Een Bijdrage tot de Geschiedenis der Mechanica von Aristoteles tot Newton*, Groningen, 1924.

Dugas, R. *Histoire de la mécanique*, Neuchatel, 1950.

Duhem, P. *L'évolution de la mécanique*, Paris, 1905.

Duhem, P. *Les origines de la statique*, vols. 1–2, Paris, 1905.

Duhem, P. *Le mouvement absolu et le mouvement relatif*, Montligeon, 1907.

Ichak, F. *Das Perpetuum Mobile*, Leipzig, 1914.

Kisch, Br. *Scales and weights, a historical outline*, New Haven, 1965.

Mach, E. *Die Mechanik in ihrer Entwicklung historisch-kritisch dargestellt*, Leipzig, 1883 and many later editions.

Moody, E. A. and Clagett, M. *The mediæval science of weights*, Madison, Wisc., 1952.

9. ALCHEMY AND CHEMISTRY

Berthelot, M. P. E. *Introduction à l'étude de la chimie des anciens et du Moyen Âge*, Paris 1889, reprint Bruxelles, 1966.

Berthelot, M. P. E. *Collection des anciens alchimistes grecs*, 3 vols. Paris, 1887, reprinted London, 1963.

Berthelot, M. P. E. *La chimie au moyen âge*, vols. 1–3, Paris, 1893.

Berthelot, M. P. E. *Les origines de l'alchimie*, Paris, 1885.

Federmann, R. *Die königliche Kunst. Eine Geschichte der Alchemie*, Vienna, 1964.

Ferguson, J. *Bibliotheca chemica*, vols. 1–2, Glasgow, 1906, reprint London, 1954 (a very comprehensive bibliography of alchemical and chemical literature).

Forbes, J. *Short history of the art of distillation*, Leiden, 1948.

Furley, D. J. *Two studies in the Greek atomists: I, Indivisible Magnitude; II, Aristotle and Epicurus on Voluntary Action*, Princeton, 1967.

Ganzenmüller, W. *Die Alchemie im Mittelalter*, Hildesheim, 1967 (reprint of the Paderborn 1938 ed.).

Hammer-Jensen, I. *Die älteste Alchymie*, Copenhagen, (Vid. Selsk.) 1921.

Hartlaub, G. F. *Der Stein der Weisen. Wesen und Bildwelt der Alchemie*, München, 1959.

Holmyard, E. J. *Alchemy*, Edinburgh, 1957.

Kopp, H. *Geschichte der Chemie*, vols. 1–4. Braunschweig, 1843–47.

Kopp, H. *Die Alchemie in älterer und neuerer Zeit*, vols. 1–2, Heidelberg 1886, reprinted 1962.

Laminne, J. *Les quatre éléments. Histoire d'une hypothèse*, Bruxelles, 1904.

Lasswitz, K. *Geschichte der Atomistik vom Mittelalter bis Newton*, Leipzig, 1890, reprint Darmstadt, 1963.

Lieben, F. *Vorstellungen vom Aufbau der Materie im Wandel der Zeiten*, Wien, 1953.

Lindsay, J. *The origins of alchemy in Graeco-Roman Egypt*, London, 1970.

Lippmann, E. O. von, *Entstehung und Ausbreitung der Alchemie*, vols. 1–2, Berlin, 1919–31 (standard work).

Mabilleau, L. *Histoire de la philosophie atomistique*, Paris, 1895.

Melsen, A. van, *From atomos to atom. The history of the concept atom*, Pittsburg, 1952.

McMullin, E. (ed.) *The concept of matter*, Notre Dame, Ind., 1953.

Partington, J. R. *A short history of chemistry*, London, 1960.

Partington, J. R. *A history of chemistry*, vols. 1–4, London/New York, 1961–70.

Read, J. *Prelude to chemistry. An outline of alchemy, its literature and relationships*, London, 1936.

Stillman, J. *The story of alchemy and early chemistry*, New York, 1960.
Taylor, F. Sherwood, *The Alchemists*, London, 1951.

10. TECHNOLOGY

Daumas, M. (ed.) *Histoire générale des techniques*. I, *Les origines de la civilisation technique*, Paris, 1962.
De Camp, L. S. *The ancient engineers*, Garden City, New York, 1963.
Derry, T. K. and Williams, T. I. *A short history of technology*, Oxford, 1961.
Diels, H. *Antike Technik*, Leipzig, 1920.
Drachmann, A. G. *The mechanical technology of Greek and Roman Antiquity*, Copenhagen, 1963.
Drachmann, A. G. *Grosse griechische Erfinder*, Zürich, 1967.
Feldhaus, F. M. *Die Technik der Antike und des Mittelalters*, Potsdam, 1931.
Forbes, R. J. *Studies in ancient technology*, vols. 1–9, Leiden, 1955–64.
Kretzschmer, Fr. *Bilddokumente römischer Technik*. 3. Aufl., Düsseldorf, 1967.
Neuburger, A. *Die Technik des Altertums*, Leipzig, 1919.
Rathgen, B. *Das Geschütz im Mittelalter*, Berlin, 1928.
Semenov, S. A. *Prehistoric technology: An experimental study of the oldest tools and artefacts from traces of manufacture and wear*, New York, 1964.
Singer, C., Holmyard, E. J. and Hall, A. R. (ed.) *A history of technology*, vols. 1–5, Oxford 1954–1958 (vols. 1–3 treat Antiquity and the Middle Ages until about 1750).
White, L. Jr. *Mediæval technology and social change*, Oxford, 1962.

Biographical Appendix

A number of standard works are referred to only by the name of the author, according to the following list:

Burnet   John Burnet, *Early Greek philosophy*, 3rd ed., London, 1920.

Clagett   Marshall Clagett, *The science of mechanics in the Middle Ages*, Madison, Wisconsin, 1959.

Crombie   A. C. Crombie, *Robert Grosseteste and the origins of experimental science*, Oxford, 1953.

De Vaux   Carra de Vaux, *Les penseurs d'Islam*, **I–V**, Paris, 1921.

Dreyer   J. L. E. Dreyer, *A history of astronomy from Thales to Kepler*, 2nd ed., New York, 1953.

Duhem   Pierre Duhem, *Le système du monde* **I–X**, Paris, 1914–1959.

Gilson   Étienne Gilson, *La philosophie au moyen âge*, 3rd ed., Paris, 1947.

Haskins   Charles Homer Haskins, *Studies in the history of mediæval science*, Cambridge, Mass., 1924.

Heath   Thomas L. Heath, *A history of Greek mathematics*, **I–II**, Oxford, 1921.

337

Kennedy	E. S. Kennedy, A survey of Islamic Astronomical Tables; *Transactions of the American Philosophical Society*, New Series, **XLVI**, part 2, Philadelphia, 1956, pp. 123–177.
Kirk and Raven	G. S. Kirk and J. E. Raven, *The presocratic philosophers*, Cambridge, 1962.
Lippmann	E. O. von Lippmann, *Die Entstehung und Ausbreitung der Alchemie*, Berlin, 1931.
Migne	Migne, *Patrologia latina*.
Nallino	C. A. Nallino, *Raccolta di scritti editi e inediti*, **V**, a cura di Maria Nallino, Roma, 1944.
PW	August Pauly, *Real-Encyklopädie der klassischen Altertumswissenschaft*, ed. G. Wissowa, Stuttgart, 1894 (still unfinished).
Sarton	George Sarton, *Introduction to the history of science* **I–III** (in 5 vols), Baltimore, 1927–1948.
Steinschneider	Moritz Steinschneider, *Mathematik bei den Juden*, Leipzig, 1893–1899, repr. Hildesheim, 1964.
Suter	Heinrich Suter, *Die Mathematiker und Astronomen der Araber und ihre Werke*, Leipzig, 1900.
Tannery	P. Tannery, *Mémoires scientifiques*, **I-XVII**, Paris, 1912–1951.
Thorndike	Lynn Thorndike, *History of magic and experimental science*, **I–III**, New York, 1923–1958.
Wiedemann	Eilhard Wiedemann, *Aufsätze zur arabischen Wissenschaftsgeschichte*, **I-II**, Hildesheim, 1970.

Abū al-Wafā
of Būzjān, Qūhistān
c. 940–997

An astronomer and mathematician, Abū al-Wafā, worked in Baghdad. In mathematics he occupied himself with constructions with a fixed compass-span, with spherical triangles, and particularly with trigonometry, in which he introduced the theorem of tangents to spherical right triangles. In astronomy he wrote an *Almagest*, after the manner of Ptolemy, sometimes believed to contain the first mention of the inequality of the Moon. This misinterpretation was corrected by de Vaux.

de Vaux, **II**, 213; *Journal asiatique*, **XIX**, (1892), 408–71.
Dreyer, 252–56.
Kennedy, 134.
Nallino, *Scritti,* **V**, 336.
Sarton, **I**, 666.
Suter, 71.
M. Youschkevitch, *Dict. Sci. Bibl.*, **I**, 39.

Abū Ma'shar
(Albumasar) of Balkh
787–886

A Persian Moslem scholar who flourished in Baghdad where he became famous as an astrologer. His prolific writings draw upon Persian, Indian, Syriac, and Greek sources. In Latin Europe he became one of the principal sources of astrology with his *Introductiorium in astronomiam*, transl. by John of Spain, in 1133 and again by Herman of Carinthia, in 1140 (first printed 1489 in Augsburg; no modern edition). The work contains a complete exposition of astrology, including a theory of the tides. Equally popular was his plagiarism from al-Kindi, *De magnis coniunctionibus et annorum revolutionibus* in John of Spain's translation (Augsburg, 1489). An important collection of tables and other material called the *Book of the thousands* is lost.

Duhem, **II**, 369.
Kennedy, 133.
R. Leamy, *Abū Ma'shar and Latin Aristotelianism in the 12th Century*, Beirut, 1962.
Nallino, *Scritti,* **V**, 331.
D. Pingree, *The thousands of Abū Ma'shar*, London, 1968.
D. Pingree, *Dict. Sci. Biogr.*, **I**, 32.
Sarton, **I**, 568.
Suter, 28.

Adelard of Bath
(Adelardus Bathoniensis)
c. 1090 to c. 1150

The foremost English mathematician and natural philosopher in the twelfth century, and one of the first great translators from the Arabic, which he had learned in Spain. Among his most important translations are Euclid's *Elements* and al-Khwarizmi's *Astronomical tables* (ed. H. Suter, Vid. Selsk. Kbh., 1914). He was the author of the first treatise in Latin on falconry, and also wrote on the abacus and the astrolabe. Of importance to physics are his *Questiones naturales* (ed. H. Gollandz, Oxford, 1920 and M. Müller, Münster, 1934) and the philosophical dialogue *De eodem et diverso* (ed. H. Willner, Münster i.W., 1903).

F. Bliemetzrieder, *Adelhard von Bath*, Munich, 1935.
Clagett, Isis **XLIV** (1953), 16–42 (Euclid transl.).
Clagett, *Dict. Sci. Bibl.* **I**, 61.

Duhem, **III**, 116.
Gilson, 294–96.
Haskins, 20, 113, 346.
Sarton, **II**, 167–69.
Thorndike, **II**, 19.

Aegidius Romanus
(Egidius Colonna)
c. 1244–1316

An outstanding Aristotelian, but anti-Averroistic philosopher, a pupil of Thomas Aquinas and teacher of Philip the Fair. In 1292 he became General of the order of St Augustine, and in 1295 Archbishop of Bourges. He wrote a criticism of Etienne Tempier and also on the classification of the sciences, commentaries on Aristotle, and a *Hexaëmeron* (about the creation of the world in relation to Genesis). He upheld the Ptolemaic system against the attacks of al-Bitrūjī and Averroës.

Duhem, **IV**, 106–119.
Gilson, 546–48, 576–78.
J. Koch, *Giles of Rome, Errores philosophorum* (text, with an English transl. J. O. Riedl, Milwaukee, 1944).
Sarton, **II**, 922–26.

Aëtios
of Antiocheia
c. A.D. 100

A Greek historian of philosophy whose partly extant *Placita philosophorum* (The opinions of the philosophers) (ed. H. Diels in *Doxographi Graeci*, Berlin, 1879) is an important source for many views of authors whose works have been lost.

It builds on Theophrastos, and is often printed along with Plutarch's works.

PW, **I** 703.

Albategnius
see Al-Battānī

Albertus de Saxonia
(Albertutius, Albertus Novus, Albert of Saxony) of Helmstedt
c. 1316–1390

With Buridan and Oresme, one of the most outstanding Paris physicists of the fourteenth century. He became Master of the Sorbonne in 1351, and Rector in 1353. He was instrumental in the establishment of the University of Vienna, of which he became the first Vice-Chancellor in 1366, at the same time becoming Bishop of Halberstadt. In logic he followed Ockham, and in mechanics Buridan; he introduced the special *resistentia intrinseca* (see p. 241) into the theory of falling bodies. He also occupied himself with geological problems (the effects of water on the surface of the Earth) as well as with purely mathematical questions. A number of older editions (from 1487 and later) exist, but no modern ones.

Clagett, 223 and 565.
Duhem deals very thoroughly with Albert of Saxonia in his *Système du Monde* (**IV**, **VII–IX**), in *Les origines de la statique*, 2 vols. Paris, 1906, and in *Études sur Léonard de Vince*, 3 vols, Paris, 1906–13.
G. Heidingsfelder, *Albert von Sachsen*, Münster i.W., 1921.
E. A. Moody, *Dict. Sci. Bibl.* **I**, 93–95.
Sarton, **III**, 1428–32.

340

Albertus Magnus
of Launingen
c. 1200–1280

The first of the great Aristotelian philosophers of the thirteenth century, and an able natural scientist. In Germany he entered the Dominican order, and in about 1241 moved to Paris, where from 1245–48 he lectured on theology and philosophy. In 1248 he established the Dominican *studium generale* (the university) at Cologne, where he taught for the rest of his life, apart from a short period from 1260–62 when he was Bishop of Regensburg. Among his pupils was Thomas Aquinas. Called *Doctor universalis* by his contemporaries, he was a very prolific author on many subjects (*Opera omnia*, 38 vols. (incomplete), ed. A. Borgnet, Paris 1890–99; new edition vol. 1 *ff.*, Münster, 1951 *ff.*), and was instrumental in changing the intellectual climate of the Middle Ages. Due mainly to him (and to Aquinas) Aristotelianism prevailed and the general principle of experience as the basis of knowledge was accepted. Albert himself made contributions to zoology, botany, and mineralogy. A number of minor works of scientific interest pass under his name, such as a valuable astronomical bibliography *Speculum astronomiae*, a *Liber de mineralibus* (transl. D. Wyckoff, Oxford, 1967), a *Libellus de alchimia* (ed. P. Kibre, *Isis*, **XXXII**, (1949), 267–300; transl. V. Heines, Berkeley, 1958) and the spurious, often printed, treatise on gynaecology *De secretis mulierum*.

Duhem, **III**, 327–45 and **V**, 412–67.
Gilson, 503–16.
G. v. Hertling, *Albertus Magnus*, Münster, 1914.
P. Kibre, *Isis*, **XXXV**, (1944), 303–16; *Osiris* **XI**, (1954), 23–29.
Sarton, **II**, 934–44.
H. C. Scheeben, *Albertus Magnus*, Cologne, 1955.
Studia Albertina (ed. B. Geyer), Münster, 1952.
Thorndike, **II**, 517–92.
W. A. Wallace, *Dict. Sci. Bibl.*, **I**, 99–103.
D. Wyckoff, *Isis* **XLIX**, (1958), 109–22.

Albumasar
see Abū Ma'shar

Alcmaion
of Croton
c. 500 B.C.

A Pythagorean philosopher, traditionally considered to be a pupil of Pythagoras himself. Wrote *De natura* (lost). Performed anatomical dissections, discovered the optic nerve and concluded that the brain is the central organ of sensory perception. He explained illness as a disturbance of the internal equilibrium of the body.

Kirk and Raven, 232.
F. Kudlien, *Dict. Sci. Biogr.* **I**, 103.
Sarton, **I**, 77.
J. Wachtler, *De Alcmaeone Crotoniata*, Leipzig 1896.
PW, **I**, 1555.

Alessandro della Spina
d. 1313

A dominican of St Catharine's monastery at Pisa, and mentioned in its chronicles as one of the first makers of spectacles.

Sarton, **II**, 1024–27 (on the history of spectacles).

Alfonso X

(Alfonso the Wise)
King of Castile and Leon from 1252
until his death in 1284

Under Alfonso's auspices a series of schools and institutions were erected, in which Arabs, Jews, and Christians cooperated on the translation of treatises by al-Battānī, al-Haitham, Ptolemy and others from Arabic into Spanish (Castilian). From 1276–77 a Great Spanish encyclopedia on astronomical instruments, *Libros del saber de astronomia* (ed. Rico y Sinobas, 5 vols. in fol., Madrid, 1863–67) was edited, and provided by the King with prefaces to the various sections. By his orders *The Alfonsine tables* were compiled in Toledo (presumably in 1272) to substitute *The Toledo tables* (several early but no modern editions).

J. L. E. Dreyer, *Monthly notices, Royal Astron. Soc.* **LXXX**, (1920), 243–62.
Duhem, **II**, 259–66; **IV**, 171–75.
E. S. Proctor, *Alfonso X of Castile*, Oxford, 1951.
Sarton, **II**, 834–42.
P. D. Thomas, *Dict. Sci. Biogr.* **I**, 122.
J. S. Viguera, *La astronomia de Alfonso X*, Madrid, 1926.
A. Wegener, *Bibliotheca Mathematica* **VI**, (1905), 129–85.

Alfraganus

see Al-Farghānī

Alhazen

see Ibn al-Haitham

Alpetragius

see Al-Bitrūjī

Anaxagoras

of Clazomene
c. 500 B.C. to *c.* 428 B.C.

Ionian philosopher who in about 480 B.C. settled in Athens as the first professional philosopher where he taught Pericles. Later he founded a school on Lampsacos. It is difficult to reconstruct his train of thought, but in astronomy he conceived of the celestial bodies as stones hurled away from the Earth and glowing because of their motion, an opinion which he perhaps assumed after a meteoric fall at Aegospotami in 468 B.C. Only a few fragments of his writings are extant (Anaxagore, *Les fragments*, Greek text and French transl. J. Zafiropulo, Paris, 1948).

Burnet, 251–75.
F. M. Cleve, *The philosophy of Anaxagoras*, New York, 1949.
D. E. Gershenson and D. A. Greenberg, *Anaxagoras and the birth of scientific method*, Intr. by Ernest Nagel, New York, 1964.
Kirk and Raven, 362.
J. Longrigg, *Dict. Sci. Bibl.* **I**, 149.
Sarton, **I**, 86.
PW, **I**, 2076–77.

Anaximander
of Miletos
c. 610 B.C. TO c. 547 B.C.

After Thales, the second of the Ionian philosophers of nature, author of the first Greek treatise in prose. Introduced the Babylonian gnomon among the Greeks. Assumed a spherical universe with a cylindrical Earth at the centre, and tried to draw the first map of the world. He imagined the heavenly bodies as wheels filled with fire and rotating about the Earth, and from this created the first mechanical planetary theory. He considered the origin of things as the infinite or unfinished (*to apeiron*). Only a few fragments of his work are still extant.

Burnet, 52–71.
C. H. Kahn, *Anaximander and the origins of Greek cosmology*, New York, 1960.
Kirk and Raven, 99.
D. O'Brian, *Class. Quart.* **XVII** (1967), 423–32.
Sarton, **I**, 72.
P. Seligman, *The 'apeiron' of Anaximander, A study in the origin and function of metaphysical ideas*, London, 1962.
L. Taran, *Dict. Sci. Biogr.*, **I**, 150.
PW, **I**, 2085–86.

Anaximenes
of Miletos
c. 540 B.C.

The third Ionian philosopher, following Thales, and Anaximander. To him, the origin of things (*physis*) was air, which, when more and more condensed, became clouds, water, earth and stones, and by attenuation became fire. In astronomy his views constitute a retrograde step compared with those of Anaximander. He regarded the Earth as flat and the heavenly bodies as exhalations of fire from the Earth.

Burnet, 72–9.
Kirk and Raven, 143.
L. Taran, *Dict. Sci. Biogr.* **I**, 151.
Sarton, **I**, 73.
PW, **I**, 2086.

Andronicus
of Rhodes
First century B.C.

A peripatetic (Aristotelian) philosopher who worked as head of the Lyceum in Athens, where he tried to revive the now languishing Aristotelism by, among other things, editing the works of Aristotle and Theophrastus, and ordering them chronologically.

Sarton, **I**, 203.
PW, **I**, 2164–67.

Anthemios
of Tralles
d. c. A.D. 534

A Byzantine mathematician and architect. He worked in Constantinople on, among other things, the restoration of Hagia Sophia. He wrote on burning-glasses, with passages on the theory of conic sections, and indicated the construction of a parabola by means of a directrix and a focus.

Heath, **II**, 541.
Heath, *Bibliotheca Mathematica* **VII** (1907), 225.

J. L. Heiberg, *Zts. f. Math. hist. Abt.*, **XXVIII** (1883), 121–29.

G. L. Huxley, *Anthemius of Tralles; a study in later Greek geometry*, Cambridge, Mass., 1959.

Sarton, **I**, 427.

PW, **I**, 2368.

H. G. Zeuthen, *Die Lehre v.d. Kegelschnitten im Altertum*, Copenhagen, 1886, 375–80.

Apollodoros
c. 180 B.C. to c. 120 B.C.

A Hellenistic historian and geographer; in about the year 146 B.C. he left Alexandria, to work in Athens where he remained until his death. He is the author of a (lost) *Chronica* for the period 1184 B.C. (the fall of Troy) to 144 B.C. the fragments of which are an important source of the history of philosophical schools (F. Jacoby, *Apollodors Chronik*, Berlin, 1902). In addition, he wrote on mythology (Apollodorus Atheniensis, *The Library;* English transl. J. G. Frazer, 2 vols, London, 1921), religion and geography.

Sarton, **I**, 198.

PW, **I**, 2855–86.

Apollonios
of Perga (Pamphylia)
c. 262 B.C. to c. 190 B.C.

One of the greatest mathematicians of Hellenistic times, Apollonios was educated in Alexandria by the successors of Euclid. He wrote on regular polyhedra, on an approximate determination of π etc., but his fame is first and foremost due to his *Conica*, in which he developed the theory of conic sections to a high degree of perfection; moreover, it is to him we owe the names ellipse, parabola, and hyperbola. *Conica* contained eight books, of which four are extant in Greek, and three in Arabic translation by Thabit ben Qurra (826–901). The astronomer Halley edited the work (Oxford, 1710) with a reconstruction of the eighth book. In astronomy Apollonius investigated eccentric and epicyclic models, and found the method of determining stationary points used in the *Almagest*.

Opera, ed. Heiberg, 2 vols, Leipzig 1891–93. *Conica*, ed. P. ver Eecke, *Les coniques d'Apollonius de Perge*, Bruges, 1923; reprinted Paris, 1963 (Greek text with French translation). English transl. T. L. Heath, *Apollonius of Perga, treatise on conic sections*, Cambridge, 1896; reprinted Cambridge 1961.

Heath, **II**, 126.

Sarton, **I**, 173–75.

G. J. Toomer, *Dict. Sci. Biogr.* **I**, 179.

PW, **II**, 151–60.

H. G. Zeuthen, *Die Lehre v.d. Kegelschnitten im Altertum*, Copenhagen, 1886, Reprinted Hildesheim 1966.

Aquinas
see Thomas Aquinas

Archimedes
of Syracuse
c. 287–212

The greatest mathematician of Antiquity, and one of the founders of mathematical physics. Born in Syracuse, the son of the astronomer Phidias, he went to study in Alexandria along with Eratosthenes and Conon, but returned to his native city, where he remained until he died at the hands of a Roman soldier when Marcellus took Syracuse in 212 B.C. He soon became a rather legendary figure as the inventor of the water screw, the endless screw, and various machines of war used against the Romans during the siege of Syracuse. He was also remembered for his famous assertion: 'Give me a place to stand, and I shall move the Earth', for his determination of the gold content of King Hieron's crown, and for his construction of a kind of planetarium which Cicero saw in Rome 200 years later.

Most of Archimedes' writings are known. In 1906 in Constantinople Heiberg found the lost treatise *On method* (J. L. Heiberg and H. G. Zeuthen, *Bibl. Math.*, **VII** (1907), 321). Here Archimedes explains how he found many of his geometrical results by means of a 'mechanical' method, that is by thought experiments in statics. In his other works he proceeds in a strictly mathematical way, using an axiomatic-deductive method. In *The Measurement of the Circle* π is found to lie between $3\frac{10}{71}$ and $3\frac{1}{7}$ by means of inscribed and circumscribed regular polygons with 96 sides. In *The sandreckoner* he shows how one may express very large numbers (like the number of sand grains filling a spherical universe of a given size) by using 10^8 as a basic unit. In *On the sphere and the cylinder* and in *The quadrature of the parabola* he applied Eudoxos's method of exhaustion to the determination of areas and volumes, and came closer than any other ancient mathematician to modern integration. In the book *On spirals* he studied the curve $r = a\phi$.

In the field of mechanics Archimedes wrote *On the equilibrium of plane figures* which is the first (extant) treatise on statics, just as *On floating bodies* is the foundation of hydrostatics. He wrote nothing on dynamics.

Later, Archimedes became the great and almost inimitable genius of creative imagination, stringent argument, and clear exposition. Time and again progress in mathematics and mathematical physics can be traced back to a renewed interest in his work. His mechanical works were translated in the twelfth century, leading to the flourishing of Mediæval statics. In the thirteenth century the rest of the works were translated from the Greek by William of Moerbeke (*c.* 1269) as a great stimulus to scholastic mathematics, just as the first printed editions (by N. Tartaglia, Venice, 1543 and Geschauff, Basel, 1544) inspired scientists all over Europe.

Opera omnia, 3 vols, ed. J. L. Heiberg (critical edition of the Greek text with Latin translation), 2nd ed., Leipzig 1910-15; T. L. Heath, *The Works of Archimedes*, Cambridge, 1897 with supplement (*The method of Archimedes*), 1912; reprinted New York, 1912; 1953; Ver Eecke, *Oeuvres complètes d'Archimède* Bruxelles, 1921; reprinted Paris, 1960.

M. Clagett, *Archimedes in the Middle Ages*, **I**, Madison, Wisc. 1964.

M. Clagett, *Dict. Sci. Biogr.*, **I**, 213; *Mélanges Alexandre Koyré*, Paris, 1964, **I**, 40.

E. J. Dijksterhuis, *Archimedes*, Copenhagen, 1956.

A. G. Drachmann, *Centaurus*, **VIII** (1963), 91, ibid. **XII** (1967), 1.

Sarton, **I**, 169.

W. Stein, *Quellen und Studien*, **I B** (1929-31), 221.

PW, **II**, 507.

Archytas
of Tarentum,
fl. 400–365 B.C.

One of the foremost Pythagorean philosophers, Archytas was a personal acquaintance of Plato, who was influenced by
him. Archytas is often said to have
founded mechanics. In acoustics he
determined the numerical ratio of the
monoharmonic, the chromatic, and the
diatonic scales. In mathematics he distinguished between harmonic, geometrical, and arithmetical progressions,
and gave a solution of the 'Delian'
problem on the duplication of the cube.

Fragments ed. by A. E. Chaignet,
*Pythagore et la philosophie pythagori-
cienne*, 2 vols, Paris, 1873, **I**, 259–331.
E. Frank, *Plato und die sogenannten
Pythagoreer*, Halle, 1923; reprinted Darm-
stadt, 1962.
Heath, **I,** 215
K. von Fritz, *Dict. Sci. Biogr.*, **I**, 231.
Sarton, **I**, 116.
Tannery, **III**, 105.
B. L. v.d. Waerden, *Hermes*, **LXXVIII**
(1943), 184; *Math. Ann.* **CXXX** (1948),
127.
PW, **II**, 600.

Aristarchos
of Samos
c. 310 B.C. *c.* 230 B.C.

The most eminent Greek astronomer of
the third century B.C., Aristarchos is
remembered for his heliocentric planetary system in which the Earth rotates
about its own axis at the same time as it
moves on a circular orbit about the Sun.
Only one of his writings is extant, *On
the sizes and distances of the Sun and Moon*

(ed. T. L. Heath, *Aristarchus of Samos*,
Oxford, 1913; German transl. by A.
Nokk, 1854).

T. W. Africa, *Isis*, **LII**, (1961), 403.
Dreyer, 136.
Duhem, **I**, 418 and **II**, 17.
O. Neugebauer, *Isis* **XXXIV** (1942), 4.
Sarton, **I**, 156.
W. H. Stahl, *Dict. Sci. Biogr.* **I**, 246.
Tannery, **I**, 371.
PW, **II**, 873.

Aristotle
of Stagira
384–322 B.C.

One of the greatest philosophers of
Antiquity, Aristotle's influence on the
development of science, right down to
the present has been enormous. He was a
son of Nicomachus, the Macedonian
court physician, spent a part of his child-
hood at the court at Pella and arrived at
Plato's Academy in Athens at the age of
17. He remained there until Plato's death
in 347 B.C. when he set out on a journey,
first to Assus, then to Mitylene and back
to Macedonia, where from 343–340 B.C.
he taught Alexander the Great. In 335
B.C. he returned to Athens and established
his own school (Lyceum, the Peripatetic
School). Here he started a library, and
other scientific collections. Alexander the
Great supported Aristotle, and ordered
hunters and fishermen of his empire to
send him interesting and rare finds.
Aristotle taught in Athens for twelve
years and educated pupils who came to
occupy prominent positions in Greek
science in the following period (as, for
example, Theophrastos, Eudemos, Aristoxenos). When Alexander died in 323

B.C. Aristotle left Athens, where anti-Macedonian feelings were now dominant, and went to Chalcis, where he died the following year. His school in Athens was carried on by Theophrastos.

In his writings, which are nearly all lecture notes, Aristotle deals with practically all aspects of Greek thought and science, establishing a series of new disciplines. Among the authentic works we have a series of writings on logic (*Analytica priora et posteriora, Topica*, and others) where for the first time formal logic together with a philosophy of science are systematically developed. Moreover, he left a number of important writings on physics and philosophy of nature: *Physica, De caelo et mundo, De generatione et corruptione, Meteorologica*, and *De anima*, besides the important *Metaphysica* which deals with fundamental epistemological questions, the theory of causation, the doctrine of hylemorphism, etc. In *Historia animalium* and *De partibus animalium* more than 500 species of animals are described and classified, thus establishing zoology as a science. Ethical and sociological questions are dealt with in an *Ethics* (found in several versions), a *Politics*, and a work on *The Constitution of the Athenians* to which is added a long series of minor writings. In the form known to us they have, without exception, been edited by Andronicus of Rhodes. Finally, works like the *Mechanica Problemata*, etc., have, mistakenly, been ascribed to Aristotle, and must be counted among the sources of the Aristotelian traditions.

Editions: *Opera omnia*, 3 vols, ed. A.I. Bekker, Berlin, 1831, together with Bonitz, *Index Aristotelicus*, 1870. Provided with Greek text and English translations, most of them are found in Loeb's Classical Library.

H. Carteron, *La notion de force chez Aristote*, Paris, 1924.

Duhem, **I**, 126.

J. Düring *Aristoteles. Darstellung und Interpretation seines Denkens*, Heidelberg, 1966.

L. Elders, *Aristotle's cosmology, a commentary on the De Caelo*, Assen, 1966.

M. Grabmann, *Die lateinischen Aristoteles-Übersetzungen im XIII. Jahrhundert*, Münster i.W., 1916.

M. Grene, *A portrait of Aristotle*, Chicago, 1963.

O. Hamelin, *Le système d'Aristote*, Paris, 1920.

T. L. Heath, *Mathematics in Aristotle*, Oxford, 1949.

W. Jaeger, *Aristoteles*, Berlin, 1923. English transl. by R. Robinson, Oxford 1934, reprinted 1962.

M. Manquat, *Aristote naturaliste*, Paris, 1932.

A. Mansion, *Introduction à la physique aristotelicienne*, Louvain, 1913, 1945

M. Mignucci, *La teoria Aristotelica della scienza*, Florence, 1965.

J. Owens, D. M. Balme, J. Wilson, L. Minio-Paluello, *Dict. Sci. Biogr.* **I**, 250.

M. D. Philippe, *Aristoteles*, Bern, 1948 (*Bibliographische Einführungen in d. Studium d. Philosophie*, No. 8).

W. D. Ross, *Aristotle*, London, 1923.

Sarton, **I**, 127–36 (very detailed bibliography).

M. Schramm, *Die Bedeutung der Bewegungslehre des Aristoteles für seine beiden Lösungen der zenonischen Paradoxie*, Frankfurt, 1962.

F. Solmsen, *Aristotle's system of the physical world. A Comparison with his Predecessors*, New York, London, 1960.

F. van Steenberghen, *Aristote en occident*, Louvain, 1946 (English transl., *Aristotle in the West*, Louvain, 1946).

PW **II**, 1012–54.

Aristoxenos
of Tarentum
Fourth century B.C.

Pupil of Aristotle, philosopher, mathematician and the foremost theorist of music of Antiquity. Of his 435 writings, by far the greater number has been lost, but *The principles and elements of harmonics* is extant (ed. H. S. Macran, *The harmonics of Aristoxenus*, Oxford, 1902, Greek text with English translation).

L. Laloy, *Aristoxène de Tarente et la musique dans l'antiquité*, Paris, 1904; 1924.
Sarton, **I**, 142.
PW, **II**, 1057–65.
R. P. Winnington-Ingram, *Dict. Sci. Bibl.*,**V**, 281.

Āryabhaṭa I
of Kusumapura (near modern Patna)
c. A.D. 500

An Indian astronomer who in A.D. 499 wrote one of the first Sanskrit works on mathematics, spherical and theoretical astronomy, called the *Āryabhaṭīya* (English transl. W. E. Clark, Chicago, 1930, and P. C. Sengupta, Calcutta, 1927). In contrast to other Indian astronomers, he assumed the daily rotation of the Earth around its axis.

D. Pingree, *Census of the Exact Sciences in Sanskrit*, **AI**, 50, Philadelphia, 1970; *Dict. Sci. Biogr.* **I**, 308.
Sarton, **I**, 409.
P. C. Sengupta, *Journ. Calcutta Univ. Dept. Letters*, **XVIII**, (1929), 1.

Arzachel
see al-Zarqālī

Augustine
of Tagaste
354–430

Aurelius Augustinus, born in Tagaste (in present Tunisia), trained as an orator and practised as such in Rome and Milan until his conversion to Christianity in 386. From 395 until his death he was Bishop of Hippo (Tunisia). The most important of the Latin Fathers of the Church, whose positive view of Christianity became the cultural foundation of the early Middle Ages. Influenced by Neoplatonism, he used its philosophical concepts to attempt a clarification of the content of the Christian faith. Down to the year 1200 he is quoted by Mediæval thinkers more frequently than any other author. To the philosophy of nature particularly his commentary on the Book of Genesis, *De genesi ad litteram*, became of importance. His epistemology, according to which the soul (possibly stimulated by the senses) is illuminated by a divine light and sees the 'eternal grounds' or 'ideas' of things in God himself, dominates the thought of the Middle Ages until, in the thirteenth century, the principle of experience is again introduced.

Opera omnia in Migne, *Patrologia latina* vols **XXXII-XLVII**
Numerous translations into the popular languages.
Duhem, **II**, 410–18.
Gilson, *Introduction à l'étude de Saint Augustin*, Paris, 3. ed. 1949.
H. I. Marrou, *Saint Augustin et le fin de la culture antique*, Paris, 4th ed., 1959.
E. Portalié, *Augustin*, in *Dictionnaire de théologie catholique*, **I**, Paris, 1902 (the best exposition of the philosophy of Augustine).
M. F. Sciacca, *Augustinus*, Bern, 1948 (*Bibliographische Einführung in d. Studium*

d. Philosophie, No. 10).
Sarton, **I**, 383–84.
Thorndike, **I**, 504–22.

Autolycos
of Pitane
c. 310 B.C.

A Greek astronomer who criticized the
system of Eudoxos because it did not
explain the variable distances of the
planets from the Earth. He wrote 'The
Small Astronomy' (as opposed to
Ptolemy's 'The Great Astronomy') in
two parts: *On the movable sphere* which is
the first, fully extant mathematical
textbook treatment of the circles of the
celestial sphere, and similar spherical-
astronomical questions; and. *On risings
and settings* (both parts ed. F. Hultsch,
Leipzig, 1885, and J. Mogenet, Louvain,
1950).

Duhem, **I**, 400–3.
G. L. Huxley, *Dict. Sci. Biogr.* **I**, 338.
J. Mogenet, *Archives internat.*, **V** (1948),
139.
Sarton, **I**, 141.
Tannery, **II**, 225–55.
PW, **II**, 2602.

Avendeut
see Johannes Hispalensis

Averroës
see Ibn Rushd

Avicenna
see Ibn Sīnā

Azarquiel
see al-Zarqālī

Bacon
see Roger Bacon

Al-Battānī
(Albategnius) of Harrān
c. 850–929

One of the very greatest Moslem astronomers, who for the major part of
his life worked at al-Raqqa on the
Euphrates. His chief astronomical work
is his *zīj* or collection of astronomical
tables with canons for their use, and is
one of the very few works of its kind
published in a modern, critical edition
(Nallino, *Albatenii Opus astronomicum*
(Arabic text with Latin translation), 3
vols, Milan 1899–1907). In the Latin
Middle Ages it was known as *De scientia
stellarum*, from a (now lost) translation by
Robert of Chester and another by Plato
of Tivoli (printed Nuremberg, 1537).
The theory behind this *zīj* is mainly
Ptolemaic, but a number of Ptolemy's
numerical parameters were improved
from new observations. Discarding
Thābit's theory of trepidation, al-
Battānī accepted his actual rate of pre-
cession (1° in 66 years).
de Vaux, **II**, 208.
Duhem, **II**, 230.
W. Hartner, *Dict. Sci. Bibl.* **I**, 507.
Kennedy, 154.
Nallino, *Scritti*,**V**, 334.
Sarton, **I**, 602.
Suter, 45.

Bede
the Venerable of Yarrow
A.D. 673–735

An English Benedictine of Yarrow near Durham, the most outstanding scholar in Europe around the year A.D. 700, and England's first historian. His scientific treatise *De natura rerum* (ed. Migne, PL 90, 187) builds on Pliny and Isidore of Seville. *De temporum ratione* (ed. C. W. Jones, *Bedae opera de temporibus*, Cambridge, Mass., 1943) is a great and excellent textbook of the calendar system and time reckoning which came to serve as a model for the Mediæval literature on this subject (the *Compotus* literature). It contains, too, original observations on tidewater.

Complete Works, 12 vols, ed. J. A. Giles, London, 1843–44.
Migne, PL 90.
Duhem, **III**, 16–20.
C. W. Jones, *Isis*, **XXIV** (1936), 397–99; ibid. **XXVII** (1937), 430–40; *Dict. Sci. Biogr.*, **I**, 564.
Sarton, **I**, 510.
A. H. Thompson, *Bede, his life, times, and writings*, Oxford, 1935.
Thorndike, **I**, 634.

Bernard Walter
1436–1504

A prosperous citizen of Nuremberg. In 1471 he erected an astronomical observatory, along with an instrument workshop and a printing press, for Regiomontanus to use. This was the first scientific research institution in Europe outside Spain independent of the universities. After the death of Regiomontanus in 1476 Bernard continued the work with observations later utilized by Copernicus and Tycho Brahe. He is said to be the first astronomer to take atmospheric refraction into account and to use mechanical clocks for observational purposes.

Thorndike, *Science and thought in the 15th century*. New York, 1929, 146–47.
E. Zinner, *Leben und Wirken des Johannes Müller*, München, 1938 (reprinted Osnabrück, 1968).

Bessarion
see Johannes of Trebizont

al-Bīrūnī
of Khwārizm (Khiva)
973 to *c*. 1050

One of the foremost Mediaeval scientists within the Islamic world. For religious reasons he nourished anti-Arabian feelings, lived mostly in India, and probably died in Afghanistan. He was one of the great promoters of contacts between East and West, and his *India* (transl. C. E. Sachau, London, 1910) is a mine of information on Indian science and civilization in the Middle Ages. Among his very numerous writings is *The chronology of ancient nations* (transl. C. E. Sachau, London, 1879) and he has left also a number of important works on astronomy, such as *Instruction in the elements of the art of astrology* (ed. and transl. R. R. Wright, London, 1934), *The determination of the co-ordinates of cities* (transl. J. Ali, Beirut, 1966), and *On transits* (transl. Mohammed Saffouri and Adham Ifram, Beirut, 1959). His great astronomical collection *Qānūn al-Mas'ūdī*

(Hyderabad, 1954 *ff.*) deals with all the main problems of astronomy. In particular, he is remembered for his contributions to trigonometry, his new method of measuring the size of the Earth, and his (inconclusive) discussion of its daily rotation. In geology he explained artesian wells by the law of communicating vessels, and the Indus Basin as an ancient sea filled up with alluvial soil. He also measured the specific weights of 18 gems and metals.

Al-Bīrūnī Commemoration Volume, Calcutta, 1951.
de Vaux, **II**, 75 and 215.
Kennedy, 133 and 157, *Dict. Sci. Biogr.,* **II**, 147.
Nallino, *Scritti,* **V**, 339.
Sarton, **I**, 707.
C. Schoy, *Die trig. Lehren des . . . al-Bīrūnī,* Hanover, 1925.
Suter, 98.
Wiedemann, **I**, 776, 822; **II**, 1, 215, 474, 516

al-Bitrūjī
(Alpetragius) of Cordoba
c. 1100 to *c.* 1190

A Spanish-Arabian astronomer who worked in Cordoba and Seville, sometimes regarded as the founder of a 'new astronomy' through his (only extant) work on the motion of celestial bodies, which must be considered as an attempt to revive a planetary theory with concentric spheres. This work represents the height of an anti-Ptolemaic current within Arabian astronomy, but the theory was never so elaborated as to become a serious rival of the Ptolemaic theory. The treatise was translated into Latin by Michael Scotus in 1217 under

the title *De motibus celorum* (ed. F. J. Carmody, Berkeley, 1952).

F. J. Carmody, *Isis,* **XLII** (1951), 121–30 (Alpetragius and Regiomontanus).
de Vaux, **II**, 230–36.
Dreyer, 264–67.
Duhem, **II**, 146–56.
L. Gauthier, *Journal asiatique,* **XIV** (1909), 483–510.
E. S. Kennedy, *Speculum,* 1954, 246–51.
Nallino, *Scritti,* **V**, 343.
Sarton, **II**, 399–400.
Suter, 131.

Boethius
(Anicius Manilius Severinus B.)
c. 480–524

One of the most eminent figures of the transition period between Antiquity and the Middle Ages, Boethius is often called the Last Roman. A high official of the Roman Gothic Kingdom he was executed at Pavia in 524. Boethius translated several Greek scientific works into Latin and wrote commentaries to the logical works of Aristotle (ed. L. Minio-Paluella, in *Aristoteles Latinus,* 5 vols, Bruges, 1961 *ff.*). A translation of the Almagest is lost. Boethius coined the term 'quadrivium', and wrote a number of manuals on its disciplines. Extant are his algebra (*De institutione arithmetica,* ed. G. Friedlein, Leipzig, 1867) and his book on music (ibid, German transl. O. Paul, Leipzig, 1872). A geometry is also ascribed to him. In prison, he wrote a philosophical treatise *De consolatione philosophiae,* which became immensely popular in the Middle Ages and was translated into English by King Alfred and Chaucer (transl. V. E. Watts, London, 1969).

351

Opera omnia, in Migne, PL, 63–64.
K. Dürr, *The propositional logic of Boethius*, Amsterdam, 1951.
M. Folkerts; *Boethius Geometrie*, Göttingen, 1967.
L. Minio-Paluello; *Dict. Sci. Biogr.*, **II**, 228.
Sarton, **I**, 424.
H. F. Stewart; *Boethius*, Edinburgh, 1891.
Tannery; *Bibl. Math.*, **I** (1900), 39.
PW, **III**, 596–600.

Bolos
of Mendes (Egypt)
Second or third century B.C.

One of the first alchemical and astrological authors, often known by the name Democritos.

E. J. Holmyard, *Alchemy*, London, 1957, 23 *ff*.
PW, **III**, 676–77.
Sarton, **I**, 89.
J. Stannard, *Dict. Sci. Biogr.*, **II**, 256.

Bradwardine
see Thomas Bradwardine

Brahmagupta
c. 598 to *c.* 670

Indian astronomer who worked in Ujjain. In about A.D. 628 he wrote the *Brahmasphuta-Siddhānta*, based on the *Sūrya-Siddhāntā* and Āryabhaṭa. Another astronomical treatise, the *Khaṇḍakhādyaka* (English transl. by P. C. Sengupta, Calcutta, 1934; and (more complete) by B. Chatterjee, Calcutta, 1970) followed later.

D. Pingree, *Dict. Sci. Biogr.*, **II**, 416.
Sarton, **I**, 474.
R. Sewell, *The Siddhāntas and the Indian calendar*, Calcutta, 1924.

Buridan
see Johannes Buridanus

Calcidius
First half of the fourth century A.D.

Platonic author; translated and commented upon the greater part of Plato's *Timaios* (*Plato Timaeus; a Calcidic translatus*, ed. P. J. Jensen and J. H. Waszink, London and Leiden, 1962) thus enabling the scholars of the early Middle Ages to become acquainted with Plato and his cosmology, and with the motion of Venus and Mercury around the Sun (a geo-heliocentric system).

Duhem, **II**, 417; **III**, 47.
Sarton, **I**, 352.
W. H. Stahl, *Dict. Sci. Biogr.*, **III**, 14–15.
J. H. Waszink, *Studien zum Timaioskommentar des Calcidius*, **I**, Leiden, 1964.
J. C. M.v. Winden, *Calcidius on Matter*, Leiden, 1959.
PW, **III**, 2042.

Callippos
of Cyzicus
c. 370 B.C. to *c.* 300 B.C.

An important astronomer, a pupil of Aristotle, who stimulated him to develop the planetary theory of Eudoxos by adding more spheres (Aristotle, *Metaphysics*, 1073 b).

Duhem, **I**, 123–28.
Heath, *Aristarchos*, Oxford 1913, 212 ff.
J. S. Kieffer, *Dict. Sci. Biogr.*, **III**, 21–22.
Sarton, **I**, 141.
PW, **X**, 1662–64.

Campanus

of Novara (Giovanni Campano de Novara)
c. 1240 to *c.* 1300

An Italian astronomer and mathematician whose life history is rather obscure. He was chaplain to Urban IV (1261–1264) and connected with the University of Paris. He edited a Latin version of Euclid's *Elements* which was used for the first printed edition (Venice, 1482). In astronomy he wrote a number of treatises among which a *Theorica planetarum* introduced the first equatorea in Latin astronomy (Edited with an English transl. by G. J. Toomer, *Campanus of Novara and Medieval Planetary Theory*, Madison (Wisc.), 1971).

F. J. Benjamin, *Osiris*, **XI**, (1954), 221.
Duhem, **III**, 317 and **IX**, 129.
Sarton, **II**, 985.
G. J. Toomer, *Dict. Sci. Biogr.*, **III**, 23–29.

Chaucer, Geoffrey

c. 1340–1400

The most famous English author of the Middle Ages, probably born in London, where he lived as a courtier and diplomat. His *Canterbury tales* give a vivid picture of English life in the fourteenth century. Among his translations were the *Romaunt of the Rose* (from French) and the *De consolatione philosophiae* by Boethius.

In astronomy he has left the first English *Tretis of the astrolabie* (from 1391) derived from Māshāllāh (ed. W. W. Skeat, *The complete works of Geoffrey Chaucer*, Oxford, 1912 (reprinted 1954) 396–418; facsimile edition by P. Pintelon, Antwerp, 1940; also ed. by R. T. Gunther, *Early Science in Oxford*, **V**, Oxford, 1929). The Peterhouse MS 75.1 contains a Middle English treatise on an equatorium written in his hand (ed. D. J. Price and R. M. Wilson, *The equatorie of the planetis*, Cambridge, 1955).

W. C. Curry, *Chaucer and the Mediæval sciences*, New York, 1926; London, 1960.
R. D. French, *A Chaucer handbook*, New York, 1927 and later ed.
F. M. Grimm, *Astronomical lore in Chaucer*, Lincoln, Nebr., 1919.
H. Lange and A. Nippoldt, *Quellen und Studien*, Part 4 **V**, (1936), 38.
J. D. North, *Review of English Studies*, New Series **XX** (1969) 129–154, 257–283, 418–444.
Sarton, **III**, 1417.

Chrysippos

of Soli
c. 280–207

A Stoic philosopher who in about the year 260 B.C. came to Athens, and in about 232 B.C. succeeded Cleanthes as head of the Stoic school, of which he can be said to be the second founder. Only a few of his numerous works are extant (J. ab Arnim, *Stoicorum veterum fragmenta* **III**, Leipzig 1903).

E. Brehier, *Chrysippe*, Paris, 1910, 1951.
Sarton, **I**, 169.
PW, **III**, 2502.

Cleanthes
of Assos
331–232 B.C.

A stoic philosopher, and Zeno's successor. Instigator of the campaign against Heracleides of Pontus and his astronomy. Most of his scientific writings have been lost.

A. C. Pearson, *The fragments of Zeno and Cleanthes*, London, 1891.
Sarton, **I**, 152.
G. Verbeke, *Kleanthes van Assos*, Brussels, 1949.
PW, **XI**, 558–74.

Cleomedes
Probably 1st century A.D.

A Stoic astronomer, difficult to date. Author of a treatise *De motu circulari corporum caelestium* (Greek text with Latin translation by H. Ziegler, Leipzig, 1891. German transl. A. Czwalina *Die Kreisbewegung der Gestirne*, Leipzig, 1927 (Ostwalds Klassiker N° 220) largely based on Poseidonios, but also containing original contributions to optics, for example the discovery of atmospheric refraction.

D. R. Dicks, *Dict. Sci. Biogr.*, **III**, 318–320.
Duhem, **I**, 310, 470.
Sarton, **I**, 211.
PW, **XI**, 679–94.

Cleostratos
of Tenedos
Second half of the sixth century B.C.

An Ionian philosopher, perhaps a disciple of Thales, who according to tradition, died on Tenedos. He is supposed to have introduced the zodiac and its twelve signs to the Greeks, following Babylonian models.

J. K. Fotheringham, *Journal of Hellenic studies*, **XXXIX** (1919), 164–184.
Sarton, **I**, 72–73.

Conon
of Samos
Third century B.C.

An Alexandrian astronomer and mathematician, a disciple of Archimedes. Wrote on astronomy in seven books containing Babylonian observations of eclipses. Hipparchos was influenced by him. Conon introduced the constellation of the Locks of Berenice. He studied conic sections; the fourth book of the *Conica* by Apollonius is based on him.

I. Bulmer-Thomas, *Dict. Sci. Biogr.*, **III**, 391.
Sarton, **I**, 173.
Tannery, **III**, 353.
PW, **XI**, 1338–40.

Copernicus, Nicolaus
1473–1543

A Polish astronomer, born in Torun (Thorn) as the son of a wealthy merchant who died in 1483; he was then protected by his uncle Lucas Watzelrode who became Bishop of Ermland in 1498. In 1491 Copernicus matriculated at the University of Cracow, where through the teaching of Albertus de Brudzewo he came to know the tradition of Regiomontanus and Peurbach. In 1496 he began studying law in Bologna where he made his first observations and learned astronomy with Maria de Novara. In the jubilee year of 1500 he was in Rome, where he gave lectures, and 1501 he returned to Poland in order to take over a canonry, at the cathedral church of Frauenberg, a post which he had accepted in 1497. In 1503 Copernicus took his doctor's degree in canon law at Padua, and following this went back to Poland. Until 1510 he lived chiefly in Heilsberg and acted as a secretary to his uncle; after that year he moved to Frauenberg and was occupied by administrative tasks in the diocese. In 1521 Copernicus was sent as ambassador to the King of Poland to complain of the exploitation by the German Order, and in 1523 he became general administrator of the diocese during a vacancy. From about 1530 he devoted himself more and more to purely scientific matters.

The work of Copernicus shows his multifarious interests. In Cracow he issued (1509) a translation of a Greek author. In 1517 he worked out the first draft of a treatise on coinage. The treatise was published in Latin in 1528, under the title *Monetae cudendae ratio*, and came to be the foundation of a monetary reform, in the same way as did Oresme's earlier writing on the same subject in France. In 1524 came his first astronomical writing, a letter criticizing the German astronomer Werner's book *De motu octavae sphaerae* (Nuremberg, 1522). The letter circulated in copy, but was not printed until the nineteenth century (English transl. in Edw. Rosen, *Three Copernican treatises*, New York, 1959, 91–106). Another minor writing was also circulating in copy from about 1530: *Nicolai Copernici de hypothesibus motuum coelestium a se constitutis commentariolus*, generally called *Commentariolus* (English transl. in E. Rosen, op.

cit., 59–90). Through this treatise the learned world first became acquainted with the new Copernican system. In 1533 Pope Clement VII was informed of it by one Widmanstad, and 1536 Cardinal Nicolaus von Schönberg urged Copernicus to give a full account of the system. This did not happen, however, until the year of his death, at the request of the Wittemberg mathematician Rheticus. The work was entitled *De revolutionibus orbium coelestium* (Nuremberg, 1543, Basel, 1566, Amsterdam, 1617. Modern facsimile editions 1927, Turin, 1932 Amsterdam, 1943, München, 1944, English transl. *On the Revolutions of the Heavenly Spheres* by C. G. Wallis, Chicago, 1952, in *Great Books of the Western World*, **XVI**. German transl. *Ueber die Kreisbewegungen der Weltkörper*, Thorn, 1879, reprinted Leipzig, 1939).

Edition: *Gesamtausgabe von Nikolaus Kopernikus*, Band I (facsimile of a manuscript of *De revolutionibus*), München, 1944, **II** (critical edition by M. C. Zeller), München, 1949. The first complete edition of Copernicus' works will be published by the Polish Academy of Science from 1973 onwards.

A. Armitage, *Copernicus, founder of modern astronomy*, London, 1938.

A. Armitage, *Sun, stand thou still: the life and work of Copernicus*, New York, 1947 (more popular than the above).

H. Baranowski, *Bibliografia Kopernikowska* 1509–1955, Warszawa, 1958.

H. Blumenberg, *Die kopernikanische Wende*, Frankfurt a.M., 1965.

Brunet, *Archeion*, **XIX** (1937), 371 (On Copernicus's nationality).

Dreyer, 305–44.

G. Eis, *Lychnos* (1952), 186 (Copernicus as a physician.)

A. R. Hall, *The scientific revolution 1500–1800*, London, 1954, 51–72, and Appendix A.

G. Harig, *Die Tat des Kopernikus: Die Wandlung des astronomischen Weltbildes im 16. und 17. Jahrhundert*, Leipzig ,Jena Berlin, 1962.

A. Koyré, *La révolution astronomique*, Paris, 1961.

T. S. Kuhn, *The Copernican revolution. Planetary astronomy in the development of western thought*, Cambridge, Mass., 1957.

A. Müller, *Nikolaus Kopernicus, der Altmeister der neueren Astronomie*, Freiburg i. Br., 1898 (biographical).

O. Neugebauer, *Vistas in Astronomy*, **X** (1968), 89.

D. J. Price, in *Critical Problems in the History of Science*, ed. Clagett, Madison (Wisconsin) 1959, 197–218.

L. Prowe, *Nicolaus Copernicus*, 2 vols., Berlin, 1883–84. Reprinted Osnabrück, 1967.

J. R. Ravetz, *Astronomy and Cosmology in the Achievement of Nicolaus Copernicus*, Warszawa *1965*.

E. Rosen, *Three Copernican Treatises*, New York, 1939, reprint (Dover) 1959 3rd ed. 1971 (introduction, a good bibliography of the Copernicus literature); *Dict. Sci. Biogr.*, **III**, 401–411.

F. Rossmann, *Naturwissenschaften*, **XXXIV** (1947), 65 (on the *Commentariolus*).

E. Rybka, *Four hundred years of the Copernican heritage*, Cracow, 1964.

Studia Copernicana, **I** *ff*, Warszawa 1970 *ff*. (Several volumes of collected papers by A. Birkenmajer, M. Markowski, B. Bieńkowska, and others).

E. Zilzel, *Journal of the history of ideas*, **I** (1940), 113 (Copernicus and mechanics).

E. Zinner, *Die Entstehung und Ausbreitung der Copernicanischen Lehre*, Erlangen, 1943 (essential).

Cusanus
see Nicolaus Cusanus

Dante Alighieri
1265–1321

A Florentine poet and philosopher who held high office in his native city until the year 1302 when he was exiled as a member of the Guelf party. For the rest of his life he wandered all over Italy, and also went to Paris. He has earned a place in the history of world literature with his historical and philosophical Italian poem *La Divina Commedia*, existing in over 500 manuscripts and over 600 printed editions. The poem discusses almost every aspect of the Mediæval universe, and reveals the author's sound knowledge of science; the poem must be studied as an expression of the general cosmological ideas of an educated but non-specialist fourteenth century writer. Also, a *Quaestio de aqua et terra* (ed. and transl. C. L. Shadwell, Oxford, 1909) from 1320 is ascribed to Dante.

I. Capasso, 'L'astronomia nella Divina commedia—IV', *Physis*, **VII** (1965), 75–106 and 129–201; **VIII** (1966), 23–98.

E. Gilson, *Dante et la philosophie*, Paris, 1939. English transl. *Dante the philosopher* by D. Moore, London, 1948.

M. A. Orr, *Dante and the early astronomers*, London, 1914. Revised, 1956.

D. L. Roberts, *The scientific knowledge of Dante*, Manchester, 1914.

Sarton, **III**, 479.

Demetrios
of Phaleron
c. 350 to *c.* 270

Athenian statesman and author, disciple of Aristotle. In about 280 B.C. he prompted Ptolemaios Soter to found the Museion in Alexandria to provide a counterweight representing the new Hellenistic culture against the old schools in Alexandria. He himself became a librarian in Alexandria.

PW, **IV**, 2817–41.

Democritos
of Abdera
c. 460–370

One of the most eminent thinkers of the fifth century B.C. His biography contains many legendary features. It is certain, however, that he was a rich citizen of Abdera and that he embraced the atomic theory of Leucippos and opposed the doctrine of elements of Empedocles. He seems to have had as widespread a range of interests as Aristotle, who greatly valued his work, although their opinions were strongly divergent. 'No one but Democritos reached further than to the surface of any question.' Plato, on the other hand, never mentions him, and if we are to believe the rumour in Aristoxenes, Plato wanted to burn all the works of Democritos. Of his seventy-odd writings only a few fragments are extant, an indication of the small support for the atomic theory in Antiquity. The views of Democritos constitute the basis of the later Epicurean philosophy, and of all later materialism.

C. Bailey, *The Greek Atomists and Epicurus*, Oxford, 1928.
Burnet, 330*ff.*
G. B. Kerfeld, *Dict. Sci. Biogr.*, **IV**, 30–35.
Kirk and Raven, 400.

S. Luria, 'Die Infinitesimaltheorie der antiken Atomisten,' *Quellen und Studien*, **IIB** (1932–33), 106–85.
L. Mabilleau, *Histoire de la philosophie atomistique*, Paris, 1895.
Sarton, **I**, 88.
PW, **V**, 135.

Dietrich von Freiberg
(Theodoricus Teutonicus) of Freiberg
c. 1250 to *c.* 1311

A German Dominican and Master in Paris in 1297. In 1304, at the general chapter in Toulouse, he was ordered to publish the optical works, the most important of which is *De iride et radialibus impressionibus* (ed. J. Würschmidt, *Über den Regenbogen*, etc., Münster i.W., 1914) on meteorological optics. It builds partly on Alhazen, but contains, too, the first correct theory of the rainbow as a phenomenon of refraction. In philosophy Dietrich was an eclectic, but strongly influenced by Neoplatonism. In many ways he reminds one of Eckhart.

Crombie, *Grosseteste*, 233–59.
E. Krebs, *Meister Dietrich*, Münster i. W., 1906.
Sarton, **III**, 704.
W. A. Wallace, *The scientific methodology of Theodoric of Freiberg*, Fribourg, 1959.
W. A. Wallace, *Dict. Sci. Biogr.* **IV**, 92–95.

Digges, Thomas
c. 1545–*c.* 1595

Digges was inspector of the fortress in Dover, and the most important English mathematician and astronomer in the reign of Queen Elizabeth. He was an eager advocate of the Copernican system and was interested in verifying it by means of better astronomical observations, for example, in *Alae seu scalae mathematicae*, 1573, which also contained observations of Tycho Brahe's 1572 nova. In *A perfit description of the coelestiall orbes*, 1576 (new ed. F. R. Johnson, *Huntington Library Bulletin*, April 1934, 69–117) he edited an English translation of large sections of the first Book of Copernicus *De revolutionibus*, at the same time supposing the solar system to be an element of an infinite universe, completely filled with stars.

J. B. Easton, *Dict. Sci. Biogr.*, **IV**, 97f.
F. R. Johnson, *Osiris*, **I** (1937), 390–410.

Diodoros Siculus
of Agyrium
c. 50 B.C.

The author of a great but not particularly valuable *Biblioteke*, of the history of the world from the fall of Troy until Caesar's Gallic War (ed. C. H. Oldfather London, 1933).

Sarton, **I**, 231.
PW, **V**, 663–704.

Diogenes Laërtius
probably third century A.D.

A Greek historian of philosophy whose extant work in ten books, *The lives of the philosophers* (Greek text with English transl. R. D. Hicks, Loeb No. 184–85) is one of the most important sources of the older Greek science and philosophy.

Sarton, **I**, 318.
PW, **V**, 738–63.

Dominicus Gundissalinus
(Domingo Gundisalvo)
Twelfth century

Archdeacon of Segovia, a most studious translator from Arabic into Latin in collaboration with the Jew John of Spain (Johannes Hispalensis, Avendeut). He wrote various philosophical treatises and a book on the classification of the sciences, *De divisione philosophiae* (ed. L. Baur, München, 1903).

Duhem, **III** 179–82.
Sarton, **II**, 172–73.

Eckhart
of Hochheim (Gotha)
c. 1260–1327

A German Dominican, Master in Paris, in 1302, and the first of the great mystic theologians of the late Middle Ages to explore a path other than that of scholasticism. In 1326 he was accused of heresy, and in 1329 Johannes XXII declared a series of his theses to be unorthodox. Eckhart did not occupy himself with scientific questions, but he greatly influenced Cusanus, and hence, the Renaissance conception of the infinity of the universe.

K. Weiss (ed.) Meister Eckhart, *Die lateinischen Werke*, 6 vols, Stuttgart, 1936–55.
J. M. Clark, *Meister Eckhart: an introduction to the study of his works*, London, 1957.
Sarton, **III**, 568.

H. Wackerzapp, *Der Einfluss Meister Eckharts auf die ersten philosophischen Schriften des Nikolaus von Kues*, Münster, 1962.

Ecphantos
of Syracuse,
c. 400 B.C.

A Pythagorean and advocate of the atomic doctrine and of the theory of the daily rotation of the Earth. Possibly a disciple of Hicetas.

Sarton, **I**, 118.
Tannery, **VII**, 249–57.
PW, **V**, 2215.

Empedocles
of Agrigentum
c. 484 B.C. *c.* 424 B.C.

A Sicilian statesman, poet and philosopher. He systematized the doctrine of the four elements. He is often—probably mistakenly—considered to be the originator of the theory. His astronomical views were primitive; he explained day and night by supposing that a bright and a dark hemisphere rotated about the Earth. To him, the Sun was merely the light of the day-hemisphere reflected from the Earth. Empedocles is also supposed to be the founder of Sicilian medicine. Of his didactic poem *De natura* only 350 verses are extant.

J. Bidez, *La biographie d'Empédocle*, Gand, 1894, Burnet, 197–250.
G. H. Clark, *Empedocles and Anaxagoras in Aristotle's De anima*, Philadelphia, 1929.
Kirk and Raven, 320.

A. Mourelatos, *Dict. Sci. Biogr.*, **IV**, 367–369
D. O'Brian, *Empedocles' cosmic cycle*, Cambridge, 1969.
Sarton, **I**, 87.
F. Solmsen, *Phronesis*, **X** (1965), 109–48.
PW, **V**, 2507.
I. Zafiropulo, *Empédocle d'Agrigente*, Paris, 1953 (trans. of fragments).

Epicuros
of Athens
c. 341–270 B.C.

A Greek philosopher who had schools in Mytilene and on Lampsacos until he settled in Athens, where he founded a philosophical community of men and women, in 306 B.C. Nearly all his numerous works have been lost, including the poem *De natura*, in which he embraced the atomic theory by introducing into the motion of atoms a moment of arbitrariness, so that to him the theory implied no pure determinism. Some fragments have been preserved (*Epicurea*, ed. H. Usener, Leipzig, 1887; Greek text with English transl. in C. Bailey, *Epicuros*, Oxford, 1926, repr. New York, 1970).

C. Bailey, *The Greek atomists and Epicurus*, Oxford, 1928.
A. J. Festugière, *Épicure et ses Dieux*, Paris, 1946; English transl. *Epicurus and his Gods*, Oxford, 1955.
D. J. Furley, *Dict. Sci. Biogr.*, **IV**, 381–82
A. Goedemeyer, *Epikurs Verhältnis zu Demokrit*, Strassbourg 1897.
Sarton, **I**, 137.
PW, **VI**, 133–55.
N. W. de Witt, *Epicurus and philosophy*, Minneapolis, 1954.

Eratosthenes
of Cyrene
c. 275 B.C. to *c.* 194 B.C.

Eratosthenes studied in Athens and later became a librarian in Alexandria. He was the most wide-ranging scientist of his time and the first to call himself a 'philosopher'. In mathematics he studied prime numbers (Eratosthenes' sieve), the doubling of the cube, etc. He was the first Greek systematic geographer. His treatise *On the measuring of the Earth* (lost) describes the first determination of the circumference of the Earth. The account of this has been preserved in Cleomedes. Also extant are some geographical fragments (Eratosthenes, *Die geographischen Fragmente*, ed. H. Berger, Leipzig, 1880, reprinted Farnham, 1964).

D. R. Dicks, *Dict. Sci. Biogr.*, **IV,** 388–93.
Heath, **II,** 104.
T. L. Heath, *A manual of Greek mathematics*, Oxford, 1931 reprinted New York, 1963.
Sarton, **I,** 172.
Tannery, **III,** 358.
PW, **VI,** 358–88.
E. P. Wolfer, *Eratosthenes von Kyrene als Mathematiker und Philosoph*, Groningen, 1954.

Etienne Tempier
d. 1279

Bishop of Paris; in 1270 and 1277 he checked the Aristotelianism at the University there by publishing a long list of Aristotelian and Averroistic propositions which it was forbidden to teach as true (ed. Denifle and Chatelain, *Chartularium Universitatis Parisiensis*, **I,** Paris, 1889, 541 *ff.*)

Duhem, **VI,** 20–29.
J. Koch, *Mélanges Mandonnet* **II,** Paris. 1930, 305 *ff.*
P. Mandonnet, *Siger de Brabant et l'averroïsme Latin*, **I,** Louvain, 1911, 196 *ff.*

Euclid
c. 300 B.C.

If not the most important, certainly the most well-known, mathematician of Antiquity, working under Ptolemaios I in Alexandria. His fame is chiefly due to his great synopsis of Greek mathematics *Stoikeia*, later called Euclid's *Elements*, throughout the ages the most consulted textbook of geometry. It was translated from Arabic into Latin about 1120 by Adelard of Bath, and later, in the thirteenth century, by Johannes Campanus, whose translation was used for the first of innumerable printed editions (Venice, 1482). A series of other mathematical writings is no longer complete. Further, Euclid wrote on astronomy in *Phaenomena*, on optics and perspective in *Optica* and *Catoptrica* (French transl. P. Ver Eecke, Bruges, 1938), as well as on the theory of music.

Editions: *Euclidis Opera omnia*, 8 vols, ed. J. L. Heiberg and G. Menge, Leipzig, 1883–1916.
T. L. Heath, *The thirteen books of Euclid's Elements* **I–III** Cambridge, 1926.
I. Bulwer-Thomas and J. Murdoch, *Dict. Sci. Biogr.*, **IV,** 414–59.
Clagett, *Isis*, **XLIV,** (1953), (Latin transl.).
A. G. Kapp, *Isis*, **XXII** (1934), 150–72; **XXIII** (1935), 54–99; **XXIV** (1935), 37–79 (Arabic translations).
A. Lejeune, *Euclide et Ptolemée. Deux*

stades de l'optique géométrique grecque,
Louvain, 1948.
Sarton, **I**, 153.
C. Thomas-Stanford, *Early Editions of
Euclid's Elements*, London, 1926.
L. Vanhée, *Isis*, **XXX** (1939), 84 (Euclid
in Chinese).
PW,**VI**, 1003–52.

Euctemon
Second half of fifth century B.C.

A Greek astronomer and geographer
who with Meton in Athens in 432
determined the solstitial points and the
length of the year.

D. R. Dicks, *Dict. Sci. Biogr.*, **IV**, 459–60.
Duhem, **I**, 108.
Sarton, **I**, 94.
PW,**VI**, 1060.

Eudemos
of Rhodes
Second half of fourth century B.C.

A disciple of Aristotle, and later principal
of his own school, presumably on the
island of Rhodes. He wrote historical
expositions of Greek mathematics, astronomy and religion. Only fragments of
these are extant(*Fragmenta*, ed. L. Spengel,
Berlin, 1866), but a large part of the
contents has been preserved through
passing into later works by Theon,
Pappos and others.

I. Bulwer-Thomas, *Dict. Sci. Biogr.*, **IV**,
460–65.
Sarton, **I**, 140.
Tannery, **I**, 168.
PW,**VI**, 895.

Eudoxos
of Cnidos
c. 408–355 B.C.

One of the greatest mathematicians of
Antiquity, Eudoxos was a pupil of
Archytas. As a young man he studied in
Egypt from 381–380, and afterwards
founded a school in Cyzicos which, in
368, he combined with Plato's Academy
in Athens. In mathematics he evolved
both the treatment of the doctrine of
proportions known from Euclid's fifth
and sixth Books, and the method of
exhaustion for determining areas and
volumes. In astronomy he was the first
to prepare a mathematical theory of
planetary motion on the basis of a
geometrical model of the solar system
(the Eudoxian system with concentric
spheres), and can thus be regarded as the
true founder of theoretical astronomy.

L. Bechler, *Centaurus*, **XV**, (1970), 113.
Dreyer, 87 *ff.*
Eudoxus, *Die Fragmente des Eudoxus von
Knidos*. ed. transl. und commentary
by François Lasserre, Berlin, 1966.
Heath, **II**, 322.
J. Hjelmslev, *Centaurus*, **I**, (1950), 2.
G. L. Huxley, *Dict. Sci. Biogr.*, **IV**, 465–67.
O. Neugebauer, *Scripta Mathematica*
XIX (1953) 225 *ff.* (the *Hippopede*).
G. de Santillana, *Isis*, **XXXII** (1949),
248–62 (Eudoxos and Plato).
G. Sarton, **I**, 117.
G. Schiaparelli, *Abhandl. z. Gesch. d.
Mathematik*, 1877, 101 (fundamental).
H. N. Swenson, *Amer. Journ. Physics*,
XXXI (1963), 456.
Tannery, **III**, 366.
PW,**VI**, 930–50.

Eugenius
of Palermo
c. 1160

A Greek-Sicilian official under William the First (1154–66). He translated Ptolemy's *Optics* from Arabic into Latin (ed. A. Lejeune, *l'Optique de Claude Ptolémée*, Louvain, 1956) and contributed to the so-called 'old Sicilian' translation of the Almagest.

Haskins, 171–78.
Sarton, **II**, 346.

al-Farghānī
of Farghānā (Transoxania)
c. 800 to *c.* 870

One of the best astronomers at the time of al-Ma'mūn. In 861 he organized the Nilometre at Fustat. He wrote two (unpublished) treatises about the astrolabe. Under the name Alfraganus he was known in the Middle Ages as the author of a widely read book, *Compilatio astronomica*, also called *Liber 30 differentiarum*, transl. Gerard of Cremona and Johannes Hispalensis (Ferrara, 1493, later printed by Melanchthon, Nuremberg, 1537), which in the thirteenth century was translated into Hebrew by Jacob Anatoli. He also wrote about the construction of sundials.

Duhem, **II**, 204–14.
Nallino, *Scritti,* **V**, 332.
A. I. Sabra, *Dict. Sci. Biogr.*, **IV**, 541–45.
Sarton, **I**, 567.
Suter, 18–19.

Galenos
of Pergamon
A.D. 129–199

A Greek physician who worked in Rome and performed valuable anatomical and physiological studies, the latter partly of an experimental character. He applied the physiological theory of elements to a psychological theory of the four temperaments. His writings were widely circulated in Latin translation, and highly esteemed until the introduction of modern anatomy in the sixteenth century (Vesalius, 1543).

Opera omnia, 20 vols, Leipzig 1821–33 (Greek text and Latin transl.).
Sarton, **I**, 301–07; *Galen of Pergamon*, Kansas, 1954.
PW, **VII**, 578–91.

Geber
see Jābir

Geminos
of Rhodes
about 70 B.C.

A Stoic philosopher and author of, among other volumes, a great exposition of the classification and principles of the mathematical sciences, and of an elementary textbook of astronomy from Hipparchos's standpoint (ed. with German transl. Manitius, *Gemini elementa astronomiae*, Leipzig, 1898).

Sarton, **I**, 212.
PW, **VII**, 1026–50.

Georg Peurbach
1423–1461

An Austrian astronomer and one of the pioneers of the rise of astronomy in the fifteenth century. He was born in Peurbach near Linz, and excelled as professor in Vienna, where Regiomontanus was among his pupils. In addition, he was astronomer and astrologer to King Ladislas of Hungary. His main work is a *Nova theorica planetarum* (Augsburg, 1482) which became widely used as a textbook of planetary theory. He taught himself Greek and at the request of Cardinal Bessarion began a new translation of the Almagest, continued later by Regiomontanus (*Epitome in Almagestum Ptolemæi*, Venice 1496). His biography has been written by Pierre Gassendi.

Dreyer, 288 *ff.*
Duhem, **X**, 351.
P. Gassendi, *Tychonis Brahei . . . vita. Accessit N. Copernici, G. Peurbachii et Joannis Regiomontani vita*, Paris, 1654.
Thorndike, *Science and thought in the 15th century*, New York, 1929, 142.
E. Zinner, *Leben und Wirken des Johannes Müller*, München, 1938 (reprinted Osnabrück, 1968).

George
of Trebizont
1395–1484

A Greek philosopher who, from about 1430, worked as a teacher in Italy and later as papal secretary. He translated numerous Greek works into Latin, including Aristotelian and Platonic writings, and Ptolemy's *Almagest* and *Tetrabiblos*. In the debate between Platonists and Aristotelians he supported the Aristotelians against Bessarion, with for example, the treatise *Comparison between Plato and Aristotle*, which led to a huge controversy.

Duhem, **X,** 358*ff.*

Gerard
of Bruxella
First half of the thirteenth century

A Flemish physicist, the author of a treatise *De motu* (ed. Clagett, *Osiris*, **XII** (1956), 73–175) which is the first purely kinematical work of the Middle Ages. It played an important part in the emancipation of kinematics in the fourteenth century. Apart from this work, the author is practically unknown.

Clagett, 163 *ff.*
Sarton, **II**, 629.

Gerard
of Cremona
c. 1114–1187

An Italian scholar who went to Toledo in 1134 eager to become acquainted with the *Almagest*. In Toledo he became the twelfth century's most studious translator from Arabic into Latin. With his collaborators he translated about 90 writings by Aristotle, Themistios, Ptolemy, and others.

A. A. Björnbo, *Bibl. Math.*, **VI** (1905), 239.
Clagett, *Isis*, **XLIV** (1953), 27*ff.*
Haskins, 104.
Sarton, **II**, 338.
K. Sudhoff, *Archiv. f. Geschichte der Medizin*, **VIII**, (1914), 73.

363

Gerardus
de Sabionetta
c. 1260

An Italian astronomer and astrologer, often confused with Gerardus Cremonensis. He is sometimes wrongly supposed to be the author of that *Theorica planetarum* which in the Middle Ages was the commonly used textbook for Ptolemy's planetary theory. It is preserved in hundreds of manuscripts and early editions (the best: Venice, 1478). His *Geomantia* enjoyed a wide popularity as an astrological textbook.

Haskins, 257.
O. Pedersen, *Classica et Mediaevalia*, **XXIII** (1962), 225.
Sarton, **II**, 987.

Gerbert
of Aurillac
c. 940–1003

A Benedictine monk, who became a teacher at the school of the court, abbot in Bobbio, archbishop in Rheims, and Pope under the name of Sylvester II (999–1003). As a young man he studied the quadrivium in Spain and became one of the first intermediaries of Moslem science in Europe. In his writings (ed. N. Bubnov, *Gerberti opera mathematica*, Berlin, 1899, and J. Havet, *Les lettres de Gerbert*, Paris, 1889) he treated of the astrolabe and the abacus, and he is the first Christian author to describe ghubār numerals (without the zero).

Duhem, **III**, 163 *ff.*
Sarton, **I**, 669–71.
Thorndike, **I**, 697–718.
H. Weissenborn, *Gerbert*, Berlin, 1888.

Giovanni de' Dondi
of Chioggia
1318–1389

An Italian physicist and astronomer, a friend of Petrarca, who was attached to the universities of Florence and Padua. He was a son of the physician Jacopo de' Dondi, and like his father became particularly known as the maker of mechanical clocks with verge escapement, which spread over Europe in the late Middle Ages. The family adopted the surname of dall'Orologio. Giovanni described the clock in his treatise *Tractatus astrarii* (ed. Città del Vaticano, 1960, with introduction by E. Morpurgo).

S. A. Bedini, Fr. R. Maddison: *Mechanical universe; the Astrarium of Giovanni de 'Dondi*. Transl. Am. Phil. Soc., 1966.
E. L. Edwardes, *Weight-driven chamber clocks of the Middle Ages and Renaissance*, Altringham, 1965.
D. J. Price, *On the origin of clockwork, perpetual motion devices, and the compass*, United States National Museum Bulletin No. *218* (1959), 81.
Sarton, **III**, 1676.
Thorndike, **III**, 386–97; *Archeion*, **XVIII** (1936), 308–17; *Isis*, **X** (1928), 360.
E. Zinner, *Aus der Frühzeit der Räderuhr*, München, 1954.

Grosseteste
see Robert Grosseteste

Hecataeos
of Miletos
c. 550 B.C. to c. 475 B.C.

An Ionian geographer and cartographer whose work *Circuit of the Earth* was important to Herodotos, who looked upon

Hecataeos as the originator of geography. The work is only partly preserved (R. H. Klausen, *Hecataei Milesii fragmenta*, Berlin, 1831.

Diels, *Hermes*, **XXII** (1887), 411–44.
Sarton, **I**, 78.
PW, **VII** (1922), 2667.

Henricus Aristippus
c. 1100–1165

A translator and archdeacon of Catania on Sicily who brought Greek manuscripts from Constantinople to Sicily. He translated Aristotle's *Meteorologica*, Plato's *Meno* and *Phaidon* and other works.

Haskins, 165–93.
Sarton, **II**, 346–47.

Henricus de Hassia
(Heinrich von Langenstein)
1325–1397

A German astronomer, educated in Paris and who was from 1383 to his death professor in the University of Vienna (founded 1365), being one of the driving forces behind its re-foundation. He wrote on many different subjects, but is mainly remembered for a series of anti-astrological writings, for example a *Questio de cometa* 1368 and the *Tractatus contra astrologos*, 1373 (ed. H. Pruckner, *Studien zu den astrologischen Schriften des H. von L.*, Berlin, 1933). Another treatise from 1364 *De reprobatione eccentricorum et epiciclorum* is unpublished, but extant in several MSS. The work is an attack on Ptolemaic astronomy and ends with an alternative theory using only concentric spheres.

Duhem, **VII**, 569 and 585.
O. Hartwig, *Henricus de Langenstein*, Marburg, 1859.
C. Kren, *Isis*, **LIX** (1968), 269 (Homocentric astronomy).
Sarton, **III**, 1502.
Thorndike, *Isis*, **XXXI** (1939), 68; **XXXII** (1940), 417.

Heracleides
of Pontos
c. 390 B.C. to *c.* 310 B.C.

A Greek philosopher from Pontos (Heraclea) on the Black Sea. He went to Athens and became a pupil of Plato and his deputy at the Academy. When, after Plato's death, he was refused as the head of the Academy in favour of Xenocrates, he returned to his native city, where he opened his own school. None of his scientific writings has been preserved, but it is known that he tried to work out a sort of molecular theory which was to surmount some of the difficulties of Democritos's atomic theory, just as he maybe on the basis of the Pythagorean Ecphantos—elaborated a planetary system in which Venus and Mercury orbit about the Sun which itself orbits about the Earth. It has also been maintained that he placed the Sun at the centre of the Universe, as Aristarchos and Copernicus did later, but this is not certain.

Heath, *Aristarchus of Samos*, Oxford, 1913, 249–83.
Sarton, **I**, 141.
H. Staigmüller, *Archiv f. Gesch. d. Philosophie*, **XV** (1902), 141–65.
PW, **VIII**, 472–84.

Heraclitos
of Ephesos
c. 500 B.C.

An Ionian philosopher who taught that fire is the principle of all things. Maintaining that 'all things are in process and nothing stays still' he raised philosophical problems which the atomists and the supporters of the doctrine of elements later on tried to solve in different ways. Only fragments of his work remain (Heraclitus, *Cosmic Fragments*, ed. G. S. Kirk, Cambridge, 1954; Heraklit, *Fragmente griechisch und deutsch* von B. Snell, 5. Aufl. München, 1965).

A. Brieger, *Hermes*, **XXXIX** (1904), 188-223 (On the physics of Heraclitos).
Burnet, 130–68.
P. Jeannière, *La pensée d'Héraclite d'Ephèse, avec la trad. intégrale des fragments*, Paris, 1959.
Kirk and Raven, 182.
Sarton, **I**, 85.
P. Wheelwright, *Heraclitus*, Princeton, 1959.
PW, **VIII**, 504.

Hermannus Contractus
(H. the Lame; H. of Reichenau)
1013–1054

Abbot of the Benedictine Monastery of Reichenau and one of the best eleventh century proponents of the quadrivium, wrote on the use of the abacus, on a mathematical game called 'rithmomachia', and on theory of music (*Musica*, ed. L. Ellinwood, Rochester N.Y., 1952). Best known are two works on the astrolabe patterned on Arabian models, *De mensura astrolabii*, and *De*

utilitatibus astrolabii (ed. J. Drecker, *Isis*, **XVI** (1931), 200–19). In addition, he wrote a world chronicle up to 1054.

Opera omnia, Paris, 1853, Migne, P. L. 143.
Duhem, **III**, 165.
Sarton, **I**, 757.
Thorndike, **I**, 701.

Hermannus Dalmata
(H. of Carinthia; H. Secundus)
c. 1140

Educated at Chartres and Paris, Herman lived in Spain in about 1138–1142 and translated various works from Arabic into Latin.

A. A. Björnbo, *Bibl. Math.*, **IV** (1903), 130.
Haskins, 43.
Sarton, **II**, 173.

Herodotos
of Halicarnassos
c. 484 B.C. to *c.* 425 B.C.

The first great Greek historian and geographer, who on his extensive journeys in the Orient gathered material for his great *History* in nine books (4 vols, ed. Godley, London, 1921–24). A series of important information on Egyptian and Babylonian science and technology is contained in this work.

J. L. Myres, *Herodotos, father of History*, Oxford, 1953.
Sarton, **I**, 105.
PW, Suppl. 2, 205–520.
J. Wright, *Scientific monthly*, **XVI** (1923), 638 (on the science of Herodotos).

Heron
of Alexandria
1st century A.D.

An Alexandrian polytechnologist. With his *Mechanics* (preserved in Arabic translation), his *Pneumatics*, and his *Automata* he continued the mechanical-pneumatic tradition of Ktsebios and Philon. His efforts mainly involved the construction of various machines, and the examination of the mode of operation of the five simple machines. *Catoptrica* deals with optics, and *Dioptra* describes a geodetic measuring instrument in which a diopter is adjusted by means of a worm gear. His *Metrica* treats the subject of surveying and, among other things, contains 'Heron's formula' giving the area of a triangle in terms of its sides.

Opera omnia 5 vols, ed. C. G. Schmidt, *et al.* (Greek text with German transl.), Leipzig, 1899–1914. Heron, *Les mécaniques*, ed. de Vaux, *Journal asiatique*, **I**, 386–476; **II**, 152–269 and 420–574, Paris, 1893.
M. Boas, *Isis* **XL** (1949), 38 (Pneumatica).
H. Diels and E. Schramm, *Herons Belopoiika* (Schrift von Schützbau, Greek text with German transl.), Berlin, 1918.
A. G. Drachmann, *Ktsebios, Philon and Heron*, Copenhagen, 1948.
A. G. Drachmann, *Centaurus*, **I** (1950) 117; **VIII** (1963), 91.
Heath, **II**, 298.
O. Neugebauer, *Über eine Methode zur Distanzbestimmung Alexandria-Rom bei Heron* 2 vols, Copenhagen, (ed. V. Selsk.) 1938–39.
Sarton, **I**, 208.
PW, **VIII**, 992–1080.

Heron
of Byzantium
First half of the tenth century

Heron flourished in Constantinople in about 938 and wrote on surveying and war machines in the tradition of Heron of Alexandria. His *Peri dioptras* describes geodetic instruments (ed. A. J. H. Vincent, *Notices et Extraits*, **XIX** (1858), 348–407, Greek text with French transl.). His *Liber de machinis bellicis* was translated into Latin by Franciscus Barocius (Venice, 1572).

Sarton, **I**, 632.
PW, **VIII**, 1074–77.

Hesiod
of Ascra, Boeothia
c. 700 B.C.

One of the oldest Greek poets whose works have been preserved (English transl. H. G. Evelyn-White, London, 1914, French by Mazon, *Hesiode*, Paris, 1928). *Works and Days* describes the farmer's work during the year and reveals the astronomical knowledge of the Greeks before the scientific revolution, while *Theogony* treats of the evolution of the Universe on a mythological basis.

G. Arrighetti, *Studi Class. Orient.*, **XV** (1966), 1–60.
R. Eisler, *Isis*, **XL** (1949), 108 (metallurgy).
Sarton, **I**, 57.
F. Schwenn, *Die Theogonie des Hesiodos*, Heidelberg, 1934.
M. C. Stokes, *Phronesis*, **VII**, (1962) 1–37; ibid., **VIII** (1963), 1–34.
PW, **VIII**, 1167–1240.

Heytesbury
see William Heytesbury

Hicetas
of Syracuse
Fifth century B.C.

A Pythagorean astronomer who assumed the rotation of the Earth around its own axis, but about whom very little else is known.

Dreyer, 49.
Sarton, **I**, 94.
PW, **VIII**, 1597.

Hipparchos
of Nicaea
c. 190 B.C. to 120 B.C.

The most outstanding of all Greek astronomers, Hipparchos combined an extensive knowledge of the early Babylonian and Greek astronomy with great skill in instrument making, as he was an expert observer and theorist. His writings have been lost but for a single, less important one (*In Arati et Eudoxi phaenomena commentariorum libri tres*, ed. Manitius Leipzig, 1894, Greek text with German transl.) and some geographical fragments (ed. D. R. Dirks, 1960), and his efforts are known mainly from Ptolemy, whose *Almagest* is partly based on Hipparchos. Among his most important achievements were the discovery of the precession, the compilation of a catalogue of the fixed stars, the calculation of lunar eclipses for a period of 600 years with a view to the determination of geographical length, and a theory for the Sun's movement on an eccentric circle, which was the first

Greek astronomical theory providing a relatively satisfactory model of an astronomical phenomenon.

A. Aaboe, *Centaurus*, **IV** (1955), 122 (Babylonian presuppositions).
F. Boll, *Bibl. Math.*, **II** (1901), 185–195.
Heath, **II**, 253.
V. M. Petersen and O. Schmidt, *Centaurus*, **XII** (1967), 73–96 (Solar theory).
Sarton, **I**, 193.
G. J. Toomer, *Centaurus*, **XII** (1967), 145–50 (Lunar theory).
PW, **XVI**, 1666–81 (fundamental).

Hippocrates
of Chios
c. 470 B.C. to c. 400 B.C.

A Greek mathematician who compiled the first survey of the elements of mathematics. The treatise has not been preserved. Hippocrates worked on the Delian problem (the duplication of the cube) and on the determination of the area of figures (lunulae) confined by two circular arcs.

Heath, **I**, Oxford, 1921, 182–202.
Sarton, **I**, 91–92.
PW, **VIII**, 1780–1801.

Hippocrates
of Cos
Second half of fifth century B.C.

Hippocrates of Cos was the first great representative of Greek medicine. He endorsed the doctrine of elements and applied it to physiology and pathology. The numerous 'Hippocratic' writings are not authentic, although many of them can be traced back to Hippocrates' own time.

368

Hippocrates. Greek text with an English transl. by W. H. S. Jones, vols. 1–4, London 1923–1931.
Oeuvres complètes, 10 vols, ed. M. Littré, Paris, 1839–61 (Greek text with French transl.).
Sarton, **I**, 96–102.
PW, **VIII**, 1801–52.

Hippolytus
First half of the third century A.D.

One time Bishop of Portus Romanus, Hippolytus was later deported. He wrote a great work in 10 books, *Philosophumena* often referred to as *Refutatio omnium haeresium* (English transl. F. Legge, 2 vols, London, 1921) against heretics. It is an important source of the philosophy, magic, astrology, etc. of Antiquity.

Sarton, **I**, 320–21.
Thorndike, **I**, 466–69.
PW, **VIII**, 1873–78.

Homer
Eighth century B.C.

Homer is the oldest Greek author. His epic poems the *Iliad* and the *Odyssey* give a glimpse of the primitive, pre-scientific Greek knowledge of nature (astronomy, botany, medicine) and technology (metallurgy, shipbuilding).

G. Arrighetti, *Studi Class. Orient.,* **XV** (1966), 1–60.
F. Buffière, *Les mythes d'Homère et la pensée grecque,* Paris, 1956.
C. Mugler, *Les origines de la science grecque chez Homère,* Paris, 1963.
Sarton, **I**, 53.
PW, **VIII**, 2188–2247.

Iamblichos
of Chalcis
c. 300 A.D.

A Neoplatonic philosopher, Iamblichos was the author of a survey of the doctrine of the Pythagoreans, consisting of nine books; five of these books have been preserved, as an important source of the history of Pythagorean number mysticism. The section of his work on Pythagoras must be regarded as a novel summing up of the ancient tradition of this somewhat legendary figure. (*De vita pythagorica,* etc., ed. T. Kiessling, 2 vols, Leipzig, 1815–16 (Greek text with Latin transl.; English transl. T. Taylor, London, 1818, reprinted, London, 1965; *Iamblichos, Pythagoras. Legende, Lehre, Lebensgestaltung.* Griechisch und deutsch, ed. Michael von Albrecht, Zürich Stuttgart, 1963).

Sarton, **I**, 351.
PW, **IX**, 645.

Ibn al-Haitham
of Basra
c. 965–1039

In Latin Alhazen, the greatest physicist of the Moslems, Ibn al-Haitham did his best work in Egypt under al-Hakim (996–1020). His most important achievement deals with spherical mirrors, the refraction of light, which is examined experimentally, and other optical subjects. His *Optics* (ed. Baarmann, *Abhandlung über das Licht von Ibn al-Haitham,* Arabic text with German transl. *Zts. Deuts. Morgenl. Ges.,* **XXXVI** (1882), 195–237; older Latin transl. ed. Risner, *Alhazeni Opticae Thesaurus,* Basel, 1572, together with Vitelo's *Optics*) forms a starting

369

point for Vitelo and was still of importance in the sixteenth century. As an
astronomer, Alhazen was very critical of
Aristotle, stressing the necessity for a
'physical' astronomy of material spheres.

de Vaux, **II**, 243.
Duhem, **II**, 119.
D. Lindberg, *Isis*, **LVIII** (1967), 321.
Nallino, *Scritti*, **V**, 339.
S. Pines, *Proc. 10th Int. Congress Hist. Sci.
Ithaca 1962* (1964), 547.
A. I. Sabra, *Journ. Hist. Phil.*, **IV** (1966),
145.
Hakim Mohammed Said (ed.), 'Ibn al-
Haitham, Proceedings of 1000th anni-
versary,' *Hamdard* **XIII** (1970) Nos 1–2
(with many of al-Haitham's minor works
in English transl.).
Sarton, **I**, 721.
M. Schramm, *Ibn al-Haythams Weg zur
Physik*, Wiesbaden, 1963.
Suter, 91.
Wiedemann, **I**, 379, 519 *ff.*, 597; **II**, 69,
87, 739, 756.
H. J. J. Winter, *Centaurus*, **III** (1954), 190.

Ibn al-Muthannā
probably tenth century A.D.

A Moslem astronomer about whom very
little is known, except that he worked in
Spain and left a *Commentary on the astronomical tables of al-Khwārizmī*, extant in a
Hebrew version (ed. and transl. B. R.
Goldstein, New Haven, 1967) as well as
in a Latin translation by Hugo Sanctal-
lensis (ed. E. Millas-Vendrell, *El com-
mentario de ibn al-Mutanna*, Barcelona-
Madrid, 1963).

J.–M. Millas-Vallicrosa, *Isis*, **LIV** (1963),
114.

Ibn ash-Shātir
1304–1376

An Arabian astronomer who lived in
Damascus. He worked out a planetary
theory which in many ways fore-
shadowed that of Copernicus, as it discarded Ptolemy's equant circles. His
lunar theory is identical to that of Copernicus. One cannot, however, infer any
direct influence, as his treatise was never
translated into Latin; and Ibn ash-Shātir
does not follow Copernicus in transferring the centre of the universe to the
Sun.

Victor Roberts, *Isis*, **XLVIII** (1957), 428–
32; ibid., **LVII** (1966), 208–19.
Kennedy, 162.
E. S. Kennedy and Victor Roberts, *Isis*,
L (1959), 227–35.
Suter, 168.
Wiedemann, **II**, 729.

Ibn Rushd
of Cordoba
1126–1198

In Latin Averroës, the last important
Moslem philosopher and the most
Aristotelian of all Arabian commentators on Aristotle. Ibn Rushd departs
from the traditional Arabian attempt to
interpret Aristotle in terms of neo-
Platonism and through his triple set of
commentaries (a brief, a medium, and
an advanced one) exerted great influence
upon the Aristotelianism of the thirteenth
century. He was known simply as the
'Commentator'. His Latin followers, or
'Averroïsts', were the most outspoken
Aristotelians of the later Middle Ages.
In astronomy most of them advocated
the 'physical' planetary system, rather
than that of Ptolemy.

Corpus commentarium Averroës in Aristotelem, ed. H. A. Wolfson *et al.* Cambridge, Mass., 1949–.

de Vaux, **IV**, 65.

Duhem, **I**, 234–39; **II**, 133–39; ,511–**IV** 20, 532–76.

L. Gauthier, *Averroës*, Paris, 1949.

P. Mandonnet, *Siger de Brabant et l'averroïsme latin du XIIIe siècle*, Fribourg, 1899.

E. Renan, *Averroës et l'Averroïsme*, Paris, 1852 (many later editions).

Sarton, **II**, 355–61.

Suter, 127.

de Vaux, **IV**, 65–83.

Ibn Sīnā
of Bukhārā
980–1037

Islam's most famous scientist, Ibn Sīnā's fame reached far beyond the Arabic-speaking world. In the Latin Middle Ages he was known as Avicenna, and his reputation as a medical author was as great as that of Galen. His large encyclopedia *Qānūn* (*Canon*) was translated by Gerard of Cremona, and during the fifteenth century was frequently reprinted. In the East new editions still appear today. His great composite philosophical work *Kitab al-shifā* (German transl. Horten, *Das Buch der Genesung der Seele*, Halle, 1907–09) contains many original scientific views. He advanced a corpuscular theory of light, thus postulating a finite speed for light. He also investigated specific gravity but differs from all contemporary alchemists in denying the possibility of transmutation of metals. His treatise on minerals is the main source for Mediaeval mineralogy and geology.

S. M. Afnan, *Avicenna, His life and works*, Hemel Hempstead, 1968.

C. de Vaux, *Avicenna*, Paris, 1900.

C. de Vaux, **II**, 263 and **IV**, 18.

Duhem, *Études*, **II** (1909), 302–09 (mineralogy).

A. M. Goichon, *La philosophie d'Avicenne et son influence en Europe médiévale*, Paris, 1944; 2nd ed. 1951.

J. Ruska, *Isis*, **XXI** (1934), 14–51.

Sarton, **I**, 709–13.

Suter, 86–90.

G. M. Wickens, *Avicenna, scientist and philosopher, A millenary symposium*, London, 1952.

Wiedemann, **I**, 142, 146; **II**, 437, 650.

Ibn Yūnus
c. 940–1009

One of the foremost Arabian astronomers, Ibn Yūnus lived in Cairo, where he compiled the important *Hakemite Tables* (Arabian text with French transl. M. Caussin, *Notices et extraits*, **VII** (an **XII** = 1804), 16–240) containing observations of eclipses and conjunctions, important astronomical constants, and an account of the geodetic measurements of al-Ma'mūn. In addition, he contributed considerably to the development of trigonometry.

Kennedy, 126.

Nallino, *Scritti*, **V**, 338.

Sarton, **I**, 716.

C. Schoy, many articles in various periodicals.

Suter, 77.

Isḥāq ibn Ḥunain
of Baghdad
c. 850–910

One of the most important translators of Greek scientific works into Arabic, he was the author of the standard translation of the *Almagest*, revised by Thābit ibn Qurra.

Sarton, **I**, 600.
M. Steinschneider, *Bibl. Math.*, **VI** (1892), 53.

Isidore of Seville
c. 560–635 A.D.

A native of Cartagena or Seville, Isidor was Bishop of Seville from about 600. He was the first of the great Mediæval encyclopaedists who tried to make essential parts of Ancient science accessible by presenting them in a lexicographic form. Throughout the Middle Ages his main work, *Etymologiarum libri xx* (ed. W. M. Lindsay, Oxford, 1911) was a much used work of reference, while the shorter *De natura rerum* (ed. with French transl. J. Fontaine, *Traité de la nature*, Bordeaux, 1960) served as a model for many similar expositions of the elements of astronomy and physics.

Opera, in Migne, PL 81–84.
Duhem, **III**, 1–12.
J. Fontaine, *Isidore de Seville et la culture classique*, Bordeaux, 1959.
Sarton, **I**, 471–72.
PW, **IX** 2069–80.

Jābir ibn Afflaḥ

Spanish-Arabic astronomer, of whose life very little is known. He probably worked in Seville, about the middle of the twelfth century, and had a considerable influence on Mediæval astronomy and trigonometry through his great critical paraphrase of the *Almagest*, translated by Gerard of Cremona as *De astronomia libri IX* (printed as an appendix to Petrus Apianus, *Instrumentum primi mobilis*, Nuremberg, 1534). His trigonometry is more advanced than Ptolemy's (proof of the spherical relation $\cos A = \cos a . \sin b$). In astronomy he was a fierce critic of Ptolemy on many points of detail, although he retains the main features of his astronomy.

Duhem, **II**, 172.
Nallino, *Scritti*, **V**, 342.
Sarton, **II**, 206.
Suter, 119.

Jābir Ibn Hayyān
Second half of the eighth century

An Arabian alchemist (named Geber in Latin) who lived in Kufa. Much material is, more or less correctly, attributed to him. Most of these works are unpublished and the majority of the Latin treatises circulating during the Middle Ages as 'Geber's Works' (E. J. Holmyard, *The Works of Geber*, Anglicised by Richard Russell (1678), London, 1928) are not authentic, although they are very important to the historian of Mediæval chemistry. A book on poisons is translated into German (Jabir, *Das Buch der Gifte*, übers. A. Siggel, Wiesbaden, 1958).

M. Berthelot, *La chimie au moyen âge*, Paris, 1893.
de Vaux, **II**, 369.
E. J. Holmyard, *Alchemy*, London, 1957, chapter 5.

P. Kraus, *Jābir ibn Hayyān: essai sur l'histoire des idées scientifiques dans l'Islam*, 2 vol., Paris, Le Caire, 1935–43.
Sarton, **I**, 532.
Many articles by J. Ruska in various periodicals.

Jacob Anatoli
Probably of Marseille
c. 1195 to *c.* 1250.

A Provençal Jew who lived in Narbonne and later at the court of Frederic II on the island of Sicily, Jacob translated, among other works, the *Almagest* with Averroës' summary from Arabic into Hebrew, and was one of the most important intermediaries between the Arabic, the Jewish, and the Latin cultures in the Middle Ages; thus his translation of al-Farghānī was made from the Latin version.

Sarton, **II**, 565.

Jacob ben Mahir
of Marseilles
c. 1236 to *c.* 1304.

A Jewish astronomer, in Latin known as Profatius Judaeus. He worked mostly in Montpellier as a diligent translator from Arabic into Hebrew, and as a constructor of the so-called *quadrans novus* which replaced older and more primitive quadrants. It is possibly his name which is attributed to the Jacob's Staff, which was in fact constructed by Levi ben Gerson.

Duhem, **III**, 298–312.
Gunther, *Early science in Oxford*, Oxford, 1923, **II**, 163.
Sarton, **II**, 850–53.

Johannes Buridanus
Jean Buridan of Bethune
c. 1290 to *c.* 1360.

A French philosopher and natural scientist. He is first mentioned in 1328 as Rector of the University of Paris, and was one of the founders of the Parisian School in the fourteenth century. In the field of logic he was a nominalist, and commented on Ockham. In his *Questiones* on the *Physics* of Aristotle (J. Buridan, *Kommentar zur Aristotelischen Physik*, Frankfurt, 1964) he applied this notion to sublunar as well as to celestial motions, just as in astronomy he took the Ptolemaic standpoint against Aristotle: *Johannis Buridani Quaestiones super libris quattuor De caelo et mundo*, ed. E. A. Moody, Cambridge, Mass., 1942. His ideas greatly influenced contemporary physics (Nicole Oresme) and the late Middle Ages as a whole (Nicolaus Cusanus).

Clagett, 532 *ff.*
Duhem, **IV**, 124.
Duhem, *Études sur Léonard de Vinci*, **II**, 279 and 420; **III**, 1 and 350.
Sarton, **III**, 540.

Johannes Dumbleton
c. 1310 to *c.* 1360

John Dumbleton, Fellow of Merton College in 1331, was one of the Merton philosophers. Very little is known of his life, and his *Summa de logicis et naturalibus* was never published in its entirety.

Clagett, 204 *ff.*
Sarton, **III**, 564–65.
J. A. Weisheipl, *Isis*, **L** (1959), 439–54.

Johannes Fusoris
of Giraumont in the Ardennes
c. 1365–1436

A French instrument maker who, from 1404 to 1416, was successively Canon of Reims, Paris, and Nancy until, in 1416, he was accused of conspiracy with the English and sentenced to 'exile' at Mezières-sur-Meuse. In 1423 he constructed the great astronomical clock of Bourges Cathedral, having already made a number of other clocks, sundials, spheres, quadrants, and astrolabes. His most interesting achievement made in 1414 resulted, following several years of work, in an equatorium, in which he solved the old problem of making an astronomical computer for the determination of planetary longitudes without any auxiliary tables. However, the computer needed a separate instrument for each planet, and was thus less simple than the previous instruments (with tables) of Peter Nightingale and Chaucer. The instrument itself has not survived but, along with other treatises by Fusoris, the text describing its construction and use has been found and published by E. Poulle, *Un constructeur d'instruments astronomiques au XVᵉ siècle: Jean Fusoris*, Paris, 1963.
G. Beaujouan, *J. Savants*, April–June 1964, 140–7.

Johannes Hispalensis
(John of Spain, J. of Seville, Avendeut)
c. 1135–1153

A Spanish-Jewish translator who worked with Dominicus Gundissalinus (see al-Farghānī, Māshāllāh, Abū Ma'shar). He left a few original works and compilations, such as a short astronomical treatise

on the theory of precession (ed. J. M. Millás y Vallicrosa, *Osiris* I (1936) 451–75).

Duhem, **III**, 177 and 198.
Sarton, **II**, 169.
Thorndike, **II**, 73 and 94.

Johannes de Lineriis
(Jean de Linières)
c. 1300 to c. 1350

A French astronomer best known for his *Canones* to the Alfonsine tables, written in 1320 for the meridian of Paris, and revised in 1322. A *Theorica planetarum* of 1335 defends Ptolemaic astronomy, while his catalogue of 48 fixed stars (based on personal observations) of 1350 is an attempt to improve the catalogue of the *Almagest*. He also described the construction of an equatorium (*Compositio equatorii*, ed. D. J. Price, *The equatorie of the planetis*, Cambridge, 1955, 188–96).

Curtze, *Bibl. Math.*, **I** (1900), 390, (Canones).
Sarton, **III**, 649.
Thorndike, **III**, 253.

Johannes Müller
(Regiomontanus)
1436–1476

Johannes Müller of Königsberg in Franconia (Johannes de Monte Regio), was a German astronomer and pupil of Peurbach, whose chair in Vienna he took over in 1461. In 1462 he went to Italy with Cardinal Bessarion. There he was taught Greek, by among others, George of Trebizont, and also continued the translation of the *Almagest* which Peurbach had started (*Epitome in Almagestum*, Venice, 1496). After some years of wandering he settled in Nuremberg, in 1471 where he organized a large astronomical institute paid for by Bernhard Walter, who became his pupil. In 1475 Pope Sixtus IV called him to Rome, to work on calendar reform, but he died in the following year having previously been appointed Bishop of Regensburg. In 1472 he determined the parallax of a comet to be 3° (sic) (*De cometa*, Nürnberg, 1544), but his most important works fell within the sphere of trigonometry (*De triangulis planis et sphaericis*, Basel, 1561. English transl. *On triangles*, *De triangulis omnimodis*, B. Hughes Madison, Wisc., 1967) and calendar calculation (*Das deutsche Kalender*, Nuremberg, 1474, and numerous later editions).

F. J. Carmody, *Isis*, **XLII** (1951), 121–30.
Dreyer, 289 *ff*.
Duhem, **X**, 351 *ff*.
L. Thorndike, **IV**, 440 *ff*. and **V**, 332 *ff*.
L. Thorndike, *Science and Thought in the 15th Century*, New York, 1929, 142–50.
E. Zinner, *Leben und Wirken des Johannes Müller*, München, 1938 (reprinted Osnabrück, 1968).

Johannes de Muris
(Jean de Meurs)
c. 1300 to *c.* 1360

A French mathematician and astronomer, who from about 1320 to about 1340 was connected with the University of Paris. He wrote a number of elementary and much-used textbooks on arithmetic (anticipation of decimal fractions), mechanics, and music. In 1318 he measured (at Evreux) the obliquity of the ecliptic, using an instrument with a radius of 15 ft. He also published astronomical tables, and proposed a calendar reform by omitting leap years for a period of 40 years.

Duhem, **IV**, 38; *Études sur Léonardo de Vinci*, **III**, 47 and 300.
Sarton, **III**, 652.

Johannes Philoponos
of Caesarea
First half of the sixth century A.D.

A philosopher, who lived in Alexandria, and was later on known as John the Grammarian. Having been well-grounded in neo-Platonism, he specialized in original commentary on Aristotle and, in particular, criticized the doctrine of motion and the theory of fall (*In physica Aristoteles*, ed. H. Vitelli, Berlin, 1887–8). In his work one finds the first hint of the impetus theory. In about 520 he was converted to Christianity, and eased the introduction of Aristotelian ideas into Christian thought. He is the author of the only extant ancient treatise on the astrolabe (*De usu astrolabii*, ed. H. Hase, Bonn, 1839, German translation by J. Drecker, *Isis*, **XI** (1928), 15–44).

Joh. Philoponos, *Ausgewählte Schriften*, übers. v. W. Böhm, München, 1967.
Duhem, **I**, 313; **II**, 108, 494.
M. Grabman, *Sitz. Ber. Bayer. Akad. Phil.—Hist. Abt.* München, 1929.
E. Grant, *Centaurus*, **XI** (1965), 79.
A E. Haas, *Bibl. Math.*, **VI** (1905), 337 (physics).
Sarton, **I**, 421 *ff*.
Tannery, **IV**, 240 (astrolabe).

PW, **IX,** 1764.
E. Wohlwill, *Phys. Zts.,* **VII** (1906), 23 (dynamics).

Johannes Sacrobosco
probably of Halifax
c. 1190 to *c.* 1255

John of Holywood, an English mathematician and astronomer, educated in Oxford, but who from about 1230 lived in Paris. His writings were among the most used textbooks of the Mediæval universities, and were continually reprinted down to the 17th century. His *Tractatus de sphaera* (ed. L. Thorndike, *The sphere of Sacrobosco,* Chicago, 1949, Latin text with English transl. and a series of Mediæval commentaries) was an elementary textbook of spherical astronomy. His *Algorismus* (ed. I.O. Halliwell *Rara arithmetica,* London, 1841) was also a much used textbook of mathematics and was commented upon by, among others, the Dane Peder Nattergal about 1300 (ed. M. Curtze, *Petri Philomeni de Dacia in Algorismum vulgarem Johannis de Sacrobosco commentarius,* Copenhagen, 1897). His last work was a textbook on the calendar or *Compotus,* which was also translated into Icelandic (Codex Arnemagn., 1812).

Duhem, **III,** 238–40.
Sarton, **II,** 617–19.

Johannes of Trebizont
(Bessarion)
c. 1403–1472

A Greek philosopher, patriarch of Constantinople, and later a cardinal. He spent the greatest part of his life in Italy, and was one of the most important intermediaries between Greek culture and Renaissance Europe. His great collection of manuscripts became the nucleus of the San Marco Library in Venice. He translated works by Aristotle, Theophrastos, Xenophanes, and others, and made Peurbach undertake a new translation of the *Almagest.* In the controversy between Aristotelians and Platonists he took the part of the Platonists and his writing *In calumniatorem Platonis* (1469), was specifically directed against George of Trebizont.

L. Mohler, *Die Wiederbelebung d. Platonstudiums in d. Zeit d. Renaissance durch Kardinal Bessarion,* Görres-Gesellsch., 1921.
L. Mohler, *Cardinal Bessarion,* Paderborn, 1923.
R. Rocholl, *Bessarion. Studie zur Geschichte der Renaissance,* Leipzig, 1904.

Jordanus Nemorarius
of Nemore
First half of the thirteenth century

Also called Jordanus of Saxonia or Jordanus Teutonicus, he is perhaps the same person as the first general master of the Dominican order (1222–1237). According to Duhem he must, however, be placed in the twelfth century. The foremost representative of mathematical statics in the Middle Ages, in whose writings (ed. E. A. Moody and M. Clagett, *The Mediaeval science of weights,* Wisconsin, 1952) the problem of the inclined plane is solved correctly by means of a method akin to the principle of virtual displacements. 'Jordanus's axiom' is the first germ of the notion of work in the sense of force times distance.

M. Clagett, 69–108.

Duhem, *Les origines de la statique*, **I**, 98–155.

B. Ginzburg, *Isis*, **XXV** (1936), 341–62.

O. Klein, *Nuclear physics*, **LVII** (1964), 345.

Sarton, **II**, 613.

al-Kāshī

Jamshīd al-Kāshī of Kashan

d. 1429

A Persian astronomer and mathematician who first worked in his native city but who later became the first director of the new Ulug Beg observatory in Samarkand. He is particularly noted for his equatorium (E. S. Kennedy, *The planetary equatorium of Jamshid Ghiyath al-Din Al-Kāshī*, Princeton, 1960, text in facs. with English transl. and commentary). A mathematical work contains valuable contributions to the theory of root extraction (Abdul-Kader Dakhel, *Al-Kashi on root extraction*, Beirut, 1960, text with English transl. and commentary). A large *zīj* is extant in MS.

E. S. Kennedy, *Isis*, **XXXVIII** (1947–48), 56; ibid. **XLI** (1950), 180; ibid., **XLIII** (1952), 42.

Kennedy, 127 and 164.

al-Khwārizmī

of Khwārizm (Khiva)

d. 860 A.D.

A Persian mathematician and astronomer who worked in Baghdad and presumably took part in the measurements of a degree of the meridian under al-Ma'mūn. Using Hindu numerals he wrote an arithmetical treatise known as the *Algorismus* (ed. K. Vogel, Aalen, 1963)—the title is a perverted form of his name—and an *Algebra* translated into Latin by Robert of Chester (Latin text and English transl. L. C. Karpinsky, New York, 1915). In the field of astronomy he wrote on the astrolabe and compiled a large collection of tables, or *zīj*, which is extant in Adelard of Bath's Latin translation of a later Arabic version by Ibn al-Muthannā (ed. H. Suter, *Die astron. Tafeln des . . . al-Khwārizmī*, Copenhagen, 1914; English transl. O. Neugebauer, *The astronomical tables of al-Khwārizmī*, Copenhagen, 1962). It is important as a testimony to the non-Ptolemaic, Indian methods used in early Islamic astronomy.

de Vaux, **II**, 119.

J. Frank, *Die Verwendung des Astrolabs nach al-Chwārizmī*, Erlangen, 1922.

S. Gandz, *Osiris*, **I** (1936), 263.

Kennedy, 128 and 148, *Scripta Mathematica*, **XXVII** (1964), 55.

Nallino, **V**, 331 and 458.

Sarton, **I**, 563.

Suter, 10.

al-Kindī

Abū Yūsuf al-Kindī, of Basra

d. c. 873

The only Islamic philosopher of Arabic (Bedouin) origin. Al-Kindī had a deep knowledge of Greek science and philosophy, translated Greek works into Arabic, and wrote on many subjects in philosophy (A. Nagy, *Die philos. Abhandlungen des al-Kindi*, Münster, 1897), an *Epistle on the Concentric Structure of the Universe* (transl. H. Khatchadourian and N. Rescher, *Isis*, **LVI**, (1965) 190), a *Buch* über . . . die Destillation (Arabic text and German transl. K. Garbers, Leipzig, 1948), on the tides (German transl. E. Wiedemann, Annalen der Physik, **LXVII** (1922), 374), on optics, medicine, music, etc. Only a few of his many treatises are extant. In the Latin Middle Ages he was an influential author, due to a number of translations by Gerard of Cremona.

G. N. Atiyeh, Al-Kindi, the philosopher of the Arabs, Rawalpindi, 1966.
A. A. Bjørnbo and S. Vogl, Abhandl. Grs. math. Wiss., **XXVI**³ (1912), (optics).
N. Rescher, Al-Kindi: an annotated bibliography, Pittsburg, 1964.
Sarton, **I**, 559.
Suter, 23.
Wiedemann, **I**, 396, 660, 810.

Ktesibios
of Alexandria
c. 200 B.C.

The first of the three great Alexandrian mechanicians. His writings have not been preserved, but we know from Philon of his invention of an improved water-clock, a water organ, and a compression pump. Philon and Heron built on his work.

A. G. Drachmann, Ktesibios, Philon and Heron, Copenhagen, 1948.
Sarton, **I**, 184–85.
PW, **XI**, 2074–76.

Lātadeva
c. 505 A.D.

An Indian astronomer and a pupil of Aryabhata I. According to al-Bīrūnī he composed the Sūrya-Siddhānta (English transl. E. Burgess, Journ. Amer. Orient. Soc., **VI** (1859–60), 141–498) but he may have only commented upon it. He also wrote commentaries on the Paulisa- and Romaka-Siddhāntas.

Sarton, **I**, 387.

Leonardo da Vinci
1452–1519

An Italian painter, sculptor, architect, engineer, and inventor, and native of Vinci near Florence. Leonardo is a central figure in Renaissance art, and also an important pioneer in natural science and technology, although this aspect of his activities has been examined in far less detail than the artistic one. In science he was self-taught, and his new ideas and inventions did not spread to the extent they deserved. Most are known only from modern studies of his manuscripts, particularly the important Codex Atlanticus (facsimile ed. with transcription, 6 vols., Milan, 1894–1906).

Duhem, Études sur Léonard de Vinci, 3 vols, Paris, 1906–1913.
I. Hart, Mechanical investigations of Leonard de Vinci, London, 1925; 2nd ed. 1963.
P. Huard and M. D. Grmek, Léonard de Vinci, dessins scientifiques et techniques, Paris, 1962.
E. McCurdy, The notebooks of Leonardo da Vinci, London, 1938.
C. Truesdell, Essays in the history of mechanics, New York, 1968, 1–83.
E. Verga, Bibliografia Vinciana 1493–1930, 2 vols, Bologna, 1931, Léonard de Vinci et l'expérience scientifique au seizième siècle (Symposium), Paris, 1953.

Leucippus
of Miletos
Fifth century B.C.

An Ionian natural philosopher who with Democritos, created the notion of atoms. He is perhaps the actual originator of the atomic theory. No extant writings.

Burnet, 330–49.
Sarton, **I**, 88.
PW, **XII**, 2266–77.

Levi ben Gerson
of Bagnols
1288–1344

A Judeo-Provençal who represents the summit of Jewish science in the fourteenth century. As a philosopher he commented on Averroës, and as a mathematician he studied algebra (*Die Praxis des Rechnens*, übers. G. Lange, Frankfurt a.M., 1909), combinatorial analysis, trigonometry, and the axioms of geometry. In the field of astronomy he criticized the *Almagest*, and tried in vain to elaborate a new 'physical' planetary system in accordance with the general wishes of the Averroists. In 1320 he compiled astronomical tables for Orange; but he is best known as the inventor of the Jacob's staff, a simple instrument for measuring arcs on the heavenly sphere. For centuries it remained a much-used instrument of navigation (his treatise on the instrument has been edited by M. Curtze, *Bibl. Math.*, **XII** (1898) 97–112). He also spread the knowledge of Alhazen's invention, the *camera obscura*, which he used to determine the variations of the apparent diameters of the Sun and Moon. There are no modern editions of his works.

Duhem, **IV**, 39–41; **V**, 201–29.
S. Günther, *Bibl. Math.*, **IV** (1890), 73.
Sarton, **III**, 594.
Steinschneider, 129.

Lucretius
Titus Lucretius Carus of Rome
c. 96 B.C. to *c.* 55 B.C.

A Roman philosopher and poet whose great didactic poem *De rerum natura* (ed. H. A. J. Munro, 3 vols, London, 4th ed., 1905–10, Latin text and English transl.; ed. K. Büchner, Zürich, 1957, reprinted Wiesbaden, 1966) is the most important source for the atomic theory of Antiquity.

P. Boyancé, *Lucrèce, sa vie, son oeuvre. Avec un exposé de sa philosophie*, Paris, 1964.
C. A. Gordon, *A Bibliography of Lucretius*, London, 1962.
K. Lasswitz, *Geschichte der Atomistik*, **I**, Hamburg and Leipzig, 1890.
J. Masson, *The atomic theory of Lucretius*, London, 1884.
G. Mueller, *Die Darstellung der Kinetik bei Lukrez*, Berlin, 1959.
Sarton, **I**, 205.
A. D. Winspear, *Lucretius and scientific thought*, Montreal, 1963.
PW, **XIII**, 1659.
Lucretius, London, 1965 with Chapters by Dudley, Farrington and others.

Macrobius
Ambrosius Theodosius Macrobius of Rome
c. A.D. 400

A Roman, neo-Platonic philosopher whose commentary, *In somnium Scipionis* (ed. F. Eyssenhardt, Leipzig, 1893, transl. annotated by W. H. Stahl, *Commentary on the dream of Scipio*, New York, 1952) on Cicero's writing about Scipio's dream gives a glimpse of the natural science and mathematics of late Antiquity. The treatise was much used in the Middle Ages, and scholars of the time thus became acquainted with the geo-helio-centric planetary system of Heracleides.

Duhem, **III,** 47, 65.
Sarton, **I,** 385–86.
M. Schedler, *Die Philosophie des Macrobius und ihr Einfluss auf die Wissenschaft des christl. Mittelalters*, Münster i.W., 1916.
T. Whittaker, *Macrobius on Philosophy*, Cambridge, 1923.
PW, **XIV,** 170–98.

al-Majrītī
of Madrid
c. A.D. 950 to *c.* A.D. 1007

Maslama ibn Aḥmad al-Majrītī of Madrid was one of the first Moorish scientists. He is particularly renowned for his Arabic version of al-Khwārizmī's astronomical tables, in which he changed the chronology from the Persian to the Arabic calendar, and the standard meridian from Arim to Cordoba (where he worked). This version was translated into Latin by Adelard of Bath, and was commented upon by Ibn al-Muthannā. Also a treatise on the astrolabe, and various alchemical writings, are ascribed to him.

E. J. Holmyard, *Isis,* **VI** (1924), 293.
Sarton, **I,** 668.
Suter, 76.
J. Vernet and M. A. Català, *Al-Andalus,* **XXX** (1965), 15.

al-Maqrīsī
of Cairo
1364–1442

An Egyptian historian whose works *Histoire des Sultans Mamelouks de l'Égypte* (4 vols, transl. M. Quatremère, Paris, 1837–45) and *Déscription topographique et historique de l'Égypte* (transl. U. Bouriant, Paris, 1895) illuminate many aspects of Islamic science and civilization.

K. Brockelmann Suppl. **II,** 36.
De Vaux, **I,** 147 and **II,** 218.

Martianus Capella
of Madaura in Africa
c. A.D. 470

A Latin author who lived and worked in Carthage. He wrote a work on the liberal arts, *De nuptiis Philologiae et Mercurii* (ed. F. Eyssenhardt, Leipzig, 1866) which in the Middle Ages was much used as a scientific encyclopedia.

Duhem, **III,** 47–52.
Sarton, **I,** 407–08.
PW, **XIV,** 2003–16.

Māshāllāh
c. A.D. 740 to *c.* A.D. 815

An Egyptian Jew called Manasse who worked as an astronomer and astrologer at the court of al-Manṣūr in Baghdad. He was active as a surveyor in its foundation. He wrote an astronomical work, *De scientia motus orbis*, extant in a Latin translation by Gerard of Cremona (printed Nuremberg, 1504), but is best remembered for his treatise on the astrolabe which, in a Latin version by John of Spain (ed. with a facsimile by R. T. Gunther, *Early Science in Oxford*, **V**, Oxford, 1929, 137–92) spread over all Latin Europe as the standard work on the subject. A collection of his horoscopes has been published by E. S. Kennedy and D. Pingree, *The astrological history of Māshā'allāh*, Cambridge, Mass. 1971.

W. Hartner, *Oriens-Occidens*, Hildesheim, 1968, 290 and 312.
Sarton, **I**, 531.
Suter, 5.
L. Thorndike, *Osiris*, **XII** (1956), 49.

Nasr Mansūr, Berlin, 1936). To a large extent Ptolemy derived his spherical astronomy by means of the 'theorem of Menelas'. From the *Almagest* it is known that Menelaos made astronomical observations in Rome in A.D. 98.

A. A. Björnbo, *Studien über Menelaos' Sphärik*, Leipzig, 1902 (with edition of the text).
A. A. Björnbo, *Bibl. Math.*, **II** (1901), 196.
B. R. Goldstein, *Physis*, **VI** (1964), 205.
T. L. Heath, **II**, PW, **XV**, 834, 260.
Sarton, **I**, 253.

Menaechmos
c. 350 B.C.

A Greek mathematician and pupil of Eudoxos. The invention of the conic sections, which he used when solving the Delian problem of doubling the cube, is ascribed to him, but should, more accurately, probably be linked with the construction of sundials. Menaechmos also studied the basis of mathematics, and he defined such terms as 'element', 'theorem', 'problem', etc.

Heath, **I**, 251 *ff.*
Sarton, **I**, 139–40.
PW, **XV**, 700–01.

Meton
Second half of the fifth century B.C.

Greek astronomer who in 432, together with Euctemon, observed the equinoctial dates in Athens. He used the so-called Metonic cycle of 6940 days, almost equivalent to 19 solar years and to 235 synodic months, 110 of them with 29 days and 125 with 30 days. This gave a value for the tropical year of 365 5/19 days, only 30 minutes too much.

Duhem, **I**, 108.
F. K. Ginzel, *Handbuch der Chronologie* **II**, Leipzig, 1911, 391 *ff.*
Sarton, **I**, 94.
PW, **XV**, 1458–66.

Menelaos of Alexandria
c. A.D. 100

A Hellenistic mathematician and astronomer, who was one of the founders of spherical trigonometry with his *Libri sphaericorum 3*, of which the original is lost. It is conserved in Arabic, Hebrew, and Latin translations (M. Krause, *Die Sphärik* [. . .] *in der Verbesserung von Abū*

Nāsir al-Dīn al-Tūsī
1201–1274

A Persian astronomer and polymath from Tūs in Khūrasān who served the new Mongol rulers both before and after the sacking of Baghdad in 1258, when the caliphate was disbanded. In 1259 he became director of the new observatory at

Marāgha in north-west Iran, where a group of astronomers wrote the famous *zīj* published in his name. He published an important manual on trigonometry considered as distinct from spherical astronomy (French transl. Carathéodory *Traité du quadrilatière*, Constantinople, 1891), and several astronomical works, as well as a multitude of writings on scientific and philosophical subjects. In planetary theory he made use of a kinematic device for producing rectilinear motion by means of a circle rolling inside another circle with twice the diameter.

De Vaux, **II,** 149.
W. Hartner, *Physis*, **XI** (1969) 287 (lunar theory).
Kennedy, 125 and 161.
Kennedy, *Isis*, **LVII** (1966), 365.
Sarton, **II**, 1001.
A. Sayili, *The observatory in Islam*, Ankara, 1960.
Suter, 146; *Bibl. Math.*,**VII** (1893), 1.
Wiedemann, **II**, 25, 653, 677, 701.

Nicolaus Cusanus
of Cues
1401–1464

A German philosopher, a Cardinal in 1448 and Bishop of Brixen in 1450. He was the most outstanding philosopher in the transition period between the Middle Ages and Renaissance. At the same time he was influenced by Mediæval mysticism (Eckhart) and an advocate of the experimental method and of the use of mathematics in the perception of nature. He supported the theory of the rotation of the Earth and the infinity of the universe. He also worked towards the reform of the calendar.

Opera omnia, Heidelberg, 1932 (not finished).
Die mathematischen Schriften, German transl. J. E. Hofmann, Hamburg, 1951.
M. De Gandillac, *La philosophie de Nicolas de Cues*, Paris, 1941.
E. Hofmann, *Das Universum des Nikolaus von Cues*, Heidelberg, 1930.
K. Jaspers, *Nikolaus Cusanus*, München 1964.
J. Meurers, *Nicolaus von Kues und die Entwicklung des astronomischen Weltbildes.*, *Philosophia nat.*, **IX** (1965), 163–90.
E. Meuthen, *Nikolaus von Kues, 1401–1464;* Skizze einer Biographie, Münster, 1967.

Oresme, Nicole
c. 1320–1382

A French philosopher from Normandy, teacher at the Sorbonne and Master of the College de Navarre in Paris who had strong connections with Charles V; later Bishop of Lisieux. With Buridan, Oresme is the most outstanding representative of the mathematicians and physicists of the Parisian school in the fourteenth century. His graphical method was useful to both kinematics and the notion of function, and later became very important to the development of geometry. His French translations and commentaries on Aristotle made him one of the foremost figures in late Mediæval Aristotelian criticism. He also wrote economic, theological, and anti-astrological works. There is no collected edition of his writings.

Le livre du ciel et du monde, ed. A. D. Menut and A. J. Denomy, Madison, Wisc., 1968.
Livre de divinacions, ed. G. W. Coopland,

Nicole Oresme and the astrologers, Liverpool, 1952.

Questiones super geometriam Euclidis, ed. H. L. L. Busard, Leiden, 1961 .

De proportionibus proportionum and *Ad pauca respicientes*, ed. with introd., transl., and critical notes by E. Grant, Madison, Wisc., 1966.

Tractatus de commensurabilitate vel incommensurabilitate motuum celi, ed. and transl. in E. Grant, *Nicole Oresme and the kinematics of circular motion*, Madison, 1971. Clagett, 331.

A. D. Menut, *A provisional bibliography of Oresme's writings*, Mediæval Stud., **XXVIII** (1966), 279.

O. Pedersen, *Nicole Oresme og hans naturfilosofiske system*, Copenhagen, 1956. Sarton, **III**, 1486.

Thorndike, **III**, 398.

Nicomachos
of Gerasa
c. 100 B.C.

A Neo-Pythagorean mathematician and theorist of music, whose *Enchiridion harmonices* is our most important source of the Pythagorean theory of music. It was translated into Latin by Boethius (P. L. Migne, 63). His mathematical efforts earned him the title 'Euclid of arithmetics', his textbook being the most copious exposition of arithmetical problems (*Introduction to Arithmetic*, English transl. M. L. D'Ooge, New York, 1926).

Sarton, **I**, 253.
PW, **XVII**, 463–64.

Ockham
see William of Ockham

Oinopides
of Chios
Middle of the fifth century B.C.

A Greek mathematician and astronomer who is traditionally considered to be the first of the Greeks to determine the obliquity of the ecliptic. He introduced a period of 730 synodic months, which was made equal to 59 years. If a synodic month is supposed to be 29 1/2 days, the length of the year will be 365 22/59 days or a little less than 365^d9^h. In mathematics he is supposed to be responsible for allowing no other instruments than the ruler and the compass.

Dreyer, 38.
Duhem, **I**, 72.
T. L. Heath, *Aristarchus of Samos*, Oxford, 1913, 130–33.
Sarton, **I**, 92.
PW, **XVII**, 2258–72.

Oresme
see Nicole Oresme

Osiander, Andreas
1498–1552

A Protestant German theologian who saw the *De revolutionibus* of Copernicus through the press at Nuremberg in 1543, and wrote the anonymous preface in which the Copernican system was explained as a mere hypothesis. This was revealed by Kepler.

W. Moeller, *Andreas Osiander. Leben u. Schriften*, Elberfeld, 1870.

Pappos
of Alexandria
c. A.D. 300

The last of the great Alexandrian mathematicians, whose *Mathematical collections* are one of our most important sources of the history of Greek mathematics (ed. F. Hultsch, 3 vols, Berlin, 1876–78. French transl. P. Eecke, *Collection mathématique*, 2 vols, Paris, Bruges, 1933). Pappos also dealt with simple machines, the problem of the inclined plane, the theory of the centre of gravity, the theory of harmonics, geography, and astronomy. His *Commentaries on the Almagest* 1–6 (ed. A. Rome, Città del Vaticano, 1931–43) is important.

Heath, **II,** 354–439.
Sarton, **I,** 337.
PW, **XVIII** (ii), 1084–1106.

Parmenides
of Elea, Italy
First half of the fifth century B.C.

An Italian-Greek philosopher who arrived in Athens in about 450 B.C. Contrary to Heraclitos he asserted the permanent and unchangeable: 'That which is, is finite, spherical, unmovable, continuously filled, and there is nothing outside'. Thus he provoked natural philosophers to undertake investigations of great importance to the doctrine of elements as well as to the theory of atoms. Only fragments of his didactic poem are extant (H. Diels, *Parmenides Lehrgedicht griechisch und deutsch*, Berlin, 1897; *Testimonianze e frammenti*, trans. M. Untersteiner, Firenze 1958).

Heath, *Aristarchus of Samos*, Oxford, 1913, 62–77.

Kirk and Raven, 263 *ff.*
A. A. Long, *Phronesis,* **VIII** (1963), 90.
K. Reinhardt, *Parmenides*, Bonn, 1916.
Sarton, **I,** 85.
PW, **XVIII** (ii), 1553.

Peter Nightingale
see Petrus Philomena de Dacia

Petrus Peregrinus
Pierre of Maricourt
c. 1270

A French physicist who taught Roger Bacon. He was the author of a treatise on the astrolabe, and a dissertation of 1269 on magnetism: *Epistola de magnete* (ed. Hellmann, *Rara magnetica*, Berlin, 1898; English transl. S. P. Thompson, London, 1902) which is a brilliant experimental work, and the most important writing on magnetism before the seventeenth century.

Duhem, *Origines de la statique*, **I,** 57 (on the perpetuum mobile).
Sarton, **II,** 1030–32.

Petrus Philomena de Dacia
(Peter Nightingale)
worked *c.* 1290–1300

A Danish astronomer and mathematician who lectured at Bologna in 1291 where he wrote a commentary to Sacrobosco's *Algorismus* (ed. M. Curtze, *Petri Philomeni de Dacia in Algorismum . . . Commentarius*, Copenhagen, 1897). In 1292–93 he was in Paris, and most of his works seem to have been written there.

These writings are only extant in MS, and comprise a *Compotus* (a manual of time-reckoning), a calendar which was much used during the following two centuries, a diagram for determining the position of the Moon, a number of astronomical tables, and finally three treatises on astronomical instruments. The first of these is a Latin version of the tract on the New Quadrant by Jacob ben Mahir (Profatius Judaeus), the second a highly original tract *De semissis* describing the construction and use of a planetary equatorium. The third is on the construction of a computer for predicting eclipses. Some of his works have been ascribed to an otherwise unknown Peter of St Omer (P. de Scto Audomaro). The place and date of his birth are unknown, but he died after 1303 as a Canon of Roskilde cathedral.

Duhem, **IV**, 29.
O. Pedersen, *Vistas in Astronomy*, **IX** (1968), 3.
E. Zinner, *Archeion*, **XVIII** (1936), 318.

Peurbach
see Georg Peurbach

Pherecydes
of Syros
Middle of the sixth century B.C.

A Greek philosopher and mythologist whose cosmology bears the marks of the mythological explanation of nature. He is said to have taught Pythagoras, and to have imparted to the latter his own theory of the transmigration of souls.

PW, **XIX**, 2025–33.
M. L. West, *Early Greek philosophy and the Orient*, Oxford, 1971.

Philo
of Byzantium
First century A.D.

A Greek physicist and engineer who worked in Alexandria. He wrote a large, partly extant, work on pneumatics and applied mechanics, building on Ktesibios and Strato (*Livre des appareils pneumatiques et des machines hydrauliques*, ed. C. de Vaux, with French translation, *Notices et extraits*, **XXXVIII**, (1903)). A work on fortification and war machines also survives (*Traité de fortification*, French translation by A. de Rochas, *Mém. Soc. d'Encouragement aux Sciences du Doubs*, 1870–71).

A. G. Drachmann, *Ktesibios, Philon, and Heron*, Copenhagen, 1948.
Sarton, **I**, 195–96.
Schmidt, *Bibl. Math.*, **II** (1901), 377–383.
PW, **XX**, 53–54.

Philolaos
of Southern Italy
Second half of the fifth century B.C.

A Pythagorean philosopher, considered to be the originator of the oldest (partly extant) Pythagorean literature. Very little is known of his life. His world system, which consisted of a spherical universe divided into spheres, came to exert a decisive influence on later Greek astronomy, but his assumption of a central fire, an anti-Earth, and the daily rotation of the Earth on a circle around the centre of the world, was abandoned in the fourth century. A few fragments of his works are still conserved.

A. Boeckh, *Philolaos*, Berlin, 1819.
W. F. M. Burkert, *Weisheit und Wissenschaft*, Nuremberg, 1962.

Kirk and Raven, 307.

W. R. Newbold, *Archiv. f. Gesch. d. Philosophie*, **XIX** (1905), 176–217.

Sarton, **I**, 93–94.

Tannery, **VII**, 131–39; **III**, 233–249.

B. L. van der Waerden, Die Astronomie der Pythagoräer *Verhdl. Nederl. Akad. Wetensch.*, **XX** (1951).

Philoponos

see Johannes Philoponos

Pierre de Maricourt

see Petrus Peregrinus

Plato

427–347 B.C.

Plato was a Greek philosopher from Athens and pupil of Socrates, whose teaching influenced him throughout his life. Following the execution of Socrates in 399 Plato spent some years travelling in Egypt, Italy, and Sicily. After 387 he returned to Athens and founded his Academy where he taught for the rest of his life, and wrote his (extant) published works. It is possible to arrange Plato's *Dialogues* more or less chronologically, as they reflect three stages of his authorship.

In the first stage Socrates is the proper main figure of the *Dialogues*, each of which illuminates one aspect of Socrates' thought. The *Apology* (Socrates' defence) and *Menon* (important for the understanding of Plato's notion of mathematics and epistemology) belong to this period.

In the second stage of Plato's thought, his own ideas are more directly exposed; here he makes several attempts to construct an extensive philosophical system as, for example, in the *Republic*, where the sections on education in the ideal state show many aspects of Plato's idea of the various sciences. *Theaitetos* deals with epistemology, and the *Sophist* with metaphysics. In the third stage the synthesizing tendency is even more obvious. *Timaios* describes Plato's primitive conception of the universe, while his conception of society is dealt with in his longest dialogue, the *Laws*, which takes a more constitutional point of view than the *Republic*. The important work entitled *Epinomis* is hardly authentic, and was possibly written by Plato's pupil Philip of Opus, but it does in many ways supplement the philosophy of nature of *Timaios*.

Most of Plato's works are printed in Loeb's Classical Library with the Greek text and an English translation.

R. S. Brumbaugh, *Plato's mathematical imagination*, Bloomington, Ind., 1954.

Burnet, *Greek philosophy*, Oxford, 1914, 205–351.

F. M. Cornford, *Plato's cosmology*, London, 1937 (commentated transl. of *Timaios*).

B. Disertori, 'Il messaggio del Timeo', *Il pensiero antico*, 1st series, **IV**, Padua, 1965.

Dreyer, 52 *ff*.

Duhem, **I**, 28–102.

K. Gaiser, *Platons ungeschriebene Lehre*, Stuttgart, 1963.

Heath, **I**, 284 *ff*.

A. Koyré, *Discovering Plato*, 3rd. ed., New York, 1960.

D. Mannsperger, *Physis bei Platon*, Berlin, 1969.

C. Mugler, *La physique de Platon*, Paris, 1960.

C. Mugler, *Platon et la recherche mathématique de son temps*, Strasbourg, 1948.

C. Ritter, *Platons Stellung zu den Aufgaben der Naturwissenschaft*, Heidelberg (Ak. Wiss.), 1919.

E. Sachs, *Die fünf platonischen Körper*, Berlin, 1917.

Sarton, **I**, 113.

J. B. Skemp, *The theory of motion in Plato's later Dialogues*, Cambridge, 1942, repr. Amsterdam, 1967.

M. Stockhammer, (ed.) *Plato dictionary*, London, USA, 1963.

A. E. Taylor, *Plato, the man and his work*, London, 1926.

A. E. Taylor, *A commentary on Plato's Timaeus*, Oxford, 1928.

A. E. O. Wedberg, *Plato's philosophy of mathematics*, Stockholm, 1955.

PW, **XX**, 2342.

Plinius

(Caius Plinius Secundus) of Como

A.D. 23–79

A Roman official, philosopher and scientist, who was killed by the eruption of Vesuvius in A.D. 79. The author of a great *Naturalis historia* in 37 books (ed. with English transl. H. Rackham and W. H. S. Jones, 10 vols., London, 1938–62). During the Middle Ages the work was extensively used as a scientific encyclopedia and is still an important source of the geography of Antiquity, etc.

H. Le Bonniec, *Bibliographie de l'histoire naturelle de Pline*, Paris, 1946.

F. Dannemann, *Plinius und seine Naturgeschichte*, Jena, 1921.

W. Kroll, *Die Kosmologie des Plinius*, Breslau, 1930.

Sarton, **I**, 249.

PW, **XXI**, 271.

Plotinus

c. A.D. 205 to *c.* A.D. 270

A Hellenistic philosopher who, for most of his life, taught in Rome as the foremost representative of the neo-Platonic school. His works, the *Enneads* (English transl. S. Mac Kenna, 4th ed., London, 1969; also ed. and transl. A. H. Armstrong, *Plotinus*, London, 1966 *ff.*) describe reality as a hierarchy in which more inferior forms of being emanate from the indivisible 'One' at the top, with base and evil matter at the bottom. Sensation is caused, he said by the action of the sensory agent upon the object; the soul of man is part of a universal soul. The system influenced early Mediæval thought through the influence of St Augustine and others, and also many Arab philosophers, such as Avicenna. In the Renaissance his arguments reappeared as part of the eclectic philosophy of Cusanus and many Italian humanists.

Duhem, **II**, 309; **IV**, 376.

W. R. Inge, *The philosophy of Plotinus*, 2 vols, 2nd ed., London, 1929.

H. Oppermann, *Plotin's Leben*, Heidelberg, 1929.

Sarton, **I**, 334.

PW, **XXI**, 471.

Plutarch

of Chaironea

c. A.D. 46 to *c.* A.D. 120

A Greek author and philosopher whose large composite works *Moralia* and *Lives* (both of them edited in Loeb's Classical Library), plus the non-authentic *De placitis philosophorum*, are important sources for the history of Greek thought. His treatise on the Moon, *De facie in orbe lunae* (English transl. A. O. Prichard, 1911) is of considerable astronomical interest.

Duhem, **II**, 293 *ff.*, 359 *ff.*
Sarton, **I**, 251–52.
PW, **XXI**, 636–962.

Polemarchos
of Cyzicos
c. 340 B.C.

A Greek astronomer and a pupil of Eudoxos when the latter stayed in Cyzicos. Polemarchos taught Calippos, with whom, in about 335 B.C., he set out for Athens. His knowledge of the variations of the apparent diameters of the Sun and Moon later became one of the main objections to the planetary system of Eudoxos.

T. L. Heath, *Aristarchus of Samos*, Oxford, 1913.
PW, **XXI**, 1256–58.

Poseidonios of Rhodes
from Apameia, Syria
c. 135 B.C. to *c.* 50 B.C.

A Stoic author with some neo-Platonic ideas. He was educated in Athens, but later lived on Rhodes as astronomer, geographer and author of encyclopedias. He determined experimentally the circumference of the Earth on the same lines as Eratosthenes, but with a slightly poorer result; but his estimate of the distance of the Sun was better than that of Hipparchos. Poseidonios was also the first to draw attention to the phenomenon of spring tide and neap tide, and he explained the tidewater as a combined effect of the Sun and Moon. His commentary on *Timaios* and a world history were much used in later years.

Posidonius, *I. The Fragments*, ed. by L. Edelstein and I. G. Kidd, Cambridge, 1972.
Duhem, **II,** 274 *ff.*
Heath, **II**, 219.
F. Hultsch, *Poseidonios über die Grösse und Entfernung der Sonne*, Göttingen (Kgl. Gesell. Wiss.), Berlin, 1897.
M. Pohlenz, *Stoa und Stoiker*, Zürich, 1950.
K. Reinhardt, *Poseidonios*, München, 1921.
Sarton, **I**, 204.
PW, **XXII**, 558.

Proclos
of Lycia
c. A.D. 412–485

A neo-Platonic philosopher who taught at the Academy in Athens. He wrote a commentary on the first book of Euclid's *Elements*, of great importance to the history of Greek mathematics (ed. G. Friedlein, Leipzig, 1873; French transl. P. Ver Eecke, *Commentaire sur le premier livre des Elements d'Euclide*, Bruges, 1948, repr. Paris, 1959; English transl. T. Taylor, 2 vols, London, 1788–89). Among his many comments on Plato his commentary on the *Timaeus* is valuable (French transl. A. J. Festugière, 2 vols, Paris, 1966–67). He also wrote an *Institutio physica* on the Aristotelian theory of motion (ed. with German transl. A. Ritzenfeld, 2nd ed., Leipzig, 1912) and an exposition of astronomical theories, *Hypotyposis astronomicarum positionum* (ed. with German transl. K. Manitius, Leipzig, 1909).

T. L. Heath, **II**, Oxford, 1921, 529.
Sarton, **I**, 402.
T. Whittaker, *The Neo-Platonists*, 2nd ed.
Cambridge, 1918.
PW, **XXIII**, 186.

Profatius Judaeus
see Jacob ben Mahir

Ptolemy
c. A.D. 100–165

Claudios Ptolemaios (from Ptolemaïs in
Egypt) spent nearly all his life in Alexandria working on expositions of all the
main branches of applied mathematics.
He is more careful than other ancient
authors in referring to his sources and has,
for that reason, often been regarded as a
mere compiler. In fact, he was one of the
most gifted Hellenistic scientists, and
made original contributions to astronomy, physics, and mathematics, which
showed a clear insight into the relations
between experience, hypotheses, and
theories. The impact of his work was
felt until the seventeenth century, particularly in astronomy, where he created
the theoretical framework used until the
time of Kepler. His main work was the
'Great Collection', or 'Megale Syntaxis',
better known as the *Almagest*, in which
he continued the work of Hipparchos
and Apollonios with his own geometrical
theories of the motions of the planets
(ed. J. L. Heiberg *et al.*, 2 vols, Leipzig
1898–1903; English transl. R. C. Ta-
liaferro in *Great Books of the Western
World*, **XVI**, Chicago 1952; French N. B.
Halma 2 vols, Paris 1816 (reprint 1927);
German by M. Manitius, 2 vols, Leipzig,
1912–13 (reprint 1963)). A number of

Opera astronomica minora (ed. J. L.
Heiberg *et* Ludwig Nix, Leipzig, 1907.)
contains among other treatises a part of
his *Hypotheses* describing his physical
ideas of the universe and his calculation
of its size; another part of the same work
has been discovered and published from
an Arabic version (B. R. Goldstein, 'The
Arabic Version of Ptolemy's Planetary
Hypotheses' (with English transl.), *Trans.
Amer. Philos. Soc.*, **VII** (1967)). His
Tetrabiblos or *Liber quadripartitum* (ed.
with English transl. F. E. Robbins, London (Loeb), 1940) is a systematic exposition of astrology. Among his minor
works is the *Analemma*, dealing with
parallel projection of the heavenly sphere
and providing the fundamental theory of
sundials. Another is his *Planisphaerium* on
stereographic projection giving the
mathematical theory underlying the
astrolabe (German transl. J. Drecker,
Isis, **IX** (1927), 255–278). The 'Handy
Tables', or *Tabulae manuales*, with their
Canones, extant in a version by Theon of
Alexandria (ed. and transl. N. B. Halma,
*Commentaire de Théon d'Alexandria sur les
tables manuelles de Ptolémée*, **I**, Paris, 1822)
are a revised version of the tables of the
Almagest with smaller intervals of the
arguments. The large *Geography* is of the
utmost importance to our knowledge of
the ancient world, and to the history of
cartography (Greek text with French
transl. N. B. Halma, Paris, 1828;
English transl. E. L. Stevenson, New
York, 1932; a Latin ed. Seb. Munster,
Basle, 1540, has been reprinted in
facsimile by R. A. Skelton, Amsterdam,
1966. An edition of the Codex Urbinas
Graecus LXXXII by J. Fischer: *Claudii
Ptolemaei Geographiae*, Leiden-Leipzig
1932, is provided with an excellent
biographical introduction). Ptolemy's
Optics survived only in an Arabic translation which in the twelfth century was

translated into Latin by Eugenius of Palermo (ed. A. Lejeune, *L'optique de Claude Ptolémée* Louvain, 1956). Among his other writings were books on the theory of music, and on philosophical problems, the latter revealing Ptolemy to be an Aristotelian strongly influenced by Stoic ideas.

A. Aaboe, *Episodes from the early history of mathematics*, New York, 1964.
F. Boll, *Studien über Claudius Ptolemäus*, Leipzig 1894 (biographical).
T. Brendan, *Math. Teacher*, **LVIII**, (1965).
Dreyer, 191 *ff.*
Duhem, **I**, 466, **II**, 83.
W. Hartner, *Oriens-Occidens*, Hildesheim, 1967, 319–48.
Heath, **II**, 273 *ff.*
L. O. Kattsoff, *Isis*, **XXXVIII**, (1947), 18.
E. Narducci, *Bibl. Math.* **II** (1888), 97.
Neugebauer, *Isis*, **L** (1959), 22.
Neugebauer, *The exact sciences in Antiquity* 2nd ed., New York, 1962.
C. H. F. Peters and E. B. Knobel, *Ptolemy's catalogue of stars*, Washington, 1915.
V. M. Petersen, *Centaurus*, **XIV** (1969), 142.
V. M. Petersen, and O. Schmidt, *Centaurus*, **XII**, (1967), 73.
Sarton, **I**, 272.
Thorndike, **I**, 100.
PW, **XXIII**, 1788–1859 by v.d. Waerden, Life and Work of Ptolemy (excellent).

Pythagoras
of Samos
Second half of the sixth century B.C.

An Ionian philosopher, Pythagoras was influenced by Oriental ideas. He emigrated to Italy and in Croton founded a sort of a religio-philosophical brotherhood of men and women who sought to liberate their souls partly by means of ascetism, and partly by means of intellectual studies. Several important mathematical and scientific discoveries are ascribed to these Pythagoreans whose doctrines are, however, difficult to explain because of the esoteric character of their society. Only Philolaos and other later Pythagoreans gave written expositions of their views, which had exerted an enormous influence on Greek thought and intellectual life on the whole. In general, the understanding of the fundamental importance of mathematics to natural philosophy can be traced back to the Pythagoreans, just as Plato's whole conception of nature builds to a great extent on the Pythagorean tradition.

W. Bauer, *Der ältere Pythagoraeismus*, Bern, 1897.
W. Burkert, *Weisheit und Wissenschaft. Studien zu Pythagoras, Philolaus und Platon*, Nuremberg, 1962.
A. E. Chaignet, *Pythagore*, 2 vols, Paris, 1873.
C. J. De Vogel, *Pythagoras and early Pythagoreanism*, Assen, 1966.
Dreyer, 35 *ff.*
R. Haase, *Geschichte des harmonikalen Pythagoreismus*, Wien, 1969.
Heath, **I**, 141 *ff.*
Kirk and Raven, 217.
J. A. Philip, *Pythagoras and early Pythagoreanism*, Toronto, 1966.
Sarton, **I**, 73.
M. Timpanaro Cardini, *Pitagorici, testimonianze e frammenti* (texts with Italian transl.), 2 vols, Florence, 1958–62.
B. L. v.d. Waerden, Die Astronomie d. Pythagoräer, *Verh. Nederl. Akad.*, **XX** (1951).
PW, **XXIV**, A1.

Quṭb al-Dīn al-Shīrāzī
1236–1310

A Persian scientist from Shīrāz, and a pupil of Naṣir al-Dīn al-Tūsī, who travelled extensively in the Near and Middle East and died in Tabriz. He wrote on geometry, optics (correct explanation of the rainbow), mechanics, medicine, philosophy, and theology, but his main work was in astronomy, where he worked out a geometrical model for planetary longitudes involving a minimum of rotating vectors. The work has been investigated by Kennedy, but the actual treatises dating from 1281 and 1284 remain unpublished.

Kennedy, *Isis*, **LVII** (1966), 365.
Sarton, **II**, 1017.
Suter, 158.
Wiedemann, **II**, 644.

Regiomontanus
see Johannes Müller

Reinhold, Erasmus
1511–1553

A German astronomer and from 1536 professor in Wittenberg. He was one of the first supporters of the Copernican system, and as early as 1542 paid homage to Copernicus in his new edition of Peurbach's *Theoricae novae planetarum*. Later he used the new system as the basis of his *Tabulae prutenicae*, or *Prussian Tables* (1551). A valuable textbook of geodesy, *Gründlicher und warer Bericht vom Feldmessen*, was edited in 1574 by a son.

Dreyer, 318 and 345.
Nature, **LXVII** (1902), 42.

Remy d' Auxerre
d. A.D. 903.

Remy d'Auxerre taught at the schools in Auxerre, Reims, and Paris, and has left a number of commentaries. His *Commentum in Martianum Capellum* (ed. C. E. Lutz, Leiden, 1962) gives a good impression of the level of 9th-century mathematics and astronomy.

Gilson, 226.
E. Narducci, *Boncompagni's Bollettino*, **XV** (1882), 505.
Sarton, **I**, 595.

Rheticus, Georg Joachim
1514–1576

A German mathematician and astronomer, born in Feldkirch. In 1536 he became Professor of mathematics at Wittenberg, but from 1539 to 1541 he stayed with Copernicus in Frauenberg, where he became acquainted with the Copernican system. After this he was engaged in spreading the knowledge of it, first by this *Narratio prima de libris revolutionum N. Copernici* (Danzig, 1540; English transl. with E. Rosen, *Three Copernican Treatises*, 3rd ed., New York, 1971). After his return Rheticus superintended, with Osiander, the printing of Copernicus's main work, which was published in Nuremberg in 1543. He was Professor in Leipzig 1542–51, where his *Canon doctrinae triangulorum* appeared in 1551. He elaborated a map of Prussia and wrote a treatise upon it

391

(ed, F. Hipler, *Zts. f. Math. u. Phys., Hist.-Litt. Abt.*, **XXI** (1876), 125–50).

Dreyer, 318 *ff.*
E. Rosen, *op. cit.*, 4–6.

Richard Suiseth
Richard Swineshead, surname "The Calculator"
probably of Glastonbury
middle of the fourteenth century

From about 1340 to 1360 Richard Suiseth belonged to the circle of natural philosophers and mathematicians of Merton College, and possibly entered the Cistercian Order. His *Liber calculationum* contains, among other things, reflections on the problem of form (reduction of qualities to quantities) and the use of infinite series in mathematics, the proof of 'the Merton relation' in kinematics, and many other mathematical ideas. The work became very influential, especially after the invention of printing, and even Leibniz urged a new edition of it. There is no critical edition.

Clagett, 290.
M. Clagett, *Osiris*, **IX** (1950), 131.
M. A. Hoskin, and A. G. Molland, *Brit. Journ. Hist. Sci.*, **III** (1966), 150.
Sarton, **III**, 736.
Thorndike, **III**, 370.

Richard of Wallingford
c. 1292–1335

The most important English mathematician and astronomer at the beginning of the fourteenth century. He was educated in Oxford and later became Abbot of the Benedictine monastery of St Albans (1326), to whose reform he applied himself diligently. During a visit to France he was struck by leprosy, and from 1332 was dumb. He was the first mathematician in Europe to give an exposition of trigonometry in his *Quadripartitum de sinibus demonstratis* (ed. J. D. Bond, *Isis*, **V** (1923), 99–115). As an astronomer he was particularly interested in the construction of measuring and calculating-instruments. As early as about 1320 he constructed a clock for St Albans showing the motion of Sun and Moon, and ebb and flow of the tides. In 1326 he invented the *albion*, a sort of equatorium or astronomical calculating-machine, and in the same year the new measuring instrument called *rectangulus*, which he described in *De arte componendi rectangulum*, and *Ars operandi cum rectangulo* (ed. H. Salter, in R. T. Gunther, *Early science in Oxford*, **II**, 1923, 337–70).

M. Curtze, *Bibl. Math.* **X** (1896), 65 *ff.*
Sarton, **III**, 662–68.

Robertus Castrensis
(Rob. Anglicus, Rob. of Chester)
First half of the twelfth century

An English mathematician, astronomer and alchemist and one of the chief scientific innovators and translators of the twelfth century. He lived for some time in Spain where he in 1143 was Archdeacon of Pamplona; later he worked in London, and in 1149 compiled a set of astronomical tables for the meridian of the city. He translated treatises on the astrolabe from Arabic to Latin, and also translated Al Khwārizmī's *Algorismus*. He introduced the word *sine* for the known trigonometrical function, and is remembered in the history of religion for the first Latin translation of the Koran.

Haskins, 120–23.
Sarton, **II**, 175–77.
Thorndike, **II**, 83 and 215–17.

Robert Grosseteste
c. 1175–1253

An English philosopher, the first Chancellor of the University of Oxford. He had strong links with the Franciscan Order, and from 1235 until his death was Bishop of Lincoln (Robertus Lincolniensis). As a promoter of the studies in Oxford, and a commentator on Aristotle (*Commentarius in VIII libros physicorum Aristotelis* ed. R. C. Dales, Colorado, 1963) he perhaps taught Roger Bacon who certainly praises him. He was one of the most outstanding exponents of the experimental method in the Middle Ages and also of the mathematical description of nature, which he himself demonstrated in a series of important optical writings (ed. L. Bauer, *Die philosophischen Werke des Robert Grosseteste*, Münster i.W. 1912). He was one of the few 13th-century philosophers familiar with Greek.

L. Baur, *Die Philosophie des Robert Grosseteste*, Münster i.W., 1917.
D. A. Callus, (ed.), 'Robert Grosseteste, Scholar and Bishop', *Essays in commemoration of the seventh century of his Death*, Oxford, 1955.
A. C. Crombie, *Robert Grosseteste and the origins of experimental science*, Oxford, 1953.
Duhem, **III**, 277–87, 460–71; **V**, 341–58.
B. S. Eastwood, *Archives Intern. Hist. Sci.*, **XIX** (1966), 313 (rainbow).
Sarton, **II**, 583.

Roger Bacon
c. 1214 to *c.* 1294

An English philosopher, educated in Oxford during Grosseteste's time, and later in Paris, where he too worked as a teacher. He entered the Franciscan Order, where he had many conflicts with his superiors, but was protected by Pope Clement IV. A chronicle of the Order from 1374 maintains—probably legendarily—that he spent his last years in prison. He was an encyclopedic author who tried to mobilize the total of philosophical and scientific forces in the defence of Christianity. He argued that mathematics were essential to the description of nature and the experimental method indispensable, thus continuing Grosseteste's optical works in a way which shows his appreciation of the inductive method. His writings are full of prophecies of future inventions, and he has, wrongly, been considered as the inventor of gunpowder, which was known during his lifetime. He was a pugnacious and unsociable person, and his furious attacks on contemporary thinkers of diverging opinions prevented him from obtaining the influence he deserved on account of his original thoughts. His life story was later embellished with many legends.

There is no complete Bacon edition, but the following single treatises, *Opus maius*, ed. J. H. Bridges, 3 vols., Oxford, 1897–1900, reprint, 1964; English transl. Pennsylvania, 1928. *Opera haectenus inedita*, 16 vols ed. R. Steele and F. M. Delorme, Oxford, 1911–1940. A fragment of *Opus tertium*, ed. P. Duhem, Quarachi, 1909. *Roger Bacon Essays*, ed. A. G. Little, Oxford, 1914.
R. Carton, *L'expérience physique chez Roger Bacon*, 3 vols, Paris, 1924.

A. C. Crombie and J. D. North, *Dict. sci. biogr.*, **I**, 377.

Duhem, **III**, 411.

S. C. Easton, *Roger Bacon and his search for a universal science*, Oxford, 1952.

D. C. Lindberg, *Isis*, **LVII** (1966), 235.

Sarton, **II**, 952.

Thorndike, **II**, 616.

S. Vogl, *Die Physik Roger Bacon's*, Erlangen, 1906.

E. Westacott, *Roger Bacon*, London, 1953.

Sacrobosco
see Johannes Sacrobosco

Scot, Michael
c. 1170 to *c.* 1235

A Scots alchemist, astrologer, and Averroistic philosopher. He worked as a translator from Arabic into Latin, first in Spain and later at the court of Frederic II on Sicily, and as such was one of the main instigators of the new Aristotelism in the thirteenth century. Later on, many traditions were ascribed to him, and he became a legendary magician.

Duhem, **III**, 241–48.

Haskins, 272–98.

Sarton, **II**, 579–82.

D. W. Singer, *Isis*, **XIII** (1929), 5–15 (alchemy).

L. Thorndike, **II**, 307–37.

L. Thorndike, *Michael Scot*, London and Edinburgh, 1965.

Seleucos
of Babylon
c. 150 B.C.

A Babylonian astronomer who was influenced by Greek astronomy and who advocated Aristarchos's theory of the diurnal motion of the Earth, as well as the conception of the infinity of the universe. He demonstrated the effect of the Moon on the tides.

T. L. Heath, *Aristarchus of Samos*, Oxford 1913, 305–7.

Sarton, **I**, 183–84.

PW, 2. Reihe **II**, 1249–50.

Seneca
(Lucius Annaeus Seneca) of Cordoba
c. 4 B.C.–A.D. 65

A Roman statesman and leading Stoic philosopher. His *Questiones naturales* (ed. Hermes, 4 vols, Leipzig, 1898–1907, English transl. by J. C. Clarke, *Physical Science in the Time of Nero*, London, 1910) is a great compilation of many different natural phenomena with, for example, valuable meteorological, geological and optical observations.

Sarton, **I**, 247–49.

Al-Shīrāzī
see Quṭb al-Dīn al-Shīrāzī

Simplicios
of Cilicia
Sixth century A.D.

One of the last neo-Platonic philosophers who flourished in Persia after the closing of the School in Athens in 529. His commentaries on Aristotle considerably influenced the Arabs, and also scholars of the Latin Middle Ages (ed. M. Hayduck and others in the Berlin series of Greek commentaries on Aristotle).

Duhem, **II**, 108.
Duhem, *Études sur Leonard de Vinci*, **II**, 64 *ff.*
Sarton, **I**, 422–23.
Tannery, **III**, 119.
PW, 2 Reihe **III**, 204–13.

Socrates
of Athens
c. 470–399 B.C.

A Greek philosopher, Plato's teacher, and one of the main figures of Greek cultural life. Socrates saved philosophy from the Sophistic spinelessness and tried to create a valid foundation for the under-standing of man and society. He had an unresponsive attitude towards the earlier Greek natural philosophy, but his exercises in logic prepared the analytic method, thus becoming of importance to the development of the knowledge of the physical world. His conception of mathematics as a system of abstract, *a priori*, truths was also important. He wrote nothing and is known to us only through Plato's *Dialogues*, of which he is the central figure, and through the more critical description in Xenophon.

O. Gigon, *Sokrates, sein Bild in Dichtung und Geschichte*, Bern, 1947.
Sarton, **I**, 89–90.
PW, 2. Reihe **III**, 811–90.
See also in *Plato*.

Sosigenes
First century B.C.

An Alexandrian astronomer and Aris-totelian philosopher, who, on Caesar's orders, created the Julian calendar which was first put into force in 45 B.C.

F. K. Ginzel, *Handbuch d. Math. u. Techn. Chronologie*, **II**, Leipzig 1911, 160–293.
PW, 2. Reihe **III**, 1153–57.
Sarton, **I**, 216.

Strabo
of Amaseia
c. 63 B.C. to *c.* A.D. 20

A Stoic historian and geographer, whose *Geographika* (ed. and transl. H. L. Jones and J. R. S. Sterrett, 8 vols, London, 1917–32) is the most comprehensive geography of Antiquity, marked more by his interest in the political and philosophical description of the countries than by the mathematical geography, which was expounded more compre-hensively by Eratosthenes.

G. Aujac, *Strabon et la science de son temps*, Paris, 1966.
Sarton, **I**, 227–29.
PW, 2. Reihe **IV**, 76–155.

Strato of Lampsakos
c. 340 B.C. to *c.* 268 B.C.

A Greek Aristotelian philosopher, a pupil of Theophrastos and his successor as master of the Peripatetic school (the Lyceum) in Athens. Strato influenced the physics of Antiquity through his con-ception of space, and his insistence that a vacuum cannot possibly exist except as passing discontinuities in bodies which the surrounding molecules try at once to fill. He was considered by some to be the author of the fourth Book of Aristotle's *Meteorology*.

H. Diels, *Über das Phys. System des Straton*, Berliner Akad. 1893.

395

H. B. Gottschalk, 'Strato of Lampsacus, Some texts, edited with a commentary', *Proc. Leeds Phil. Lit. Soc.*, **XI** (1965), part 6, 95–182.
Sarton, **I**, 152.
PW, 2. Reihe **IV**, 278.

Suiseth
see Richard Suiseth

Swineshead
see Richard Suiseth

Terpander
of Lesbos,
c. 700 B.C.

According to legend, Terpander was the father of Greek music. He founded the first school of music in Sparta, improved the seven-stringed lyre, arranged a musical scale, and developed the art of singing to the accompaniment of instruments.

Sarton, **I**, 61.
PW, 2. Reihe **V**, 785–86.

Thābit ibn Qurra
of Harran
c. A.D. 826–901

A Mesopotamian mathematician and astronomer, who lived for the most part in Baghdad. He was one of the most industrious translators of Greek and Syriac texts into Arabic, and the author of a long series of works of mathematical, astronomical or medical content

(ed. F. J. Carmody, *The Astronomical Works of Thabit B. Qurra*, Berkeley, 1960). In the field of astronomy he was responsible for the introduction of the later widespread idea of the *trepidatio* or oscillatory motion of the equinoctial points, which made precession a complicated function of time. In statics his writing on the balance *Liber Karastonis* (Latin text with English transl. in E. A. Moody and M. Clagett, *The Medieval science of weights*, Madison Wisc., 1952, 87–117) came to play an important role.

F. J. Carmody, *Isis*, **XLVI** (1955), 235.
Dreyer, 276.
Duhem, **II**, 117; 238.
B. R. Goldstein, *Centaurus*, **X** (1965), 232.
Kennedy, 136.
Nallino, *Scritti*, **V**, 338.
Sarton, **I**, 599.
Suter, 34.
Thorndike, **I**, 661.
Wiedemann, **II**, 548.

Thales
of Miletos
c. 625 B.C. to *c.* 545 B.C.

The first of the Ionian natural philosophers, possibly of Phoenician stock. Thales travelled in Egypt and after his return to Miletos became the disseminator of a certain knowledge of Egyptian mathematics among the Greeks. In natural philosophy he prepared the way for a rational description of phenomena by departing from the mythological conception of nature, trying instead to explain the evolution of the universe by a prime matter (water), from which all other substances emerge.

No writings are extant, and the effect of his contribution to science is rather uncertain.

Burnet, 39–50.
W. Hartner, *Centaurus*, **XIV** (1969), 60.
Heath, **I**, 118.
Kirk and Raven, 74.
Sarton, **I**, 72.
Tannery, *Bull. d. sci. mathématiques*, **X** (1886), 115.
PW, 2. Reihe **V**, 1210.

Theodoricus Teutonicus
see Dietrich

Theon of Alexandria
worked *c.* A.D. 360–370

A Hellenistic astronomer and mathematician who worked at the Museum in Alexandria. He edited Euclid and wrote an unfinished commentary on the *Almagest* (ed. A. Rome, *Commentaire de Pappus et de Théon d'Alexandria sur l'Almageste*, 3 vols, Rome, 1931–43; French transl. N. Halma, Paris, 1821). Ptolemy's *Handy Tables* are only extant in Theon's version (ed. N. Halma, Paris, 1822).

Heath, **II**, 526.
Sarton, **I**, 367.

Theon of Smyrna
c. A.D. 130

A Greek, Platonic philosopher who wrote *On the mathematical knowledge necessary to the study of Plato* (ed. with French transl. by Dupuis, *Ce qui est utile en Mathématiques pour comprendre Platon*,

Paris, 1892) which is an important source of the history of Greek mathematics and astronomy.

Heath, **II**, 238.
Sarton, **I**, 272.
PW, 2. Reihe **V**, 2067–2075.

Theophrastos
of Lesbos
c. 372 B.C. to *c.* 288 B.C.

A Greek philosopher from Lesbos, Aristotle's disciple and his successor as head of the Lyceum. His main achievement was the foundation of botany as a science, and he wrote two works on the classification and ecology of plants (ed. A. Hort, *Enquiry into Plants*, 2 vols, London, 1916). Most of his many other writings are lost, but his *Characters*, a collection of psychological essays are still extant.

Duhem, *Études sur Léonard de Vinci*, **II**, 473.
Sarton, **I**, 143–44.
PW, Supplement **VII**, 1354–1562.

Thomas Aquinas
1225–1274

An Italian Dominican, disciple of Albert the Great and one of the principal philosophers and theologians in the thirteenth century. Thomas Aquinas was also renowned as a teacher in Paris, Cologne, Orvieto, Viterbo, and Naples. He was less interested in natural science than his master Albert the Great, but played a considerable role in the reintroduction of the principle of experience as the foundation of natural

397

philosophy. In contradicting the Averroists he maintained the hypothetical character of astronomical theories. In physics he was (at least in some of his writings) one of the few thirteenth century supporters of the impetus theory. His principle work is the huge introduction to theology, *Summa theologica*, but his scientific opinions are to be found mainly in his long series of commentaries on Aristotle (*Commentary on Aristotle's Physics*, transl. R. J. Blackwell, R. J. Spath, and W. E. Thirlkel, introd. by V. J. Burke, New Haven, Conn., 1963) and in his collections of *Quaestiones*. His *opera omnia* are published in two modern editions, one in Rome (Editio Leonina), the other in Torino (Marietti).

F. C. Copleston, *Aquinas*, London, 1955.
R. J. Deferrari, and M. I. Barry, and I. McGuiness, *A lexicon of St Thomas Aquinas*, Washington, 1948–53.
Duhem, **III**, 348 *ff.*
Gilson, *Le Thomisme*, 5th ed., Paris, 1945.
R. Laubenthal, *Das Verhältnis des hl. Thomas von Aquin zu den Arabern in seinem Physikkommentar*, Würzburg, 1933.
T. Litt, *Les corps célestes dans l'univers de Saint Thomas d'Aquin* (*Philosophes médiévaux*, **VII**), Louvain, 1963.
Sarton, **II**, 914–21.
Thorndike, **II**, 593–615.
J. De Tonquedec, *Questions de cosmologie et de physique chez Aristote et S. Thomas*, Paris, 1950.
W. A. Wallace, *Dict. Sci. Biogr.*, **I**, 196.

Thomas Bradwardine

c. 1290–1349

One of the most important English natural philosophers of the fourteenth century. Born at Chichester, he was educated at Balliol, and then moved to Merton College. In 1337 he became Chancellor of St Paul's Cathedral, London, and in 1349 he was appointed Archbishop of Canterbury, only to die of the plague in the same year. He was the author of a large theological work, *De causa Dei contra Pelagianos* (ed. Henry Savile, London, 1618) giving an outstanding exposition of spiritual movements and tensions in his own times, and containing his personal conception of the universe as embedded in an infinite, empty space. His *Tractatus proportionum* of 1328 (ed. with a rather unreliable English transl. H. Lamar Crosby, *Thomas of Bradwardine, His tractatus de proportionibus*, Madison, Wisc., 1955) deals with the problem of motion and proposes Thomas's new relation between force, resistance, and velocity. An *Arithmetica speculativa* and a *Geometria speculativa* were widely used manuals after the invention of printing. The treatise *De continuo* is concerned with the problem of continuity in mathematics and physics.

Clagett, 199 *ff.*
S. Hahn, *Bradwardinus und seine Lehre v.d. mensch. Willensfreiheit*, Münster i.W., 1905.
J. E. Hofmann, *Centaurus*, **I** (1951), 293–308.
A. Koyré, *From the closed world to the infinite Universe*, Baltimore, 1957.
A. Koyré, *Archives d'hist. doctr. moyen. âge*, **XVII** (1949), 45.
G. Leff, *Bradwardine and the Pelagians*, Cambridge, 1957.
A. Maier, *Die Vorläufer Galileis*, 2nd. ed., Rome, 1966.
A. G. Molland, *Brit. Journ. Hist. Science*, **IV** (1968), 108.
J. E. Murdoch, *Dict. Sci. Biogr.* **II**, 390.

Sarton, **III**, 668.

E. Stamm, *Isis*, **XXVI** (1936), 13 (on *De continuo*).

al-Tūsī

see Nāṣir al-Dīn al-Ṭūsī

Ulug Beg
of Samarkand
c. 1400–1449

A grandson of Tamerlane, Ulug Beg was Governor of Turkestan until he became Emperor of the Mongol empire in 1447. Two years later he was murdered by his son. In 1420 he founded an astronomical observatory near Samarkand, which for a while became the astronomical centre of the world and was one of the last great achievements of Moslem science. From the observations made there a new collection of tables was prepared and a star catalogue (ed. E. B. Knobel, *Ulugh Beg's Catalogue of Stars*, Washington, 1917) which soon spread in Arabic versions.

W. Barthold, *Ulug Beg und seine Zeit*, Leipzig, 1935.
Kennedy, 125.
A. Sayili, *The observatory in Islam*, Ch. 8, Ankara, 1960.
V. P. Scheglov, *L'observatoire d'Ouloug-Beg à Samarkande*, Moscow, 1958.
Sarton, **III**, 1120.

Varāha Mihira
worked *c.* A.D. 505

An Indian astronomer who wrote the *Pancasiddhāntika* (ed. and partly transl.

G. Thibaut and S. Dvivedī, Benares, 1899) as a compendium of the previous astronomical *Siddhāntas*. An astrological *Brihatsaṃhitā* (transl. C. Iyer, Madura, 1884) was also written by him.

P. V. Kane, *Journ. Bombay Branch Roy. Asiatic Soc.*, 1948–49, 1–31.
Sarton, **I**, 428.

Villard de Honnecourt
c. 1200– *c.* 1260

A French engineer and architect from Honnecourt (Picardy) whose personal sketch-book (*Album de Villard de Honnecourt*, Facsimile edition, Paris, 1906) containing a wealth of drawings of buildings, machines, etc., is an important source of our knowledge of Mediæval technology.

Sarton, **II**, 1033.
H. Schimank, *Der Weg zur Erkenntnis des Energieprincips*, Dresden, 1929.

Vitello
(Vitellio, Vitellius) of Silesia
Second half of thirteenth century

A Silesian physicist and philosopher who was educated in Paris, Padua, and Viterbo. His main work was his *Perspectiva*, a huge treatise on optics of 1270 (ed. Risner, *Opticae thesaurus*, Basel, 1572, printed together with al-Haitham's optics) founded upon Ibn al-Haitham's similar work; it was later commented on by Kepler (*Ad Vitellionem paralipomena*, Frankfurt a.M., 1604). It is dedicated to the translator William of Moerbeke, whose acquaintance Vitello had made in Viterbo in 1269. It contains some of the first Mediæval descriptions of laboratory instruments in the proper sense of the word.

C. Baeumker, *Witelo*, Münster, i.W., 1908.
Duhem, V, 358–74.
Sarton, II, 1027–28.

Vitruvius
(Marcus Vitruvius Pollio)
Second half of first century B.C.

A Roman architect and engineer whose 10 books *De architectura* (ed. and transl. F. Granger, 2 vols, London, 1931–34) constitute an impressive handbook of all the architectural disciplines of his time. They were much used in the Middle Ages and during the Renaissance.

V. Mortet, *Recherches critiques sur Vitruve*, Paris, 1902–08.
PW, 2. Reihe **IX'**, 427–89.
Sarton, I, 223–25.

Walter
see Bernard Walter

William Heytesbury
c. 1300–*c.* 1380

A fellow of Merton College in 1330, and in 1371 Chancellor of Oxford University. In the field of natural philosophy he was one of the most prominent representatives of the Merton School, and much concerned with the problem of the reduction of qualities to quantities. His *Sophismata* contain 32 questions, many of them important to kinematics. In it are statements such as: 'It is impossible that anything becomes hotter unless something else becomes colder', or 'If something is expanded, something else must be compressed', the latter of which has been interpreted as a primitive form of the law of conservation of mass. There are many early printings of his work, but no critical editions.

Clagett, 199 *ff.*
Sarton, III, 565–66.
C. Wilson, *William Heytesbury: Medieval Logic and the Rise of Mathematical Physics*, Madison, 1956.

William of Moerbeke
(Guillielmus Morbecanus)
c. 1215–*c.* 1286

A Flemish Dominican who spent much of his life as a member at the papal court at Orvieto and Viterbo. From 1278 he was Archbishop of Corinth. He was a friend of Thomas Aquinas, at whose invitation he made a long series of remarkably exact translations from the Greek, including nearly all the works of Aristotle, and also those of Simplicios, Heron, Proclos, and Archimedes. The *Hydrostatics* of the latter is conserved only in William's translation, of which Tartaglia's edition of 1543 is a mere plagiarism.

Clagett, *Isis*, **XLIII** (1952), 236–42 (Archimedes translations).
Clagett, *Archimedes in the Middle Ages*, I, Madison, Wisc., 1964.
Clagett, *The Science of Mechanics in the Middle Ages*, Madison, Wisc., 1959.
Sarton, II, 829–31.

William of Ockham
c. 1300–*c.* 1350

An English Franciscan from Ockham, Surrey, educated in Oxford where he taught until 1324, when he became involved in the struggle between the Holy See and his own order as a bitter adversary of the Pope. From about 1328 he lived in Munich under the protection of the Emperor (Louis of Bavaria) and wrote a long series of anti-papal treatises. In science he exercised an important influence on logic as the leading nominalist thinker of the fourteenth century, mainly because of his *Summa totius logicae*. In his *Philosophia naturalis*, and in the *Tractatus de successivis* (the latter ed. by P. Boehner, St Bonaventure, N.Y., 1944) he employed his logical principles in a merciless attack on Aristotelian philosophy. In physics he tried to solve the problems of motion by leaving all questions of the cause of motion out of account, but this gained him few followers. There are several early and a few modern printings of individual works, but no complete editions.

L. Baudry, *Lexique philosophique de Guillaume d'Ockham*, Paris, 1958.
Gilson, 638*ff.*
R. Guelley, *Philosophie et théologie chez Guillaume d'Ockham*, Louvain-Paris, 1947.
E. A. Moody, *The logic of William of Ockham*, New York, 1935.
S. Moser, *Grundbegriffe d. Naturphilosophie bei Wilhelm von Ockham*, Innsbruck, 1932.
S. Rabade Romeo, *Guillermo de Ockham y la filosofia del siglo XIV*, Madrid, 1966.
Sarton, **III**, 549.
P. Vignaux, *Occam, Dic. Théol. Catholique*, XI, 717*ff.* and 864*ff.*

William of St Cloud
(Guilelmus de Sto Clodoaldo)
worked *c.* 1285–1296

An astronomer of unknown origin who worked in Paris and seems to have been connected with Peter Nightingale, whose calendar principles he adopted. He observed the Sun in *camera obscura*, and in 1290 determined the obliquity of the ecliptic as 23°; 34. From his observations he proposed a number of corrections to existing astronomical tables. His work is much in need of further investigation.

Duhem, **IV**, 10 and 580.
Sarton, **II**, 990.
Thorndike, **II**, 262 and 668.

Xenophanes
of Colophon
c. 570 B.C. to *c.* 480 B.C.

An Ionian philosopher who is traditionally thought to have founded the Eleatic school in Southern Italy. His cosmological ideas were primitive, but he correctly interpreted fossils to be testimonies of periodical inundations of the dry parts of the Earth.

Burnet, 112–129.
Kirk and Raven, 163–81.
Sarton, **I**, 73.

al-Zarqālī
(Arzachel, Azarquiel) of Cordoba
c. 1029–*c.* 1087)

A Spanish-Arabic astronomer, who was a good observer. His surname means 'the Engraver', and he is remembered as the inventor of a new kind of instrument (derived from the astrolabe) called the *safika* (saphea, assafea) which he described in a particular treatise (*Tractat de l'assafea*, ed. and transl. J. M. Millas y Vallicrosa, Barcelona, 1933; cfr. *Libros del saber* 3, Madrid, 1864). In the Latin Middle Ages he was famous because the (lost) Arabic original of the *Toledo Tables* and their *Canones* were ascribed to him; they were translated into Latin by Gerard of Cremona.

M. Boutelle, *Centaurus*, **XII**, (1967), 12.
Duhem, **II**, 246.
B. R. Goldstein, *Centaurus*, **X** (1965), 232.
Kennedy, 128.
J. M. Millas y Vallicrosa, *Estudios sobre Azarquiel*, Madrid-Granada, 1943–50.
Nallino, *Scritti*, **V**, 341.
Sarton, **I**, 758.
T. Schmid, *Classica et Mediaevalia*, **XV** (1954), 253.
A. Wittstein, *Zts. Math. Phys. Hist. Abt.*, **XXXIX** (1894), 41.
E. Zinner, 'Die Tafeln von Toledo', *Osiris*, **I** (1936), 747.
Wiedemann, **II**, 622.

Zeno
of Elea
5th century B.C.

Pupil of Parmenides and one of the founders of the Eleatic School. Only a few fragments of his works remain. Aristotle went to considerable length to confute Zeno's paradoxical arguments against the possibility of motion (*Physica*, **VI**, 2 *ff.*).

Burnet, 310–320.
F. Cajori, *Isis*, **III** (1920), 7.
Heath, **I**, 271*ff.*
Kirk and Raven, 286.
Sarton, **I**, 85.

Zosimos
of Panapolis (Egypt)
c. A.D. 300

An author on alchemy whose partly extant works are one of the main sources of early alchemy (P. Berthelot, *Collection des anciens alchimistes grecques*, **II**, Paris, 1888, Greek text with French transl.).
F. Sherwood Taylor, *The Alchemists*, London 1951, *passim*.

Index